Wave/Current–Structure–Seabed Interactions around Offshore Foundations

Wave/Current–Structure–Seabed Interactions around Offshore Foundations

Guest Editors

Wengang Qi
Shengjie Rui
Zhen Guo

MDPI

Basel • Beijing • Wuhan • Barcelona • Belgrade • Novi Sad • Cluj • Manchester

Guest Editors
Wengang Qi
Institute of Mechanics Chinese Academy of Sciences
Beijing
China

Shengjie Rui
Advanced Modelling
Norwegian Geotechnical Institute (NGI)
Oslo
Norway

Zhen Guo
College of Civil Engineering and Architecture
Zhejiang University
Hangzhou
China

Editorial Office
MDPI AG
Grosspeteranlage 5
4052 Basel, Switzerland

This is a reprint of the Special Issue, published open access by the journal *Journal of Marine Science and Engineering* (ISSN 2077-1312), freely accessible at: https://www.mdpi.com/journal/jmse/special_issues/Z66EWLI59V.

For citation purposes, cite each article independently as indicated on the article page online and as indicated below:

Lastname, A.A.; Lastname, B.B. Article Title. *Journal Name* **Year**, *Volume Number*, Page Range.

ISBN 978-3-7258-5901-6 (Hbk)
ISBN 978-3-7258-5902-3 (PDF)
https://doi.org/10.3390/books978-3-7258-5902-3

Contents

About the Editors

Qi Wengang

Qi Wengang is an Associate Professor at the Institute of Mechanics, Chinese Academy of Sciences. His main research focuses on fluid–structure–soil coupling mechanics and ocean engineering mechanics, involving issues such as foundation scour, structure–wave/current–seabed interactions, and the stability of marine foundations. He has presided over projects including the National Natural Science Foundation Youth B and Youth C programs, as well as two general projects. In 2021, he was selected as a member of the "Youth Innovation Promotion Association" of the Chinese Academy of Sciences. He has published over 40 SCI papers in the fields of ocean and geotechnical engineering mechanics, such as *Ocean Engineering*, *Coastal Engineering*, *JGR: Oceans*, and *Géotechnique*. Currently, he serves as a Technical Program Committee (TPC) Member of the International Society of Offshore and Polar Engineers (ISOPE), Council Member of the Ocean Engineering Division of the Chinese Society of Oceanography, Editorial Board Member of *Mechanics in Engineering*, Editorial Board Member of *Ocean Engineering* and *Applied Ocean Research*, and Youth Editorial Board Member of *China Ocean Engineering*. His research achievements on scour and bearing capacity prediction models for large-diameter monopile foundations of offshore wind turbines were awarded the 2024 Natural Science Second Prize by the Chinese Society of Theoretical and Applied Mechanics (ranked 1/2).

Rui Shengjie

Rui Shengjie is an Assistant Professor in the Department of Civil and Environmental Engineering at the National University of Singapore (NUS). Dr. Rui's research focuses on geotechnical and ocean engineering, with a particular emphasis on offshore renewable energy development and addressing complex geotechnical challenges in demanding marine environments. Dr. Rui has visited many universities and presented his work at several world-leading institutions such as the University of Oxford, Cambridge, MIT, Imperial College London, UC Berkeley and Davis and ETH Zürich.

Guo Zhen

Guo Zhen is a Professor and Doctoral Supervisor. He serves as the Assistant Director of the Center for Coastal and Urban Geotechnical Engineering Research and the Deputy Director of the Industrial Technology Transformation Research Institute at Zhejiang University. His primary work focuses on education, research, and scientific investigation in the fields of marine geotechnical engineering and materials. He is dedicated to ensuring the safety and stability of marine infrastructure through theoretical research, material development, and technological innovation. He has led over 20 national and provincial research projects, including those funded by the National Natural Science Foundation, the Innovation Special Zone, and the Provincial Key R&D Program, as well as more than 30 engineering projects. He has published over 200 academic papers, including 157 SCI-indexed articles, 5 ESI highly cited papers, and 1 hot paper, with a total citation count exceeding 3,700 and an H-index of 37. Additionally, he has authored 4 books and 1 textbook, obtained more than 60 invention patents (including 7 U.S. and Japanese patents), and registered 25 software copyrights. His research outcomes have been applied in critical projects such as offshore wind power development, deep-sea oil and gas extraction, construction of islands and reefs in the South China Sea, and airport runway repairs. His contributions have earned him awards including the First Prize of the Zhejiang Provincial Science and Technology Progress Award and the First Prize of the Hainan Provincial Science and Technology Progress Award.

Journal of
Marine Science
and Engineering

MDPI

Editorial

Wave/Current–Structure–Seabed Interactions Around Offshore Foundations

Wengang Qi [1,2,*], Shengjie Rui [3] and Zhen Guo [4]

1 Key Laboratory for Mechanics in Fluid Solid Coupling System, Institute of Mechanics, Chinese Academy of Sciences, Beijing 100190, China
2 School of Engineering Science, University of Chinese Academy of Sciences, Beijing 100049, China
3 Advanced Modelling, Norwegian Geotechnical Institute (NGI), 0480 Oslo, Norway; shengjie.rui@ngi.no
4 Zhejiang Province Key Laboratory of Offshore Civil Engineering and Materials, Zhejiang University, Hangzhou 310058, China; nehzoug@163.com
* Correspondence: qiwengang@imech.ac.cn

The rapid expansion of the offshore energy infrastructure, including wind turbines, subsea pipelines, and marine platforms, has underscored the critical need to overcome the challenges presented by harsh marine environments. These structures are exposed to complex interactions among waves, currents, seabed dynamics, and structural vibrations, which can trigger geotechnical instabilities, sediment scouring, and fatigue failure. This Special Issue, titled "Wave/Current–Structure–Seabed Interactions Around Offshore Foundations", brings together cutting-edge research to deepen our understanding of these multifaceted interactions and their implications for the design and safety of offshore engineering. The 11 articles featured here employ experimental, numerical, and theoretical approaches to investigate pivotal topics such as scour dynamics, seabed liquefaction, hydrodynamic loading, and multi-physics coupling effects. Below, we synthesize the novel contributions of each study and discuss their significance for future research and industrial applications.

Ref. [1] conducted an experimental study on local scour patterns around complex multi-pile foundations under steady currents. Employing a meticulously designed 5×5 pile group model, they systematically investigated the relationship between flow dynamics and scour depth evolution. Their results demonstrate that the maximum scour depth exhibits strong dependence on flow intensity, and the distribution of scour depths among piles differs at different flow intensities. This research elucidates scour mechanisms in pile groups by demonstrating that hydrodynamic interactions induce significant shielding effects—progressively reducing scour depths downstream—and trigger a critical transition from isolated-pile-like scour patterns to group-dominated behavior at elevated flow intensities. The findings indicate that pile group scour is governed by emergent spatial dependencies rather than superposition of single-pile responses, thereby challenging conventional extrapolation approaches in foundation design. These insights provide a mechanistic basis for advancing predictive models and optimizing risk mitigation strategies in marine infrastructure. Ref. [2] pioneered wave-current-vibration flume experiments to quantify monopile vibration effects on local scour dynamics. The equilibrium scour depths in clear-water conditions are reduced by pile vibration while triggering non-monotonic depth variations under live-bed scour conditions near the critical Shields parameter. Crucially, higher vibration frequencies predominantly suppress the scour depth, whereas increased amplitudes expand the width of the scour hole. This behavior stems from vibration-induced sediment ratcheting convection and densification, which enhance scour hole backfilling. Their predictive model—the first to integrate vibration intensity with

https://doi.org/10.3390/jmse13081595

hydrodynamic parameters (Froude number, Keulegan–Carpenter number, and velocity ratio)—provides a critical design framework for offshore wind turbine foundations under dynamic loading. Ref. [3] addressed pipeline instability challenges on "sand wave" seabeds, where natural undulations create hazardous free spans that are prone to Vortex-Induced Vibration (VIV) fatigue. Through coupled geotechnical–hydrodynamic modeling, they evaluated post-lay trenching as a mitigation strategy. A comparative fatigue assessment of trenched vs. untrenched pipelines across realistic sand wave morphologies was performed. It was shown that targeted trenching for post-lay rectification (trenching to a depth of 1 m) significantly reduces stress ranges and extends fatigue life, meeting both Ultimate Limit State (ULS) and Fatigue Limit State (FLS) requirements. The study provides a methodology for pipeline route planning and intervention design on morphologically active seabeds.

Ref. [4] presented a significant advancement in modeling wave-induced seabed liquefaction around buried pipelines by adopting a non-Darcy flow-based 2D finite element model. Their simulations reveal that nonlinear flow resistance fundamentally alters the pore pressure distribution, resulting in a shallower yet broader liquefied zone compared with conventional predictions. Crucially, the pipeline itself exerts a "liquefaction shielding effect," suppressing seabed liquefaction and constraining the extent of the liquefied zone relative to far-field conditions. These findings establish a more realistic framework for assessing pipeline flotation risks and evaluating mitigation strategies such as increased burial depth or soil densification in liquefaction-prone environments. Ref. [5] conducted pioneering numerical investigations into tidal bore-induced seabed liquefaction—an extreme but often neglected hazard. By integrating generalized Biot theory with advanced integral transform methods, they simulated the rapid propagation of a high-energy tidal bore through a shallow estuary and its destabilizing effects on sandy riverbeds. Their results demonstrate that the abrupt pressure gradient preceding the bore front induces instantaneous liquefaction in the upper seabed layer within seconds, with the liquefaction depth and intensity being governed by the sediment permeability, saturation level, and bore height. This work delivers critical predictive relationships for assessing liquefaction risks under transient hydraulic loading conditions. Ref. [6] introduced an innovative hybrid Discrete Element Method (DEM)–Finite Element Method (FEM) framework to elucidate the mechanics of lateral pipe–soil interaction on sloping seabeds. The DEM component accurately captures granular sediment behavior, including soil upheaval, particle flow, and localized densification, while the FEM efficiently models the structural pipe response. Their simulations uncover how the seabed slope dictates the failure wedge geometry and lateral resistance during pipe movement, while also revealing nuanced mechanisms such as post-breakout soil resistance evolution and the influence of pipe rotation on deeper soil mobilization. This methodology provides unprecedented insights for refining the large-scale pipeline–soil interaction models that are used in slope stability design.

Ref. [7] developed an advanced method for predicting the fatigue life of free-spanning pipelines that are subject to Vortex-Induced Vibration (VIV), integrating three key components: (1) dimensionless vibration amplitude A/D-V_r relationships for a wall-free circular cylinder with low mass damping, (2) beam-theory stress distribution analysis, and (3) high-cycle ($N > 10^7$) S-N curves for high-strength steel pipelines with cathodic protection under seawater environments. Their parametric studies revealed nonlinear fatigue life-flow velocity relationships that challenge conventional linear assumptions in pipeline design. Ref. [8] employed the novel technique of transparent soil combined with Particle Image Velocimetry (PIV) to visualize and quantify the complex coupled thermal–hydraulic processes around a buried pipeline. This non-invasive method allows for unprecedented observation of the convective flow field and temperature distribution in the soil annulus surrounding an actively heated or cooled pipeline. Key findings demonstrate that natural convection currents become the dominant heat transfer mechanism at shallow burial depths, significantly altering isotherm shapes and reducing thermal gradients compared with pure conductive models. The study quantifies the convection strength relative to the burial depth, heat flux, and soil properties, offering valuable data to improve the accuracy of numerical models predicting pipeline temperature dissipation and long-term impacts on seabed thermal regimes. Ref. [9] employed Discrete DEM simulations with realistically shaped, crushable particle models to elucidate the micromechanical mechanisms behind strength/stiffness anisotropy in calcareous sands. Their triaxial shear simulations with varying particle breakage ratios revealed that contact force redistribution from vertical to horizontal directions during particle crushing weakens deviatoric stress networks, fundamentally altering the peak strength, deformation modulus, and failure patterns. These micromechanical insights provide essential foundations for developing anisotropic constitutive models to enhance offshore foundation safety in calcareous environments.

Ref. [10] performed high-fidelity CFD simulations to unravel the complex hydrodynamics of semi-submersible platforms for floating offshore wind turbines (FOWTs). Their study reveals that submerged braces and pontoons contribute substantially to the total drag forces at high Reynolds numbers—a finding that challenges conventional models focusing solely on column effects. The research provides quantitative relationships between drag coefficients, Reynolds numbers, and platform orientation, offering critical data for optimizing structural design and mooring systems to enhance storm survivability. Ref. [11] explored the innovative synergy of offshore wind energy infrastructure with aquaculture by numerically analyzing the hydrodynamic interactions between wind turbine monopile foundations and pontoon raft aquaculture facilities (PRAFs). Using CFD modeling validated against physical experiments, they assess how the raft arrays modify wave propagation, reflection, and dissipation around the monopile and, reciprocally, how the monopile influences raft motions and mooring tensions. A major finding is that multi-row raft configurations positioned upstream of the monopile effectively act as wave dampers, reducing wave heights at the monopile by up to 20% depending on the row number, wave period, and raft submergence. Concurrently, the structural presence of the monopile provides shielding, reducing the peak mooring tension in multi-row PRAF arrangements by up to 73% compared with single-row PRAF arrangements. This research provides hydrodynamic evidence supporting the technical feasibility and mutual benefits (reduced structural loads, sheltered aquaculture conditions) of integrated offshore multi-use platforms.

This collection of studies elucidates the intricate multi-physics phenomena governing wave/current–structure–seabed interactions in offshore foundation systems, while reporting pioneering novel methodologies to enhance marine infrastructure resilience. Through systematic integration of experimental investigations, numerical modeling breakthroughs, and theoretical advancements—spanning scour dynamics, seabed liquefaction mechanisms, coupled hydrodynamic loading, and sustainable design paradigms—this Special Issue makes substantive contributions to the fundamental science of offshore engineering. The rigorous findings presented not only elevate predictive frameworks for geohazard evaluation but also propel the development of innovative engineering solutions. These advances are particularly timely and vital as the offshore industry confronts the dual challenges of expanding marine resource exploitation and increasingly stringent safety requirements in complex marine environments.

Funding: This research was funded by the Youth Innovation Promotion Association CAS (Grant No. 2021018).

Acknowledgments: We extend our gratitude to the authors and reviewers for their invaluable contributions to this work. Special thanks to the *JMSE* editorial team for their support in creating this timely Special Issue.

Conflicts of Interest: The authors declare no conflicts of interest.

References

1. Gong, M.; Pan, X.; Du, S.; Wu, G.; Lv, Y.; Sun, Y.; Ding, G.; Liang, B. Laboratory Study of Local Scour Around an Array of Pile Groups in Clear-Water Scour Conditions. *J. Mar. Sci. Eng.* **2025**, *13*, 137. [CrossRef]
2. Shi, L.; Cheng, Y.; Zheng, Y.; Xia, B.; Huang, X. Experimental Study on Local Scour at the Monopile Foundation of an Offshore Wind Turbine under the Combined Action of Wave–Current–Vibration. *J. Mar. Sci. Eng.* **2024**, *12*, 963. [CrossRef]
3. Zou, X.; Xie, B.; Bai, X.; Guo, Y.; Yuan, Q.; Zang, Z. Investigations on Free Span Rectification for a Submarine Pipeline on Sand Wave Seabeds. *J. Mar. Sci. Eng.* **2025**, *13*, 107. [CrossRef]
4. Han, S.; Zhou, M.; Zhang, D.; Qi, W.; Xue, C.; Fang, Q. Wave-Induced Instantaneous Liquefaction of a Non-Cohesive Seabed around Buried Pipelines: A Liquefaction-Associated Non-Darcy Flow Model Approach. *J. Mar. Sci. Eng.* **2024**, *12*, 373. [CrossRef]
5. Pan, D.; Li, Y. Dynamic Response Analysis and Liquefaction Potential Evaluation of Riverbed Induced by Tidal Bore. *J. Mar. Sci. Eng.* **2024**, *12*, 1668. [CrossRef]
6. Peng, Y.; Qu, L. Micromechanical Analysis of Lateral Pipe–Soil Interaction Instability on Sloping Sandy Seabeds. *J. Mar. Sci. Eng.* **2024**, *12*, 225. [CrossRef]
7. Song, Q.; Liu, J.; Gao, F. Very High Cycle Fatigue Life of Free-Spanning Subsea Pipeline Subjected to Vortex-Induced Vibrations. *J. Mar. Sci. Eng.* **2024**, *12*, 1556. [CrossRef]
8. Li, H.; Meng, Y.; Sun, Y.; Guo, L. Visualized Experimental Study of Soil Temperature Distribution around Submarine Buried Offshore Pipeline Based on Transparent Soil. *J. Mar. Sci. Eng.* **2024**, *12*, 637. [CrossRef]
9. Gao, Y.; Sun, K.; Yuan, Q.; Shi, T. Stiffness Anisotropy and Micro-Mechanism of Calcareous Sand with Different Particle Breakage Ratios Subjected to Shearing Based on DEM Simulations. *J. Mar. Sci. Eng.* **2024**, *12*, 702. [CrossRef]
10. Dhar, N.; Lloyd, C.J.; Walker, J.; Dorrell, R.M. The Influence of Structural Design on the Hydrodynamics of Floating Offshore Wind Turbine Platforms. *J. Mar. Sci. Eng.* **2025**, *13*, 248. [CrossRef]
11. Chen, D.; Lin, M.; Zhou, J.; Tang, Y.; Zhao, F.; Wang, X.; Yu, M.; Li, Q.; Kitazawa, D. Experimental Study of the Hydrodynamic Forces of Pontoon Raft Aquaculture Facilities Around a Wind Farm Monopile Under Wave Conditions. *J. Mar. Sci. Eng.* **2025**, *13*, 809. [CrossRef]

Journal of
Marine Science
and Engineering

MDPI

Article

Laboratory Study of Local Scour Around an Array of Pile Groups in Clear-Water Scour Conditions

Ming Gong [1,2,3], Xinying Pan [1,2,*], Shengtao Du [1,2,4,5,*], Guoxiang Wu [1,2], Yingxue Lv [5], Yunjia Sun [5], Guangjia Ding [5] and Bingchen Liang [1,2]

1 College of Engineering, Ocean University of China, Qingdao 266100, China
2 Shandong Provincial Key Laboratory of Ocean Engineering, Ocean University of China, Qingdao 266100, China
3 Second Institute of Oceanography, Ministry of Natural Resources, Hangzhou 310012, China
4 School of Civil & Environmental Engineering and Geography Science, Ningbo University, Ningbo 315211, China
5 Key Laboratory of Coastal Engineering Hydrodynamics, CCCC, Tianjin Port Engineering Institute Co., Ltd. of CCCC First Harbor Engineering Co., Ltd., Tianjin 300222, China
* Correspondence: xinying@ouc.edu.cn (X.P.); dushengtao@nbu.edu.cn (S.D.)

Abstract: Current-induced local scour around pile groups weakens the capacity of structures. In this paper, experimental tests of local scour around an array of 5 × 5 pile groups were conducted in a steady current in a hydraulic flume. The pile-to-pile space was five times the diameter of a single pile. All the tests were in clear-water scour conditions. The effects of upstream piles on the local scour characteristics of downstream piles, as well as the outer-arranged side piles on the inner-arranged piles, were studied within flow intensities of 0.37–1.0. Both the three-dimensional topography of bed elevation changes and the maximum temporal scour depths are discussed. The results showed that the minimum threshold of flow intensity that can induce local scour around the pile groups was 0.40. The scour holes were independent of each other, though a global scouring phenomenon occurred between piles at a flow intensity of 1.0. The temporal scour depths of the downstream piles increased slowly throughout the local scour processes. During the initial scouring stage, they accelerated rapidly. At flow intensities of 0.60, 0.80, and 1.0, the scour development then progressed gradually, resembling the behavior of a single pile. The developing scouring stage can hardly be distinguished in the case of flow intensity of 0.80. The maximum scour depths in the flow intensity of 0.60 showed irregular variations with increasing row and column numbers. The equilibrium scour depths in the central-positioned piles tended to a constant value of 0.5 times the pile diameter. In larger flow intensities of 0.80 and 1.0, they decreased linearly with pile row number, with the maximum scour depths at the piles in the first row. The local scour depths of the inner-positioned piles in the parallel arrangement showed few differences at the front and rear piles.

Keywords: pile groups; flow velocity; flow intensity; scour; maximum scour depth; bed elevation

1. Introduction

Pile groups are widely used forms of foundations for bridges crossing seas, canals, and coastal port structures in soil foundation and bedrock strata areas [1–3]. The riverine and coastal upper structures are usually supported by pile groups. Apart from the current and wave loads on these structures, local scour is considered one of the most common hazards [1,2,4–7]. When these piles are embedded in sand or soft soil bed, sediment erosion often occurs in the vicinity of the piles under waves or currents [8,9]. As a result, local scour holes appear and weaken the stability of the upper structures [10]. It is inevitable to study the local scour depths and diameters of scour holes around the piles, for instance, for a high-pile wharf-designing project in a coastal area.

A large number of papers discussing local scour around piles is available in the literature. Examples can be found in Baker [11], Raudkivi and Ettema [12], Melville [13], Sheppard [14], Oh [15], Zhao [16], and Qi [17]. The mechanisms of local scour around a single pile in steady current are summarized as follows [8]: (i) a horseshoe vortex is formed in front of the pile as the flow approaches the pile and the boundary layer separates from the sand bed; (ii) the streamlines are contracted at two sides of the pile, and the flow velocity is accelerated; (iii) Lee-wake vortexes, which are separated from the pile surface, are formed behind the pile. Consequently, both the flow turbulence intensity [18,19] and the bed shear stress [8,11] are amplified, leading to local scour around the pile. When a local scour hole is formed, additional, smaller horseshoe vortexes in the scour hole and upstream of the pile will be also produced [19]. The flow intensity of U/U_C is defined as the ratio of the approaching flow velocity (U) to the threshold flow velocity (U_C) by Melville [13]. The U_C is the minimum flow velocity below which the sediment particles cannot be motivated to move in the approaching flow bed without a pile. The ratio of U/U_C is about 0.4 for local scour around a pile [20,21], whereas it is 1.0 for general scouring in a sand bed without a pile. U/U_C = 1.0 is also employed as a threshold constant value for clear-water scour and live-bed scour conditions by Melville [13]. They found that $U/U_C \leq 1.0$ and $U/U_C > 1.0$ are in clear-water scour and live-bed scour conditions, respectively. From this perspective, the amplified shear stress near the bed can be more than 2.5 times that of the approaching flow bed in clear-water scour conditions. The simulation results of Roulund [8] found this ratio can be as much as 9. However, the above three major turbulent processes may not always be effective in a local scour process. Du [21] found that the maximum scour depth is at the upstream pile corners rather than the middle position in front of the pile when the flow intensity is less than 0.67. The flow acceleration caused by contracted streamlines contributes to the maximum scour depth. Lee-wake vortexes, however, do not affect local scour. This is evident from the absence of sediment movement behind the pile.

In local scour-protection design, the maximum scour depth and the scour hole diameters are the two main parameters considered. In general, a local scour hole is assumed to be an inverted conical hole, and the diameter of the scour hole can be obtained by the maximum scour depth and the scour hole slopes around the pile [8,22]. Therefore, most of the papers in the literature focus on the maximum scour depth and its influential factors. As the method of dimensional analysis is an effective way to study a variable that relates to several valuables, it is employed by many researchers in studying the maximum scour depth [23,24]. Furthermore, in clear-water scour conditions, the temporal maximum scour depth varies only slightly. In live-bed scour conditions, it approaches a constant value as scouring time progresses. This indicates that the equilibrium scour depth has been reached.

Extensive research has been conducted on local scour around a single pile. However, studies on local scour around pile groups are relatively scarce. This is notable, since pile groups are commonly used in hydraulic engineering foundations [1,2,9]. Apart from the differences in flow patterns between the piles, local scour characteristics including the temporal maximum scour depths, scour hole sizes, and sand deposit positions are very different from those of single piles [5,25]. According to the literature, the flow patterns and scour depths depend on the Reynolds number or flow intensity, pile arrangements, and pile-to-pile spaces [9,21,25–30]. For instance, in the case of two tandem piles, flow patterns include a single bluff body, shear layer reattachment, and vortex impinging, which depend on the Reynolds number and pile-to-pile distances. In contrast, for two parallel piles, the flow patterns are characterized by a single bluff body, a biased flow pattern, and parallel vortex streets [26]. The relative pile-to-pile space of $S/D \geq 3.94$ is a key factor in controlling scour depth. Here, S and D are the pile-to-pile space and pile diameter, respectively. However, the threshold value for this space, as reported by several researchers, varies widely, from 3.5 to 11. Above this threshold, the local scour around the piles is believed to resemble that of a single pile [30–33]. When the piles are arranged in four rows and four columns, i.e., an array of 4 × 4 pile groups, impinging flow pattern occurs with $S/D \geq 3.94$ [34]. Thus, it can be concluded that local scour around pile groups with varied pile-to-pile space may present very different characteristics.

Previous studies mainly focused on bridge pier foundations, which were usually supported by piles with $S = 1D - 2D$ [5,10,30,35] and compound piles [28,29]. The whole diameters of the pile groups or the compound piles were generally represented by an equivalent diameter to apply the equilibrium scour depth predicting equations to a single pile [2,9]. However, this method cannot be applied to pile groups when the pile-to-pile space is larger, for instance $S = 5D$. More importantly, turbulent flow interactions between the piles are rather complicated. Local scour holes do not always emerge into a scour hole around the pile groups, especially in clear-water scour conditions [25,30]. Motivated by these issues, the present paper aims to study local scour characteristics in an array of pile groups with their $S = 5D$. As U/U_c is a very important parameter in local scour around a single pile, herein, several experimental tests in an array of $R \times C$ = 5 × 5 pile groups with U/U_c ranging from 0.37 to 1.0 are studied, aiming at distinguish the effects of adjacent piles on local scour in clear-water scour conditions.

2. Experimental Setups

The experimental tests were carried out in a 68 m × 1 m × 1.2 m waves–current flume in the Key Laboratory of Coastal Engineering Hydrodynamics of China Communications Construction Corporation. As seen in Figure 1, steady current was generated from the right and flowed toward the left side before entering an external circulation pipeline. The test section was positioned in the middle areas of the flume with a length and height of 9 m and 0.3 m, respectively. Two wedge-shaped slopes were used to support the sediment particles in the test section. The piles in the first line were settled 5 m downstream of the right slope crest.

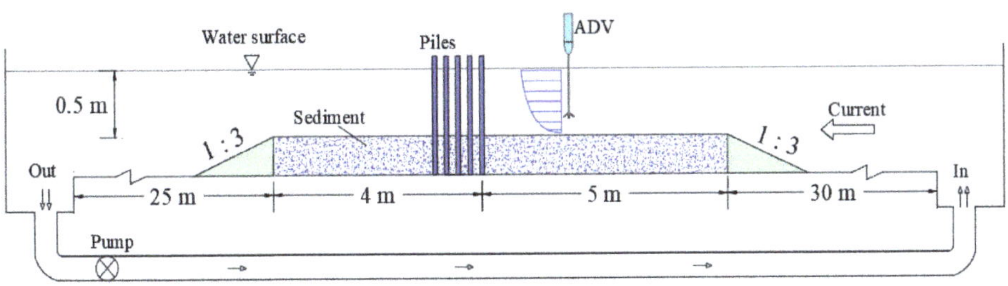

Figure 1. Experimental flume and test section arrangement.

All the piles were circular in cross-section with diameters of D = 2 cm. Each pile was labeled with a number and securely fixed at the bottom of the flume. The pile-to-pile space was $S = 5D$ in the longitudinal and horizontal directions (Figure 2). A Cartesian coordinate system was established as follows: (i) the origin of x and y coordinates was located at the center of P13; (ii) z = 0 corresponded to the sand bed surface. P_{i3} (i = 1–5) is at the center line of the flume. The ratio of the total width of piles in a row to the flume width was 1/10, which is significantly less than 1/5. As a result, the blockage effect of the piles was negligible in the experimental tests [36].

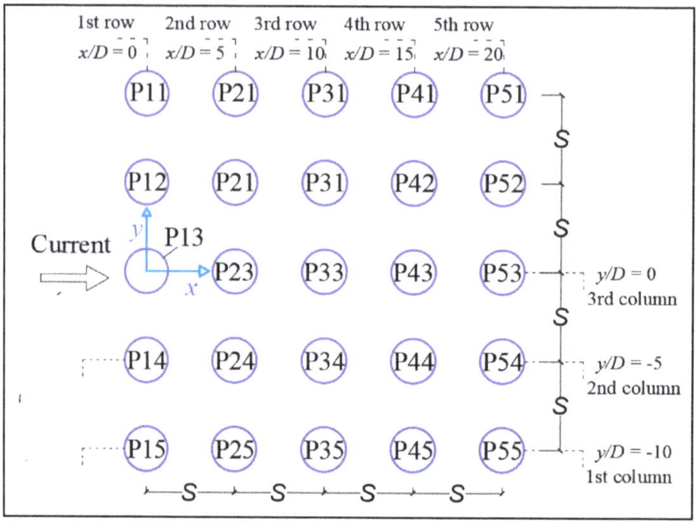

Figure 2. Array of 5 × 5 pile groups arrangement.

The sand box in Figure 1 was filled with non-cohesive sediment whose median diameter d_{50} and geometric standard deviation σ_g were 0.19 mm and 1.39, respectively. In all the local scour tests, water depth was maintained at a constant value of h = 0.5 m. The threshold of flow velocity for sediment particles moving was observed at 0.279 m/s by increasing the flow velocity gradually. Once a small amount of sediment particles was found to move as bedload, the flow velocity was considered as the threshold flow velocity U_c. Then, a Nortek acoustic Doppler velocimeter (ADV) with an accuracy of ±0.1 mm/s was used to collect flow velocity data along water depth. Sampling rate of 100 Hz and collecting duration of more than 2 min were conducted so that the collected data at each position along the water depth could be averaged with an accurate value. Consequently, the flow velocity profile along the water depth and the depth-averaged flow velocity were obtained.

For the other flow velocity values in Table 1, similar procedures were copied from the 0.279 m/s case to obtain the designed flow velocity. Note that R and C are short for row and column in Table 1. Since local scour holes around the downstream piles are usually supplied by sediment particles from the upstream local scour holes, temporal maximum scour depths of the downstream pile P_{53} was monitored using the ADV. The accuracy of ADV distance function was ±0.1 mm. It is known that the maximum scour depth can be as much as 80% of the equilibrium scour depth after a scouring duration of 2 h [16,22,37]. All the local scour tests in this paper were conducted with 2 h scouring durations. As listed in Table 1, Test A was a single pile with its $U/U_c = 1.0$, while Tests B1–B5 were square-arranged arrays of 5 × 5 pile groups in conditions of $U/U_c = 0.37 - 1.0$ and Reynolds numbers of 2080–5580. When the scouring duration reached 2 h, the current generator was turned off. Then, the flume was very slowly drained out. After the moist sand was naturally dried, which took about 3 to 5 h for each test, an iReal2E laser scanner (produced by SCANTECH in Hangzhou, China) was employed. This scanner had a minimum resolution of 0.2 mm in the xy plane and an accuracy of ±0.1 mm in the z direction. It was used to scan the sand bed, capturing data on the changes in bed elevation. The scanned data generated points cloud data through the iReal 3D (SCANTECH) software. These points cloud data were outputted as STL format file. The STL file was imported into the widely used software

Tecplot 360 2024 Release 1. Lastly, topography maps were obtained by processing the STL file.

Table 1. Test lists and parameter values in experiments.

Test	D (cm)	d_{50} (mm)	$R \times C$	S (cm)	h (cm)	U (m/s)	U/U_c	Re	t (h)
A	2	0.19	1 × 1	–	50	0.279	1.0	5580	2
B1	2	0.19	5 × 5	10	50	0.104	0.37	2080	2
B2	2	0.19	5 × 5	10	50	0.118	0.42	2360	2
B3	2	0.19	5 × 5	10	50	0.167	0.60	3340	2
B4	2	0.19	5 × 5	10	50	0.223	0.80	4460	2
B5	2	0.19	5 × 5	10	50	0.279	1.0	5580	2

3. Results and Discussion

Since Test A was employed to be a comparative test for Tests B1–B5, its local scour results will be discussed and compared with Test B5 later. The arrays of 5 × 5 pile groups were tested with a series of different flow intensities in clear-water scour conditions. Under a flow intensity of $U/U_c = 0.37$ (Test B1), no sediment scour or deposit was found around the piles. However, when the flow intensity increased to 0.42 (Test B2), very little scour holes and sediment deposition were observed. Scour could hardly be seen around the front four rows of piles, whereas a 0.05D scour depth was found in front of the fifth row of piles. The maximum scour depths and sand dune heights behind the front three rows of piles were 0.05D–0.1D and 0.1D, respectively. The very small scour holes behind these piles were not very close to the rear pile, indicating that they were induced by the wake vortexes. Additionally, no sediment deposit was observed at the rear piles in the fourth or fifth row. The two local scour tests revealed that (i) the minimum threshold of flow intensity for local scour around the pile groups was between 0.37 and 0.42; (ii) the wake vortexes were a little stronger than the horseshoe vortexes in the front four rows of piles; (iii) vortex impinging occurred at the front of the fifth rows of piles, enhancing the sediment transport rate. Herein, the threshold is considered as the averaged values of Test B1 and Test B2, that is, 0.40.

When the flow intensity of U/U_c was 0.60–1.0 (Test B3–Test B5), local scour phenomena became prominent, as shown in Figures 3a–c and 4a–c. With flow intensity increasing, the sizes of scour holes and the areas of bed elevation changes induced by wake vortexes behind the fifth row of piles grew larger. The scanned topographies of Test B3–Test B5 are presented in Figure 4a–c. No contraction scour occurred in any test, because the sand bed was not found at spaces between these piles according to the photographs and bed elevation profiles.

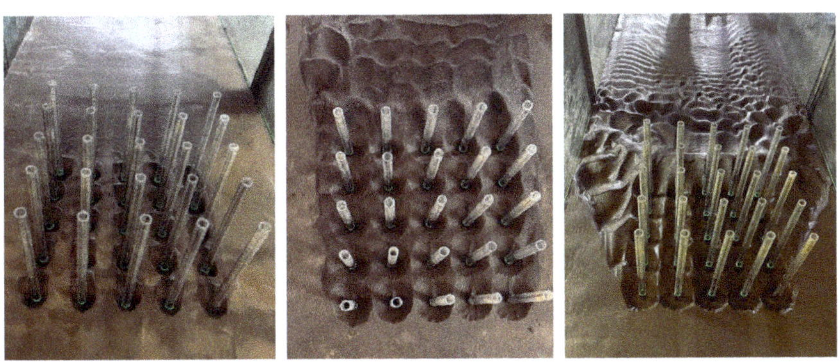

(**a**) Test B3, $U/U_c = 0.60$ (**b**) Test B4, $U/U_c = 0.80$ (**c**) Test B5, $U/U_c = 1.0$

Figure 3. Photos shot of tests B3–B5 in the flume.

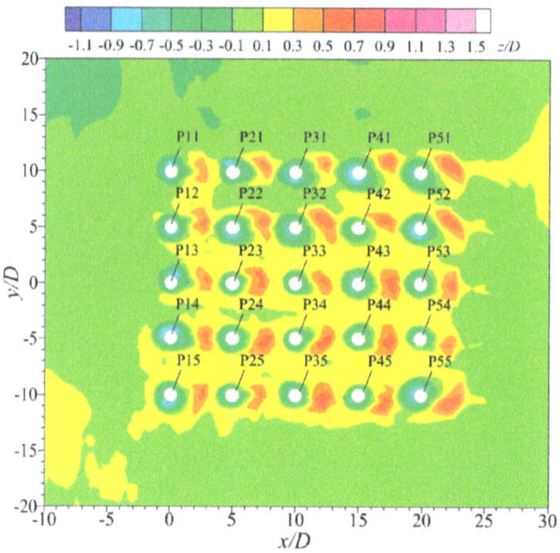

(**a**) Topography of Test B3

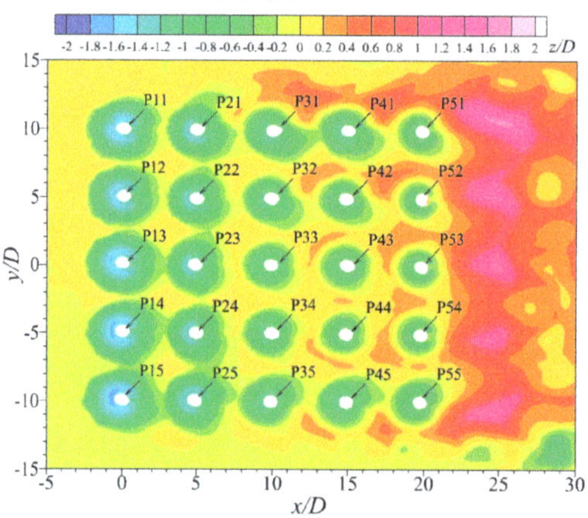

(**b**) Topography of Test B4

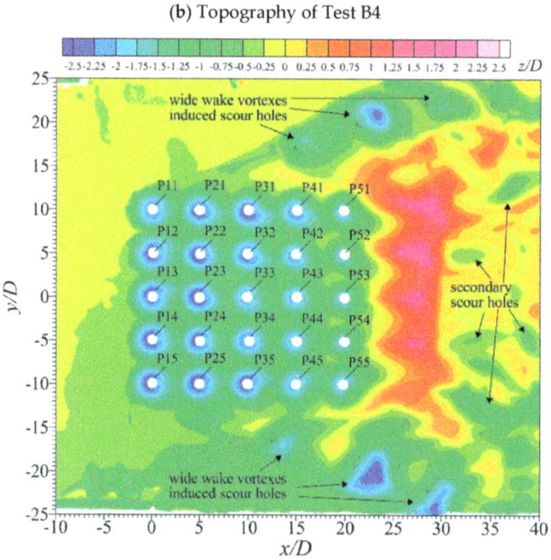

(**c**) Topography of Test B5

Figure 4. Contours of topography of tests B3–B5.

For $U/U_c = 0.60$ (Test B3), sediment particles were deposited between the piles rather than scoured by the current. From the first row to the fifth row, these deposit dune heights changed little, from 0.40D to 0.64D (Figure 5a). Most of the maximum heights of the sand dunes were about 0.60D. The sediment deposit characteristics in this test differed significantly from those in Test B2. In Test B2, sediment particles were deposited behind the piles in the front three rows, not the fifth row. It is known that the sediment particles deposited were from the scour holes upstream. Sediment particles around the downstream piles were supplied from the upstream. This led to additional sediment transport work by the turbulent flow near the downstream piles. Despite this, the maximum height of the sand dunes between these piles did not increase with the row number. This suggests that there was enhanced turbulent energy between these downstream piles. The bed elevation profiles of the first–third columns (Figure 5a) and the first–fifth rows (Figure 5b) showed that the maximum scour depths were not always in front of the pile. For example, the scour depth at the front pile of P24 was 0.46D (x/D = 4 in Figure 5a) smaller than at the pile side (y/D = −0.4 in Figure 5b). The bending of streamlines between P23 and P24 and the reattachment of wake vortexes from P13 and P14 may have contributed to this phenomenon [26,38]. Furthermore, the maximum scour depths did not present apparently decreasing or increasing trends with the increasing row number. These results are rather different from Liang [10] and Wang [33], in which the maximum scour depth decreased with the row number because of the weakened turbulence intensities [25,32]. Seeing the first column of profile y/D = −10 (Figures 2 and 5a), the square blue data show that the maximum scour depth was at the front pile of P55 (0.71D) rather than at P15 (0.40D). Moreover, in a few cases, the scour depths at the front piles were nearly identical to those at the rear piles, such as P13, P23, P33, P15, and P45. Scour holes at front piles are typically induced by horseshoe vortexes, while those at rear piles are caused by wake vortexes [8]. In clear-water scour conditions with single piles, the scour depth is usually larger at the front pile than at the rear pile [8,21,22,39]. However, the strength of the horseshoe vortexes at the front pile of P45 appears to be enhanced by the wake vortexes generated from the upstream piles. In addition, scour depths at the front sides of piles are generally found to be much larger than those at front piles, as seen in Figures 4a and 5b. The scour depths around the piles in the first to fifth rows also present irregular data, as seen in Figure 5b. For example, the scour hole slopes of piles in the first and fifth rows were not symmetric, although the piles were arranged symmetrically about the piles of the third row (center line with y/D = 0). The maximum scour depths at P11 and P15 were 0.52D and 0.92D, respectively, while they were individually 0.69D and 0.88D at P12 and P14, as shown by the blue data in Figure 5b. Although the initial sand bed was not very flat in accuracy, the maximum error should not be larger than 0.2D, as the color bar in Figure 4a shows. Thus, it cannot be considered the reason for these irregular symmetric distributions of scour holes. In fact, the computational results of [25] showed similar irregular scouring characteristics in 2 × 2 and 4 × 4 pile group cases. The irregular scour holes distribution along the central line was caused by the biased flow patterns in paralleled piles, wake vortexes impinging, and shear layer reattachment between the piles [26,38].

For the case of $U/U_c = 0.80$ (Test B4), as seen in Figures 3b and 4b, scour depths, scour hole sizes, and sand dune heights increased significantly. The sediment accumulation area was concentrated in the vicinity of the last two rows of piles. Moreover, the heights and extents of the sand dunes behind the fifth rows of piles were far larger than those upstream piles. For instance, the heights of sand dunes between the first and the fifth rows of piles were −0.2D–0.3D, whereas they were 1.2D–1.5D behind the piles in the fifth row. In contrast to Test B3 ($U/U_c = 0.60$), scour depth decreased gradually from the first row of piles to the fifth row of piles (Figure 6a). The gaps between bed profiles of y/D

= −10 (blue square data) and y/D = 0 (pink triangle data) seemed to become larger and larger with increasing row number. Furthermore, scour hole slopes in each row presented good symmetry regarding the third column (profile of y/D = 0), as seen in Figure 6b. The maximum scour depths at the front piles and upstream sides were 1.55D (Figure 6a) and 1.52D (Figure 6b), respectively. Moreover, the maximum scour depths at other rows were close to each other and did not vary much. For example, the maximum scour depths ranged from 1.4D–1.6D in the first row (square blue data for x/D = 0), while they were 0.93D–1.1D in the third row (pink triangle data for x/D = 10). Another apparent difference is that the maximum scour depth around each pile was at front of the pile in Test B4 rather than at the upstream corners as in Test B3. As scour depths at front piles are slightly larger than those at upstream pile corners, the horseshoe vortexes are thought to play more important roles in local scour processes.

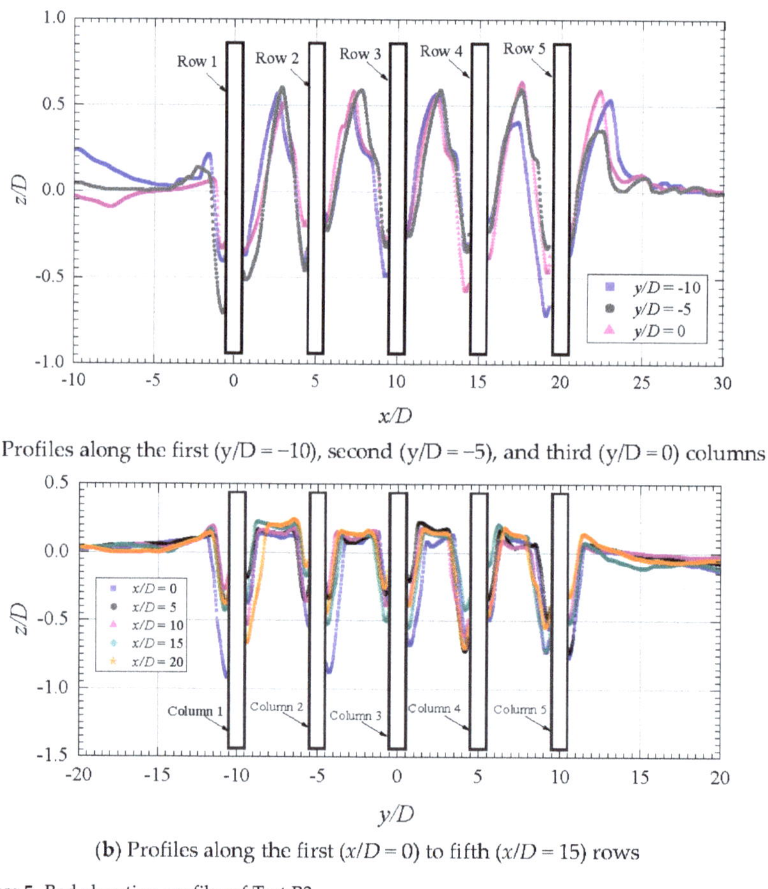

Figure 5. Bed elevation profiles of Test B3.

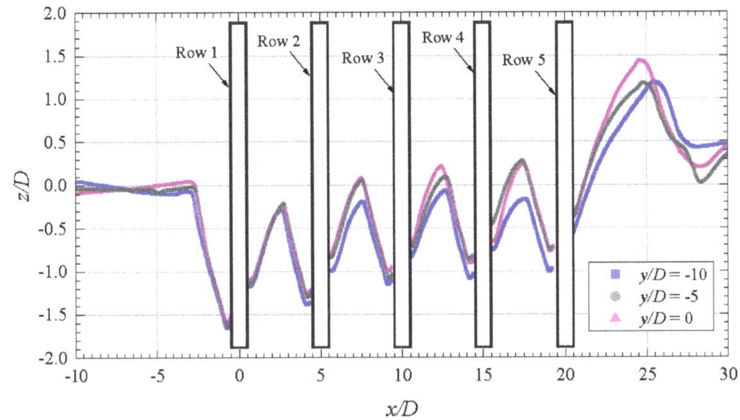

(**a**) Profiles along the first ($y/D = -10$), second ($y/D = -5$), and third ($y/D = 0$) columns

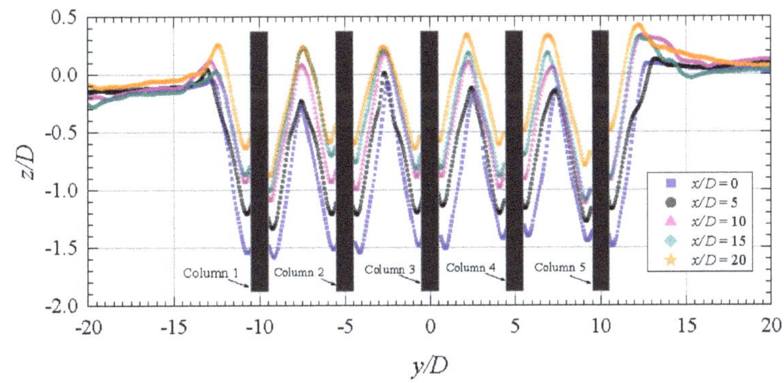

(**b**) Profiles along the first ($x/D = 0$) to fifth ($x/D = 15$) rows

Figure 6. Bed elevation profiles of Test B4.

When the flow velocity is equal to the threshold of flow velocity, i.e., $U/U_c = 1.0$ (Test B5), dramatic terrain changes occur in the vicinity of the pile groups, as seen in Figures 3c and 4c. Apart from the increased scour depths and larger sizes of the scour holes, the wide wake-induced scour holes are clearly visible, distributed along the sides of the pile groups. Furthermore, the sizes and depths of scour holes within $20 \leq x/D \leq 33$ are even larger than local scour holes around the piles. In contrast to Test B4 (Figures 3b and 4b), more and more extensive secondary scour holes behind the sand dunes are found in Test B5 (Figures 3c and 4c) due to the stronger flow intensity. Global scour between piles is obviously seen in Figure 3c, Figure 4c, and Figure 5a,b. The scour depths of global scour decrease gradually from the upstream to downstream piles. Moreover, the closer the pile is to the inner side, the smaller the global scour depth is. The minimum global scour depth is 0.25D according to the color bar and topography of Test B5 in Figure 4c. The scour depths of P15, P25, and P35 at the front piles of the first column, as indicated by the blue square data in Figure 7a at the profile of $y/D = -10$, appear to be almost identical. They are approximately 2.4D in depth. This is also depicted in Figure 4c. These illustrate that the strength of the horseshoe vortexes in front of these side piles was affected little by the piles' blockage. Due to the blockage and weakening of the front two rows of piles, scour depth at the front piles decreases gradually in the last rows of the downstream piles. The maximum scour depths of P45 and P55 are 2.05D and 1.95D, respectively. For the second (profile of $y/D = -5$) and third (profile of $y/D = 0$) columns, the maximum scour depths at the parallel front piles vary little. Furthermore, except for P34 and P33, scour depths at the rear piles of these two rows of piles are also very close to each other. In addition, the crests of scour holes as well as the slopes between piles present nearly the same values. Scour depths at the pile sides, which are thought to be induced by the contracted streamlines, show a slight gap. This gap is particularly noticeable at the crests of the scour holes between the second and third columns. As illustrated in Figure 7b, this discrepancy becomes more evident when the row number exceeds 3. For instance, the pink triangle data within $-6 \leq x/D \leq 6$ show that scour depths at P24 and P23 are 2.17D and 2.0D–2.03D, respectively. The crests of scour holes between P44 and P43 are 0.88D and 0.76D, respectively. Similar local scour characteristics around piles in the second and third columns can also be seen in Figure 6a (Test B4). Within a flow intensity of $U/U_c = 0.80$–1.0, it probably can be inferred that for pile groups with more than four columns, the local scour around the paralleled piles, excluding those in the side columns, likely exhibits similar characteristics. This includes consistent scour depths at both the front and rear piles as well as similar sizes of the scour holes.

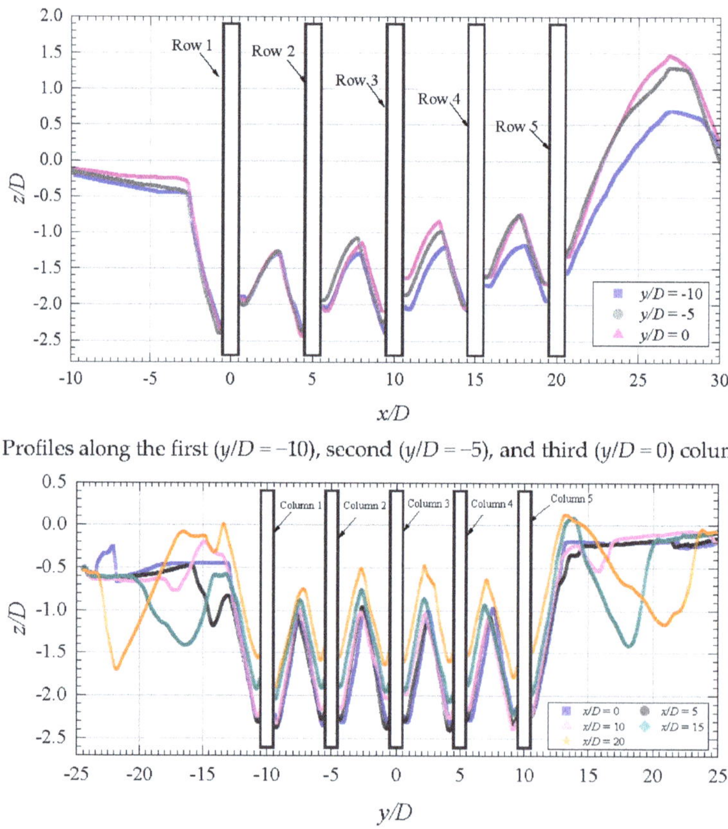

(**a**) Profiles along the first ($y/D = -10$), second ($y/D = -5$), and third ($y/D = 0$) columns

(**b**) Profiles along the first ($x/D = 0$) to fifth ($x/D = 15$) rows

Figure 7. Bed elevation profiles of Test B5.

To make a comprehensive comparison of the local scour around a single pile versus pile groups, contours and temporal scour depths from Test A are employed. Additionally, the bed elevation profile at y/D = 0 from Test A is presented. It should be noted that except for the number of piles, all the conditions in Test A were the same as in Test B5. Its topography and bed elevations profile are presented in Figure 8a,b, respectively. As seen in Figure 8b, scour depth at the rear pile is slightly smaller than at the front pile. The local scour hole around the pile and sand dunes behind the pile present very typical and representative characteristics of a single pile. The distance between the rear pile and the sand dunes is about 4.0*D*. Thus, sediment particles of these sand dunes will be accumulated upstream of the piles when the piles are in an array of groups. These situations are considered to occur in Test B5 because the pile-to-pile space is 5*D*. Consequently, these accumulated sediment particles in front of the downstream piles had to be transported by the turbulent flow. In contrast, scour depths either at the front pile or at the rear pile (Figure 8b) are smaller than those at piles in the front four rows in Test B5 (Figure 7a). The scour depths at front piles of the first and second rows in Test B5 are about 0.4*D* larger than those in Test A. From these comparisons, it can be concluded that local scour holes and sand dunes around piles in an array of arranged piles present different characteristics than those of a local scour test around a single pile. The comparisons of temporal scour depth differences will be discussed in the following text.

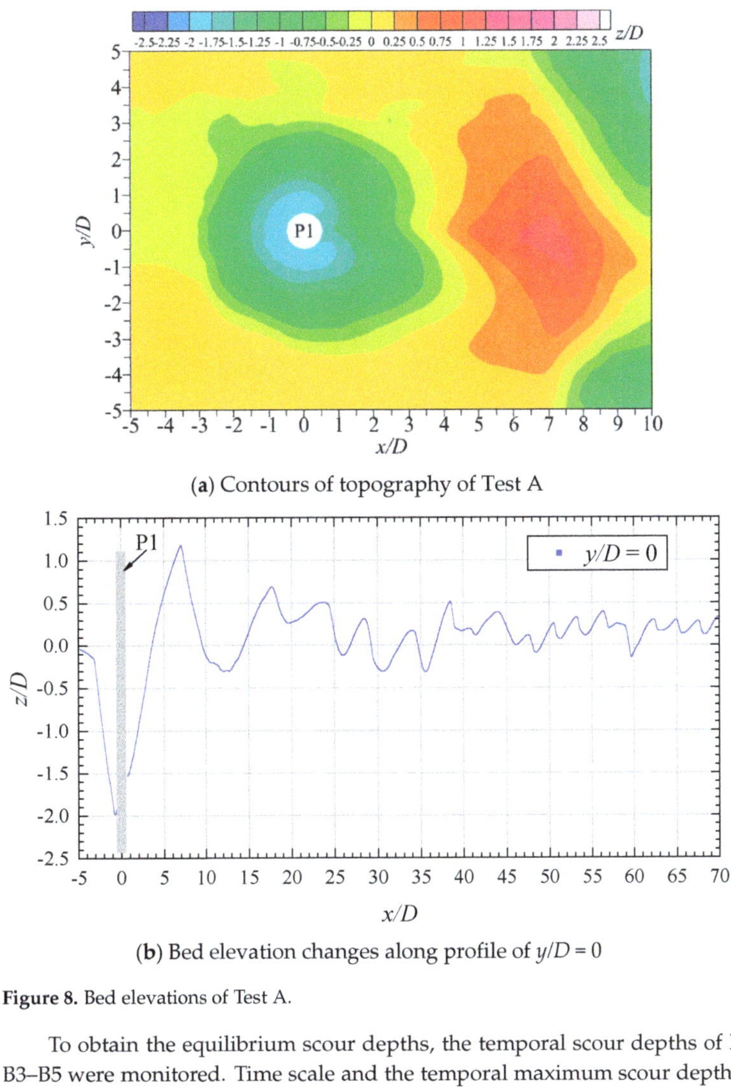

(**a**) Contours of topography of Test A

(**b**) Bed elevation changes along profile of $y/D = 0$

Figure 8. Bed elevations of Test A.

To obtain the equilibrium scour depths, the temporal scour depths of P53 in Tests B3–B5 were monitored. Time scale and the temporal maximum scour depth are written as d_s/D and $t^* = \frac{g^{1/2}}{D^{1/2}}t$, respectively, according the dimensionless analysis from Du [21]. As depicted in Figure 9, the square blue, circular pink, and triangular black data represent the temporal scour depths from Tests B3, B4, and B5, respectively. These measurements were taken at the front piles of P53, which is the pile located in the last row and third column. It is known that the maximum scour depth evolutions of a single pile can be divided into the initial, the developing, and the equilibrium scouring stages [12]. In the initial scouring stage, scour depth increases extremely rapidly. In the developing scouring stage, scour depth increases gradually with a decreased growth rate. In the equilibrium scouring stage, scour depth remains at a constant value when the flow intensity is not larger than 1.0. For Test B3 ($U/U_c = 0.60$), the relative temporal scour depth of d_s/D increased very slowly. The initial and developing scouring stages were difficult to distinguish due to the significantly smaller growth rate in the initial stage compared with the developing stage. This contrasts with the typical three scouring stages observed in local scour processes around a single pile. The evolution of temporal scour depths in this low flow intensity are very different from those results reported in the literature, for example, Sheppard [14] and Yao [20]. However, the equilibrium scouring stage can be seen when the time scale of $t^{*2} = gt/D$ surpassed 1.8×10^9. The equilibrium scour depth tended to approach a constant value of $0.47D$. When the flow intensity was 0.80, the initial and equilibrium scouring stages could be clearly seen, whereas a slight drop down of d_s/D was found within a very short time after the initial scouring stage, as shown in the circular pink data in Figure 9. When the flow intensity was 1.0, the temporal scour depth of P53 in Test B5 was the same as the typical three scouring stages of a single pile. Thus, the double exponential equation, which was proposed by Sheppard [40] and modified by Du [21], was employed to fit with the temporal scour depths of P53 in Test B5. The data were very well fitted with its $R^2 = 0.98$. The equilibrium scour depth obtained from the fitted curve was $1.11D$, while the temporal scour depth at $t^{*2} = 2.5 \times 10^{10}$ was $1.12D$. The local scour processes around the downstream piles generally take more time because the scour holes around these piles are supplied with the accumulated sediment particles from upstream scour holes. Thus, the scouring durations needed to attain the equilibrium scour depths around the piles at the last rows should be the longest. The temporal scour depths of P1 in Test A can illustrate this point well. As seen in Figure 9, the solid triangle data present the typical three scouring stages. The data are very well fitted, with $R^2 = 0.99$. The equilibrium scour depth is attained within $t^{*2} = 2.5 \times 10^{10}$, which is much shorter than that of the dotted curve of particles 3 in Test B5. Consequently, the final maximum scour depth in each local scour test can be considered the equilibrium scour depth.

As discussed above, the maximum scour depths in all the tests are considered the equilibrium scour depths. To study the effects of upstream piles and lateral piles on local scour around the downstream piles and adjacent piles, respectively, the relative equilibrium scour depths of d_{se}/D and the row number of R_N are drawn in Figure 10a–c.

The maximum and minimum scour depths are approximately $0.85D$ and $0.30D$ in the case of Test B3 ($U/U_c = 0.60$), respectively. The equilibrium scour depth at the first and fifth column piles is slightly greater than that at the second and fourth columns, with the third column in the middle exhibiting the smallest depth. Although the equilibrium scour depth generally decreases with increasing U/U_c, the complex interplay of flow fields between the piles results in asymmetric scour phenomena for symmetrically positioned piles in the same row. For example, as shown in Figure 10a–c, except for P11, P15, P51, and P55, where the scour depths are relatively close to each other, there are significant differences in scour depths between other piles, such as P21 and P25. The maximum difference is between P41 and P45 and is $0.35D$ ($0.78D$–$0.43D$), representing an 81.4% discrepancy. A similar situation is observed in the second and fourth column piles, particularly in the second and third rows with piles P21, P25, P31, and P35, as shown in Figure 10a–c. The equilibrium scour depth of the third column piles exhibits more irregular variations, as indicated by the blue data in Figure 10a–c. The maximum and minimum scour depths occur at the first row pile P13 and the second row pile P23, with scour equilibrium depths of 0.68D and 0.31D, respectively. Polynomial and linear fittings were performed on the scour equilibrium depths of the third column piles P13, P23, P33, P43, and P53 at $U/U_c = 0.60$, as shown by the solid and dashed blue lines in Figure 10c. It can be observed that the scour equilibrium depth of the third column piles fluctuates around an average value of 0.5D in Test B3.

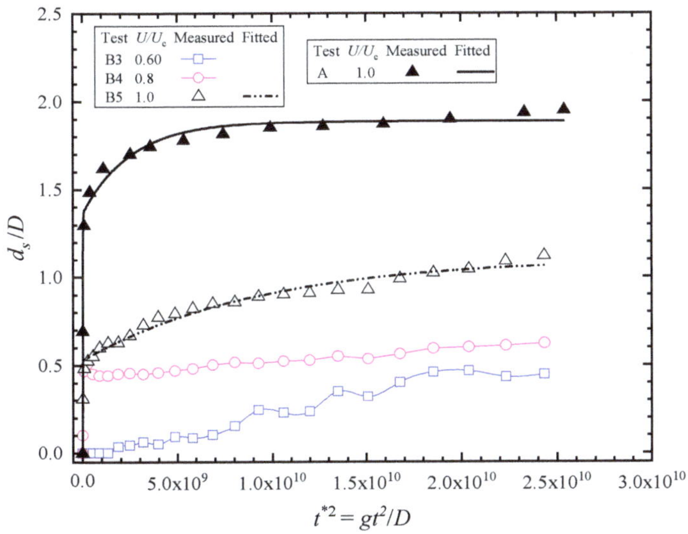

Figure 9. Temporal scour depths of P1 in Test A and P53 in Tests B3–B5.

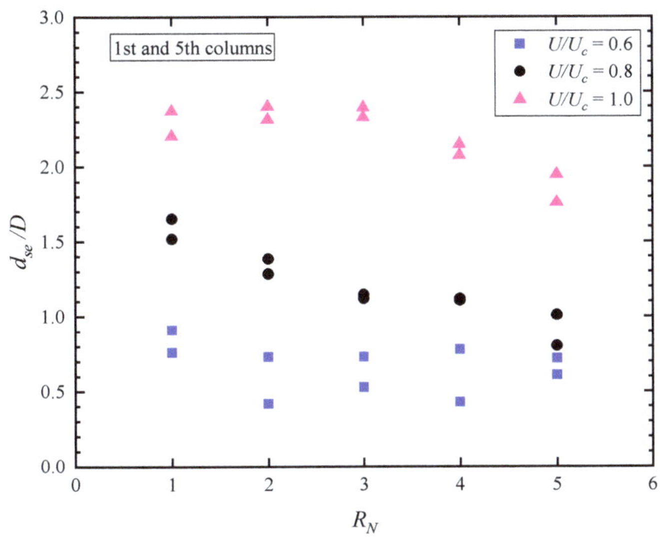

(**a**) The maximum scour depths of piles in the first and fifth columns

Figure 10. *Cont.*

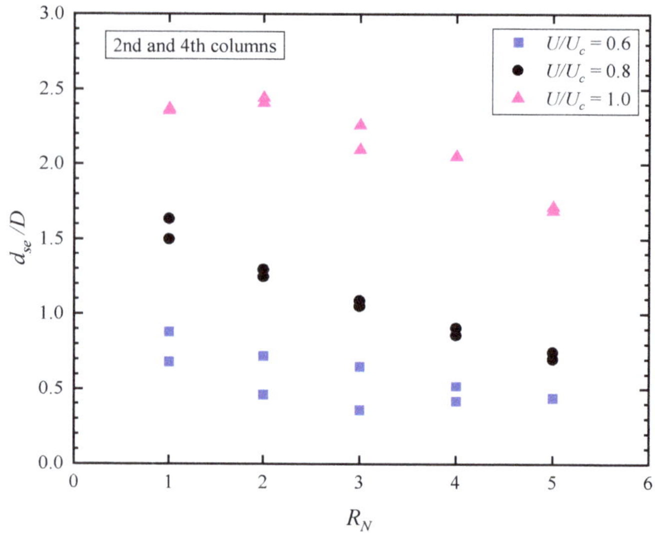

(**b**) The maximum scour depths of piles in the second and fourth columns

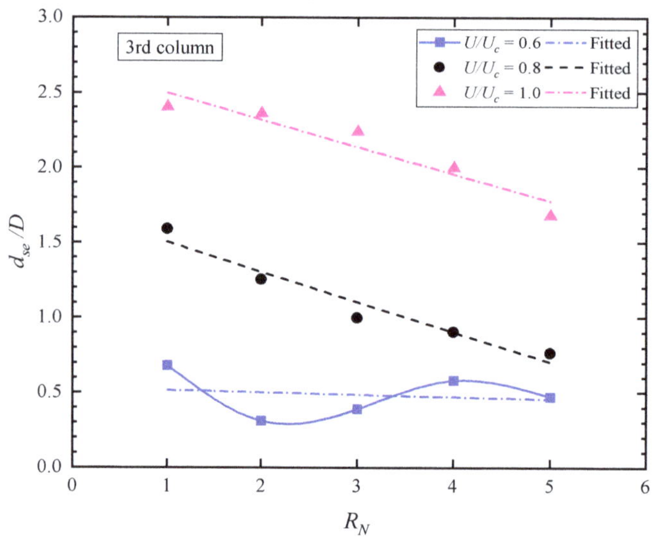

(**c**) The maximum scour depths of piles in the third column

Figure 10. Equilibrium scour depths of Tests B3–B5.

When $U/U_c = 0.80$, as shown by the black data points in Figure 10a–c, the equilibrium scour depths of the piles from the first to the fifth column decrease approximately linearly with the row number R_N. The equilibrium scour depths of the piles exhibit good symmetry, with the maximum difference being 0.198*D* (1.01*D*–0.803*D*), which is 24.7% and is located between the first column pile P15 and the fifth column pile P55. The scour depths of the other piles are very close to each other. Compared with the condition at $U/U_c = 0.60$, the scour depths of the piles are significantly increased. Taking the second row, third column pile P23 as an example, scour depth at $U/U_c = 0.80$ is about three times larger than the scour depth at $U/U_c = 0.60$.

As the flow intensity increases, at $U/U_c = 1.0$ (Test B5), the equilibrium scour depths of the piles significantly increase, as shown by the red data in Figure 10a–c. The relationship between the first and fifth column piles of d_s/D and R_N exhibits a nonlinear decreasing trend. When $R_N \leq 3$, the value of d_s/D basically remains constant at 2.35. After that, it decreases to 1.86 at $R_N = 4$ and $R_N = 5$. Observing from the first to the fifth column piles, it can be seen that the equilibrium scour depths of the second row piles differ more significantly from those of the first row piles. The maximum and minimum scour depths appear in the first and fifth rows, respectively, with values of 2.5D and 1.7D.

Knowing that the Re in Tests B3–B5 are 3340, 4460, and 5580, respectively, the flow patterns around these piles might be drawn as [1] suggested. As shown in Figure 11, Re= 2500 and Re = 3900–5000 individually for Figure 11a,b [1], the dotted lines in Figure 11 were found to coexist with the solid lines. However, vortexes represented by the solid lines dominated the flow. From this view, the differences in local scour holes around the piles in the first row in Figure 4 and in scour depths at P11 and P15 (Figure 10a–c) may be caused by the asymmetric vortexes shown in Figure 11a,b. These influences were found in most of the piles in Test B3 (the square blue data in Figure 11a,b), in which the maximum scour depths were at the piles' sides, as shown in Figure 4a. For the maximum scour depths at the side positions, specifically the first and fifth columns, the square blue data in Figure 10a show a decreasing trend. They range from 0.76D to 0.91D and eventually converge to an average value of 0.62D for the two paralleled piles in each row. However, adjacent piles within the pile groups presented different characteristics, as shown by the square blue data presented in Figure 10b. In the second and fourth columns, d_{se}/D decreased gradually from 0.68D–0.88D in the first row to 0.42D–0.72D in the fifth row. For the piles in the central line of the third column, as presented in Figure 10c, the maximum scour depths changed down and up dynamically around a nearly constant value of 0.5D. In contrast to the other columns, each pile in the third column was surrounded by two paralleled piles in each row. The maximum scour depth of P13, which was positioned at the first row and third column, was 0.68D.

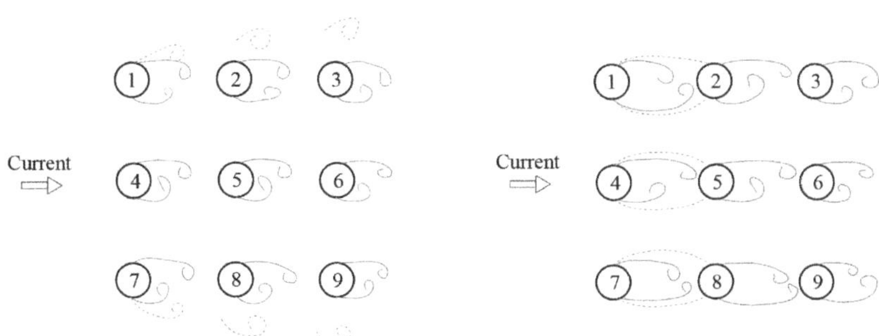

(**a**) S/D=3, Re=2500 (**b**) S/D=3, Re=3900 – 5000

Figure 11. Diagrams of flow patterns around the piles [1].

When the flow intensity is large enough to generate strong horseshoe vortexes, the influences of asymmetric flow patterns around piles in the first row weakens significantly. These can be indicated from the results of the larger flow intensities of 0.80 and 1.0. The black circular and pink triangular data in Figure 10a,b present that the maximum scour depths of piles in symmetric positions are very close to each other. Except for the maximum scour depths in the first and fifth columns in Test B5 (data of $U/U_c = 1.0$ in Figure 10a), they tend to decrease linearly in the other columns in Tests B4 and B5, as shown in Figure 10a–c. Taking the third columns of Tests B3–B5 as examples, the R^2 of the three linearly fitted curves are 0.03, 0.94, and 0.91, respectively. Moreover, the maximum scour depths for the piles in parallel arrangement within each row of the second, third, and fourth columns exhibit minimal variation. This suggests that adjacent piles in the same row have a negligible impact on the maximum scour depth of an individual pile, despite slight differences in the scour holes and the accumulation of sediment around them.

4. Conclusions

This paper experimentally investigated local scour around an array of 5 × 5 pile groups in clear-water scour conditions. To distinguish local scour around pile groups in an overall scour hole, the pile-to-pile space was arranged as five times the diameter of a single pile. Flow intensities ranging from 0.37 to 1.0 were tested. The main conclusions, which will give some new perspectives to the future studies in local scour around pile groups, are as follows:

(1) The minimum threshold flow intensity below which local scour around the pile groups would not occur was about 0.40, while the critical flow velocity in the approaching flow sand bed was 2.5 times this value. When the flow intensity increased slightly to 0.42, very small scour holes with scour depths ranging from 0.05 to 0.1 times a single pile diameter were found. The scour holes induced by horseshoe vortexes were observed exclusively in front of the piles in the fifth row, whereas scour holes caused by wake vortexes were detected behind all the piles.

(2) The temporal evolution of scour depths for downstream piles in an array differed significantly from those of a single pile, depending on the flow intensity values. The temporal scour depths in flow intensities of 0.60 presented very slow development in the initial and developing scouring stages, while they increased rapidly in the initial scouring stage for flow intensities of 0.80 and 1.0. The maximum scour depth of each pile can be considered as the equilibrium scour depth within a nondimensional square time scale of 2.5×10^{10}.

(3) The differences in scour holes and depths around piles in the flow intensity case of 0.60 were likely due to asymmetric vortexes. Adjacent piles showed varying trends, with depths in the inner arranged columns gradually decreasing. Downstream piles flanked by two adjacent piles exhibited scour depths that fluctuated around a relatively stable value. Within flow intensity of 0.80–1.0, excluding the piles in the side columns, the local scour surrounding the parallel piles exhibited similar characteristics, such as scour depths at both the front and rear piles, and sizes of scour holes, when the array of pile groups surpassed four columns.

Author Contributions: Conceptualization, M.G., X.P. and S.D.; Methodology, M.G., X.P., S.D., G.W., Y.L., Y.S., G.D. and B.L.; Software, M.G. and X.P.; Validation, X.P., S.D. and G.W.; Formal analysis, M.G. and X.P.; Investigation, X.P., S.D., Y.L., Y.S. and G.D.; Resources, X.P., G.W. and Y.L.; Data curation, M.G. and S.D.; Writing—original draft, M.G. and S.D.; Writing—review & editing, M.G., X.P., S.D., G.W., Y.L., Y.S. and B.L.; Supervision, X.P., G.D. and B.L.; Funding acquisition, S.D. All authors have read and agreed to the published version of the manuscript.

Funding: This research was funded by National Nature Science Fund (Grant No. 52301324), the Natural Science Foundation of Zhejiang Province (LQ24E090001), and the Shandong Provincial Key Laboratory of Ocean Engineering (Grant No. kloe202009).

Institutional Review Board Statement: Not applicable.

Informed Consent Statement: Not applicable.

Data Availability Statement: Data is contained within the article.

Conflicts of Interest: Author Yingxue Lv, Yunjia Sun and Guangjia Ding was employed by the company CCCC, Tianjin Port Engineering Institute Co., Ltd. of CCCC First Harbor Engineering Co. Ltd. The remaining authors declare that the research was conducted in the absence of any commercial or financial relationships that could be construed as a potential con-flict of interest.

References

1. Ma, L.L.; Wang, L.Z.; Guo, Z.; Jiang, H.Y.; Gao, Y.Y. Time development of scour around pile groups in tidal currents. *Ocean Eng.* **2018**, *163*, 400–418. [CrossRef]
2. Sarmiento, J.; Guanche, R.; Losada, I.J.; Rosendo, E.M.; Guindo, A.; Guevara, J.L. Scour processes around pile clusters of jacket foundations under steady currents. *Ocean Eng.* **2024**, *313*, 119502. [CrossRef]
3. Wu, J.; Zhang, Y.; Lin, L.; Chen, C.; Najafzadeh, M. Experiment investigation on the local scour around circular and square piles in open wharf. *Mar. Georesour. Geotech.* **2024**, *2024*, 1–12. [CrossRef]
4. Zhu, F.; Jiang, Y.; Li, W. Analysis and Calculation of Ocean Current Load and Wave Load on a Floating Crane under a Normal Working State. *J. Coast. Res.* **2015**, *31*, 48–53. [CrossRef]
5. Baghbadorani, D.A.; Ataie-Ashtiani, B.; Beheshti, A.; Hadjzaman, M.; Jamali, M. Prediction of current-induced local scour around complex piers: Review, revisit, and integration. *Coast Eng.* **2018**, *133*, 43–58. [CrossRef]
6. Gao, J.; Shi, H.; Zang, J.; Liu, Y. Mechanism analysis on the mitigation of harbor resonance by periodic undulating topography. *Ocean Eng.* **2023**, *281*, 114923. [CrossRef]
7. Gao, J.; Hou, L.; Liu, Y.; Shi, H. Influences of Bragg reflection on harbor resonance triggered by irregular wave groups. *Ocean Eng.* **2024**, *305*, 117941. [CrossRef]
8. Roulund, A.; Sumer, B.M.; Fredsøe, J.; Michelsen, J. Numerical and experimental investigation of flow and scour around a circular pile. *J. Fluid Mech.* **2005**, *534*, 351–401. [CrossRef]
9. Zhang, Z.; Du, S.; Guo, Y.; Yang, Y.; Zeng, J.; Sui, T.; Wei, R.; Li, Z. Field study of local scour around bridge foundations on silty seabed under irregular tidal flow. *Coast. Eng.* **2023**, *185*, 104382. [CrossRef]
10. Liang, F.; Wang, C.; Huang, M.; Wang, Y. Experimental observations and evaluations of formulae for local scour at pile groups in steady currents. *Mar. Georesour. Geotechnol.* **2017**, *35*, 245–255. [CrossRef]
11. Baker, C.J. The laminar horseshoe vortex. *J. Fluid Mech.* **1979**, *95*, 344–367. [CrossRef]
12. Raudkivi, A.J.; Ettema, R. Clear-water scour at cylindrical piers. *J. Hydraul. Eng.* **1983**, *131*, 338–350. [CrossRef]
13. Melville, B.W. Pier and abutment scour: Integrated approach. *J. Hydraul. Eng.* **1997**, *123*, 125–136. [CrossRef]
14. Sheppard, D.M.; Melville, B.; Demir, H. Evaluation of existing equations for local scour at bridge piers. *J. Hydraul. Eng.* **2014**, *140*, 14–23. [CrossRef]
15. Oh, S.J. Experimental Study of Bridge Scour in Cohesive Soil. Ph.D. Thesis, Texas A&M University, College Station, TX, USA, 2009.
16. Zhao, M.; Zhu, X.; Cheng, L.; Teng, B. Experimental study of local scour around subsea caissons in steady currents. *Coast. Eng.* **2012**, *60*, 30–40. [CrossRef]
17. Qi, W.G.; Gao, F.P. Equilibrium Scour Depth at Offshore Monopile Foundation in Combined Waves and Current. *Sci. China Technol. Sci.* **2014**, *57*, 1030–1039. [CrossRef]
18. Dey, S. Three-dimensional vortex flow field around a circular cylinder in a quasi-equilibrium scour hole. *Sādhanā* **1995**, *12*, 771–785. [CrossRef]
19. Guan, D.; Chiew, Y.M.; Wei, M.; Hsieh, S.C. Characterization of horseshoe vortex in a developing scour hole at a cylindrical bridge pier. *Int. J. Sediment Res.* **2019**, *34*, 118–124. [CrossRef]
20. Yao, W.D.; An, H.W.; Draper, S.; Cheng, L.; Harris, J.M. Experimental investigation of local scour around submerged piles in steady current. *Coast. Eng.* **2018**, *142*, 27–41. [CrossRef]
21. Du, S.; Wang, Z.; Wang, R.; Liang, B.; Pan, X. Effects of flow intensity on local scour around a submerged square pile in a steady current. *Phys. Fluids* **2022**, *34*, 085126. [CrossRef]
22. Liang, B.; Du, S.; Pan, X.; Zhang, L. Local Scour for Vertical Piles in Steady Currents: Review of Mechanisms, Influencing Factors and Empirical Equations. *J. Mar. Sci. Eng.* **2020**, *8*, 4. [CrossRef]
23. Oliveto, G.; Hager, W.H. Temporal evolution of clear-water pier and abutment scour. *J. Hydraul. Eng.* **2002**, *128*, 811–820. [CrossRef]
24. Dey, S.; Raikar, R.V.; Roy, A. Scour at submerged cylindrical obstacles under steady flow. *J. Hydrul. Eng.* **2008**, *134*, 105–109. [CrossRef]
25. Kim, H.S.; Nabi, M.; Kimura, I.; Shimizu, Y. Computational modeling of flow and morphodynamics through rigid-emergent vegetation. *Adv. Water Resour.* **2015**, *84*, 64–86. [CrossRef]
26. Sumner, D. Two circular cylinders in cross-flow: A review. *J. Fluid Struct.* **2010**, *26*, 849–899. [CrossRef]
27. Tsutsui, T. Experimental study on the instantaneous fluid force acting on two circular cylinders closely arranged in tandem. *J. Wind Eng. Ind. Aerod.* **2012**, *109*, 46–54. [CrossRef]

28. Yao, W.D.; Draper, S.; An, H.W.; Cheng, L.; Harris, J.M.; Whitehouse, R.J.S. Experimental study of local scour around submerged compound piles in steady current. *Coast. Eng.* **2021**, *165*, 103831. [CrossRef]
29. Yang, Y.; Qi, M.; Wang, X.; Li, J. Experimental study of scour around pile groups in steady flows. *Ocean Eng.* **2020**, *195*, 106651. [CrossRef]
30. Yang, Y.; Melville, B.W.; Macky, G.H.; Shamseldin, A.Y. Experimental study on local scour at complex bridge pier under combined waves and current. *Coast. Eng.* **2020**, *160*, 103730. [CrossRef]
31. Amini, A.; Melville, B.W.; Ali, T.M.; Ghazali, A.H. Clear-water local scour around pile groups in shallow-water flow. *J. Hydraul. Eng.* **2012**, *138*, 177–185. [CrossRef]
32. Zhang, Q.; Zhou, X.L.; Wang, J.H. Numerical investigation of local scour around three adjacent piles with different arrangements under current. *Ocean Eng.* **2017**, *142*, 625–638. [CrossRef]
33. Wang, H.; Tang, H.; Xiao, J.; Wang, Y.; Jiang, S. Clear-water local scouring around three piers in a tandem arrangement. *Sci. China* **2016**, *59*, 888–896. [CrossRef]
34. Lam, K.; Gong, W.Q.; So, R.M.C. Numerical simulation of cross-flow around four cylinders in an in-line square configuration. *J. Fluid Struct.* **2008**, *24*, 34–57. [CrossRef]
35. Ataie-Ashtiani, B.; Beheshti, A.A. Experimental Investigation of Clear-Water Local Scourat Pile Groups. *J. Hydraul. Eng.* **2006**, *132*, 1100–1104. [CrossRef]
36. Nezu, I. Open channel flow turbulence and its research prospect in the 21stcentury. *J. Hydraul. Eng.* **2005**, *131*, 229–246. [CrossRef]
37. Karimi, N.; Heidarnejad, M.; Masjedi, A. Scour depth at inclined bridge piers along a straight path: A laboratory study. *Eng. Sci. Technol.* **2017**, *20*, 1302–1307. [CrossRef]
38. Liu, M.; Huai, W.; Ji, B. Characteristics of the flow structures through and around a submerged canopy patch. *Phys. Fluids* **2021**, *33*, 035144. [CrossRef]
39. Wang, S.Y.; Qi, W.G.; Gao, F.P.; Li, B.; He, B. Time development of clear-water scour around a pile foundation: Phenomenological theory of turbulence-based approach. *Coast. Eng.* **2024**, *190*, 104511. [CrossRef]
40. Sheppard, D.M.; Odeh, M.; Glasser, T. Large scale clear-water local pier scour experiments. *J. Hydraul. Eng.* **2004**, *130*, 957–963. [CrossRef]

Journal of
Marine Science and Engineering

MDPI

Article

Experimental Study on Local Scour at the Monopile Foundation of an Offshore Wind Turbine under the Combined Action of Wave–Current–Vibration

Li Shi [1], Yongzhou Cheng [1,2,*], Yuwei Zheng [1], Bo Xia [1] and Xiaoyun Huang [1]

[1] School of Hydraulic and Environmental Engineering, Changsha University of Science & Technology, Changsha 410114, China; sl990416@163.com (L.S.); zhengyuwei7@163.com (Y.Z.); boxia@csust.edu.cn (B.X.); xiaoyun.huang@csust.edu.cn (X.H.)

[2] Key Laboratory of Water-Sediment Sciences and Water Disaster Prevention of Hunan Province, Changsha 410114, China

* Correspondence: chengyz@csust.edu.cn; Tel.: +86-13786191344

Abstract: Monopile foundations are the most widely used offshore wind turbine foundations. The experiments were conducted to investigate the influencing factors of local scour around the monopile under the action of wave–current–vibration. The study analyzed the characteristics of local scour, including the maximum scour depth, the development of scour hole shape, and the shape of the scour hole profile. The dimensionless influencing factors (vibration intensity, Froude number, Keulegan–Carpenter number, and combined wave–current parameter) are subsequently analyzed. An empirical formula is developed to predict the local scour depth of a monopile under the combined influence of wave–current–vibration. The formula provides a theoretical underpinning for engineering design.

Keywords: wave–current–vibration interaction; offshore wind turbine; monopile; local scour; empirical formula

1. Introduction

In recent years, the safety and reliability of wind turbine foundations has become a research priority due to the large-scale construction of offshore wind turbines. The environment for offshore wind turbine foundations is significantly harsher than its onshore counterpart. The challenging marine environment poses significant obstacles to the design and construction of offshore wind turbine foundations, and results in severe local scour around these structures, ultimately leading to instability and failure. As a result, local scour remains a long-standing issue in marine science research. The offshore wind turbine foundation bears both the complex coupling effect of wind, wave, and current, as well as the dynamic and static loads generated by the upper rotating blades, transferring them to the seabed soil. This transfer has a significant impact on the soil properties around the foundation [1]. Additionally, strong sediment movement caused by waves and currents is a further factor leading to local scour. According to Yu [2], offshore wind turbines experience between 10^6 and 10^8 vibration loads during their service life. These loads are primarily caused by strong winds, waves, and rotating blades. Long-term vibrations can cause particles to migrate and soil to become denser around the monopile [3]. This can not only lead to changes in mechanical properties, like soil stiffness, that can weaken the pile foundation's bearing capacity but also affect how the bed sediment moves in the eddy current environment and further affects local scouring around the pile foundation [4]. Offshore wind turbine foundations come in various types, including monopile, gravity, mono-caisson, tripod-pile/caisson, jacket-pile/caisson, high-rise pile cap, and others [5]. Monopile foundations are particularly common in offshore wind turbine construction due to their excellent applicability, simple structure, high bearing capacity, and low construction cost [6,7]. Monopile foundations have a good application prospect for offshore wind power, although many scholars have studied the design and protection of offshore wind power foundations [8–11]. However, research on the local scouring characteristics of monopile foundations under the joint action of many factors is still far from enough; especially, the sediment movement rules around the monopile considering the superimposed transverse vibration load is still not clear. And there are fewer studies on the local scour depth of monopile foundations under the joint action of multiple dynamics.

Firstly, offshore wind turbine pile foundations are unavoidably influenced by water currents and waves, inducing alterations in the surrounding conditions of the original wind turbine pile. As a result, the exposure of the offshore generator foundation may imperil the safety of the offshore wind turbine. Currently, numerous scholars have extensively researched local scour of pile foundations caused by waves and currents. Sumer [12] demonstrated that the depth of scouring under wave action changes with the Shields number, and when the Shields number is greater than the critical value causing sediment suspension on the pile side, the scour depth becomes independent of the Shields number. Qi and Gao [13] performed a local scour and pore pressure response test on a large-diameter monopile under the combined action of wave and current, using a wave–current flume. The test findings indicate that the combination of waves and current enhances the formation of local scour around individual piles. The effect on the equilibrium scour depth is significant, and it becomes more apparent under clear water scour. Zhang [14] also carried out experiments under three conditions: current-only, wave-only, and combined wave–current. The study's results indicate that scour depth is greater under combined wave–current conditions than under current-only conditions, and that this increases with flow velocity and decreasing water depth. Gautam [15] discovered that equilibrium scour depth did not notably differ for scour under waves alone, waves with weak currents, and waves with mild currents when developed in a low *KC* number environment. The equilibrium scour depth significantly increases when waves are combined with strong currents. For large *KC* numbers, the equilibrium scour depth is found to have increased with an increase in *KC* number as well as the combined wave–current parameter (U_{cw}).

Secondly, the offshore wind turbine pile foundation is subjected to numerous lateral vibration and other lateral loads from external sources, leading to reduced soil stiffness surrounding the foundation and subsequent local scour of the pile foundation. Therefore, the current research focus should be on the settlement and convection characteristics of sand surrounding the pile foundation under complex loading and traditional scour research's hydrodynamic loads. Shi [16] used numerical methods to study the unsteady flow around a monopile under the influence of waves and vibrations. The calculation results show that monopile vibrations cause a great disturbance to the unsteady flow around the pile. Yu [17] conducted an experimental investigation of a vibrating monopile inside a cohesionless saturated sand bed. The experimental results indicate that the lower the vibration frequency, the more prominent is the delayed attenuation process during the oscillation cycle. Additionally, the pore pressure amplitude decay rate increases with increasing frequency as the depth of the seabed grows, and the pile side pore pressure shows a trend of increasing and then decreasing with frequency.

Under the influence of vibration load, the soil surrounding the monopile undergoes compaction and densification, causing a change in the soil stiffness [18,19]. Currently, few studies have investigated the effects of vibration load on the scour process of monopiles. However, the local scour depth of the pile's foundation is closely associated with the cumulative settlement resulting from vibration load. Therefore, it is crucial to investigate the principles of scour growth around a monopile experiencing vibration loading. Al-Hammadi and Simons [20] analyzed the local scour process of pile foundations under vibration loading. Their research concludes that the scour hole is broader and less deep when compared to the equilibrium scour hole without vibration loading. Initially, the compaction of the sand bed results in a decrease in the scour rate during the early stages of scour, although this does not impact the equilibrium scour depth. The alternate application of water current and vibration load significantly increases the local scour depth and width of the monopile. Guan [21] conducted a study on the effects of lateral vibration loads on the local scour of monopiles. The research found that vibration loads can increase the initial scour velocity, mainly due to soil compaction and pile–soil interactions. Furthermore, raising the frequency and amplitude of vibration can lead to decreased equilibrium scour depth and slope of the scour hole. These effects are attributed to sediment ratchet convection occurring on the surface of the scour hole. Qin [22] investigated how monopile vibration, water flow, and intermittent operation of OWTs (offshore wind turbines) work together to affect the scour around the monopile foundation. The results show that coupling of a large-amplitude monopile vibration and water flow will exacerbate the scour, while the coupling of a small-amplitude monopile vibration and water flow will retard the scour.

The research results above demonstrate that physical model testing on vibration mainly focuses on the settlement and convective motion characteristics of sand around a monopile under pure vibration, and the law of tracking the trajectory of sediment particles is studied through numerical simulation tests. For the vibration-induced scour test, a local scour test of a monopile under the influence of current and vibration was conducted, providing reference significance for studying the local scour of a monopile under the influence of wave–current–vibration. However, very few studies [23] have been conducted on the existing local scour of a monopile under the influence of wave–current–vibration in a marine environment. Therefore, it is of great significance to carry out physical model test research on the local scour of a monopile under the combined action of wave–current–vibration by arranging a vibration loading device in a wave–current flume, in order to explore the local scour characteristics and scour depth development law of monopiles, and to fit the empirical formula for predicting the maximum scour depth of monopiles.

In summary, this study uses a wave flume to test the local scour characteristics of a monopile foundation subjected to combined wave–current–vibration. Additionally, a generalized test is conducted to investigate the coupling mechanism between sand movement deformation and wave–current–structure–seabed action under vibration loading. Based on the test data, an empirical formula has been developed to predict the maximum scour depth. This formula offers a theoretical foundation for the design and safeguarding of monopile structures in actual complex environments.

2. Experimental Research Program

2.1. Experimental Layout

This experiment was carried out in the wave–current flume of the Hydraulic Experiment Center of Changsha University of Science and Technology. The test arrangement is shown in Figure 1. The length $\times$ width $\times$ height of the wave–current flume is 60 m $\times$ 1.5 m $\times$ 1.8 m. The flume is emptied by arranging the slope section and the flat slope section. The length $\times$ width $\times$ height of the flat slope section is 24 m $\times$ 1.5 m $\times$ 0.6 m, the length $\times$ width $\times$ height of the slope section is 6 m $\times$ 1.5 m $\times$ 0.6 m, the slope foot is 5.71°, and the sand in the sand trough experimental area with length $\times$ width $\times$ height of 4.5 m $\times$ 1.5 m $\times$ 0.6 m is reserved between the flat slope sections. Non-cohesive sand with a median particle size of d_{50} = 0.334 mm, a relative density ρ = 2.65 $\times$ 10^3 kg/m^3, and a unit weight of γ = 14.58 KN/m^3 is laid in the sand trough test area. The model circular piles with a diameter of D = 0.09 m are buried between the sand troughs. The model piles are connected with the exciter to carry out cyclic vibration loading. One end of the flume is equipped with a push plate wave-making system and a cyclic current-making system. At the other end, a slope-type energy dissipation network is added to reduce the influence of wave reflection on the test, so as to ensure that the wave-making and current-making system can work for a long time and maintain the required wave and current types. The ratio of the distance between the side walls of the flume to the diameter of the model pile is 70.55/8.9 $\approx$ 7.927 > 3, which is in line with the wave model test specification. Referring to the suggestion of Whitehouse [24], the ratio of the width of the flume to the diameter of the model pile should be greater than 6 in the clear water scour, and the ratio of this test is $1.5/0.09 \approx 16.85 > 6$. Therefore, the side wall effect can be ignored in this test. Several pre-experiments were carried out before the formal test to observe the test results to ensure the stable operation of the wave-making and current-making system. The working water depth of this test is 0.5 m, which is greater than 4 times the pile diameter D, and the influence of water depth on local scour can be ignored [25]. The experimental wave height data were collected by a WG-50 wave height meter (RBR Ltd., Ottawa, ON, Canada). A three-dimensional Doppler profile current meter (Nortek AS, Oslo, Norway) was used to collect velocity data. A ULS-100 terrain scanner (2G Robotics Inc., Waterloo, ON, Canada) is used to collect the local scour terrain data, and the three-dimensional terrain scanner can be translated and fixed through the multi-functional measuring frame to accurately collect the terrain data. The vibration load of the monopile is loaded by an HEV-500 high energy vibrator (Nanjing Foneng Technology Industry Co. Ltd., Nanjing, China), and a D050 strain displacement sensor from Yangzhou Jingming (Yangzhou Jingming Testing Technology Co. Ltd., Yangzhou, China) is installed to collect and calibrate the frequency and amplitude of pile movement. A scale is attached around the pile body to record the development of erosion depth in real time.

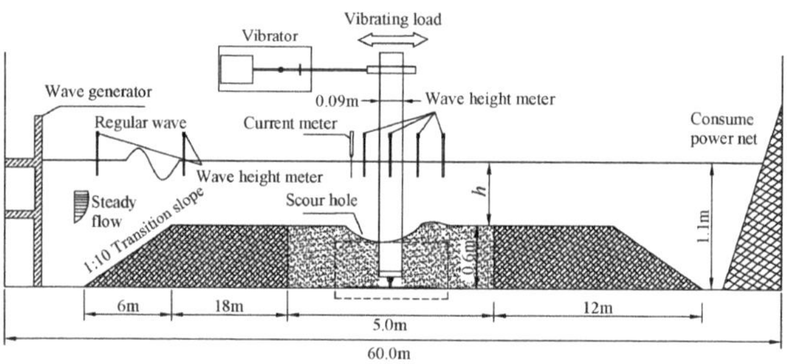

Figure 1. Experimental layout diagram.

In this paper, the wave-making system is calibrated by the wave height meter before the start of the test to ensure that the waves produced by the computer during the test are consistent with the requirements, and it is convenient to collect the wave height data during the test. During the test, a total of six wave height meters were arranged to collect the incident wave height and the wave height change around the pile.

2.2. Key Parameters of the Experimental

The main parameters such as vibration frequency f_v, amplitude A_v, wave height H, wave period T, and flow velocity U_c required for the test are mainly considered and selected from the following aspects: (1) All the above parameters are within the normal working range of the test equipment. (2) Because an offshore wind turbine is subjected to complex environmental loads, these environmental loads have the characteristics of low frequency. When the inclination angle of the offshore wind turbine pile exceeds 0.5°, the offshore wind turbine will not be able to work normally. The selected test vibration frequency and amplitude should be low frequency and low amplitude. Based on the above principles, the vibration frequency f_v = 3,6,9 Hz and the amplitude A_v = 2, 4, 6 mm are determined. The amplitude reference position is the horizontal displacement distance at the loading point of the pile top exciter. (3) Before the test, the important parameters such as the Froude number Fr, KC number, U_{cw} number, and Shields number θ of each wave–flow condition are calculated, and it is confirmed that they are within a reasonable range. In order to make the difference between each working condition obvious and facilitate data processing and analysis, the wave height H = 5.4, 8.7, 11.2, 8.3, 9.1 cm, the wave period T = 1, 1.3, 1.6 s, in order to prevent breaking during wave propagation; all wave parameters meet the following requirements of Equation (1). Moreover, the flow velocity of the reference point at 1D from the bed surface under the flow condition is U_c = 0.213 m/s. (4) Since the construction environment of offshore wind turbines is an offshore area, the water depth is much larger than the pile diameter of offshore wind turbines. Therefore, the water depth of the test is selected as a fixed value h = 50 cm, which is similar to the actual environment of offshore wind turbines.

$$H/L \leq 0.142\tanh(kh), \tag{1}$$

In the formula: H is the wave height; h is the water depth; L is the wavelength; k is the wave number.

2.3. Experimental Scheme

The test is an experimental study on the local scour characteristics of a monopile under the combined action of wave–flow–vibration. The local scour of a monopile under the action of wave–flow is a comparative test of local scour of a monopile under the combined action of wave–flow–vibration. The test has a total of five groups, and the test duration is 3.5 h. The local scour test of a monopile under the combined action of wave–current–vibration is the main research objective of this paper, which is divided into the following two parts: one group of local scour tests of the monopile under the action of current–vibration, 25 groups of local scour tests of the monopile under the action of wave–current–vibration. Before the start of the test, the flow velocity, wave height, wave period, vibration frequency and amplitude were calibrated, and the calibration files of each working condition were saved to control the exciter, wave-making, and flow-making system.

During the test, the development of scour depth was recorded by reading the scale around the pile and the critical scale of the sand bed. The scale was read every 1 min at 10 min before the start, every 5 min at 10 min~30 min, every 10 min at 30 min~120 min, every 20 min at 120 min~210 min, and every 30 min after 210 min. After the completion of the test and at the characteristic time points, shading cloth is used to cover the treatment, and then the underwater terrain scanner is used to scan and collect the sand bed terrain. Each scanning position is kept in the same position to ensure that it will not affect the scanning results. Finally, the required terrain is obtained by processing.

2.3.1. Wave–Current Test Conditions

Table 1 is the wave parameter table, and Table 2 is the working conditions table of the local scour test of a monopile under the combined action of wave and current. The wave–current tests were carried out with five wave heights, three wave periods, one flow rate, a total of six working conditions, including one group of pure flow conditions and five groups of wave–current conditions. The parameters in the table are obtained as follows:

Table 1. Table of wave parameters.

Working Condition	Wave Height H/cm	Wave Period T/s	KC	U_{wm}	θ	θ/θ_c
1	5.4	1.3	1.07	0.074	0.012	0.321
2	8.7	1.3	1.78	0.123	0.024	0.639
3	11.2	1.3	2.23	0.155	0.035	0.930
4	8.3	1.6	2.45	0.138	0.027	0.724
5	9.1	1	0.96	0.087	0.014	0.382

Table 2. Wave–current test conditions table.

Working Condition	Incidence Velocity U_c/m·s^{-1}	Wave Height H/cm	Wave Period T/s	KC	U_{cw}	Fr	S/cm	S/D	θ	θ/θ_c
1	0.213	0	0	0	0	0	7.5	0.83	0.014	0.38
2	0.213	5.4	1.3	1.07	0.74	0.28	7.7	0.86	0.026	0.70
3	0.213	8.7	1.3	1.78	0.63	0.31	7.3	0.81	0.038	1.02

Table 2. *Cont.*

Working Condition	Incidence Velocity U_c/m·s^{-1}	Wave Height *H*/cm	Wave Period *T*/s	*KC*	U_{cw}	*Fr*	*S*/cm	*S*/*D*	θ	θ/θ_c
4	0.213	11.2	1.3	2.23	0.58	0.33	7.2	0.8	0.049	1.31
5	0.213	8.3	1.6	2.45	0.61	0.32	7.5	0.83	0.041	1.1
6	0.213	9.1	1	0.96	0.71	0.29	7.7	0.86	0.028	0.96

U_c is the reference point velocity at 9 cm from the sand bed surface ($1D$ position) under the action of water flow, U_{wm} is the maximum velocity of the reference point at 9 cm from the sand bed surface ($1D$ position) under the action of wave, S is the maximum measured scour depth on the side of the monopile, and S/D is the relative scour depth. In the table, KC, U_{cw}, and Fr are dimensionless parameters. Among them, the Froude number Fr is one of the important parameters affecting the local scour of the monopile foundation. The KC number is one of the important parameters describing the local scour process of pile foundation under wave action, which shows the relative ratio relationship between viscous force and inertial force. U_{cw} is a dimensionless number reflecting the relative strength of water flow velocity and wave flow velocity under the combined action of wave and current. The calculation method of each parameter is as follows: $KC = U_{wm}T/D$, $U_{cw} = U_c/(U_c + U_{wm})$, $Fr = U_a/\sqrt{gD}$. U_a is the 1/4 period water particle velocity at the reference point under the combined action of wave and current, and its calculation formula is:

$$U_a = \frac{1}{T/4}U_a = \frac{4}{T}\int_0^{\frac{T}{4}} \left(U_c + U_{wm}\sin(\frac{2\pi t}{T})\right)d_t = U_c + \frac{2}{\pi}U_{wm}\,, \tag{2}$$

θ is the Shields number, which reflects the ratio of the force of water flow to the resistance of bed sand to movement. θ_{cr} is the critical Shields number, which is the starting parameter of sediment. It can distinguish between clear water erosion and mobile bed erosion, and its calculation formula is:

$$\theta = \frac{U_f^2}{g(\rho_s/\rho_w - 1)d_{50}}\,, \tag{3}$$

$$\theta_{cr} = \frac{0.3}{1 + 1.2D_*} + 0.055[1 - \exp(-0.02D_*)], \tag{4}$$

In the calculation formula, U_f is the maximum frictional velocity at the reference point, and D_* is defined as the dimensionless sediment particle size, which is defined as $D_* = d_{50}\left[(s-1)g/\nu^2\right]^{1/3}$, where s is the sand–water density ratio, ν is the kinematic viscosity coefficient, and d_{50} is the median particle size. The critical Shields number under the wave–current and wave action conditions of this test is calculated by Equation (4).

The wave–current test conditions in Table 2 include both clear water erosion and mobile bed erosion. When $\theta/\theta_c < 1$, only the local sediment of the monopile moves, and this condition is clear water erosion. When $\theta/\theta c > 1$, the sand bed indicates that the sediment is generally starting to move, and this condition is the mobile bed erosion.

2.3.2. Wave–Current–Vibration Test Conditions

Table 3 is the working conditions table of multi-stage vibration and wave–current load alternating development, and Table 4 is the working conditions table of monopile local scour tests under the combined action of wave–current–vibration. The wave–current–vibration tests carried out one group of local brush tests of a monopile under the combined action of current and vibration, 25 groups of local scour tests of a monopile under the combined action of wave–current–vibration, and two groups of multi-stage vibration and wave–current load alternating development test conditions, a total of 28 groups of test conditions. This test condition explores the local scour law of monopiles under wave–current–vibration, and provides data for the fitting formula of local scour depth of a monopile under wave–current–vibration.

Table 3. Multi-stage vibration and wave–current load alternating development conditions.

Working Condition	Incidence Velocity U_c/m·s^{-1}	Vibration Frequency f_v/Hz	Amplitude A_v/mm	Wave Height H/cm	Wave Period T/s
1	0.213	6	4	0	0
2	0.213	5	5	8.7	1.3

Table 4. Wave–current–vibration test conditions.

Working Condition	Incidence Velocity U_c/m·s^{-1}	Vibration Frequency f_v/Hz	Amplitude A_v/mm	Wave Height H/cm	Wave Period T/s	S/cm	S/D
1	0.213	6	4	0	0	4.1	0.456
2	0.213	3	4	5.4	1.3	5.4	0.6
3	0.213	6	4	5.4	1.3	4.5	0.5
4	0.213	9	4	5.4	1.3	4.1	0.456
5	0.213	6	2	5.4	1.3	4.4	0.489
6	0.213	6	6	5.4	1.3	3.2	0.356
7	0.213	3	4	8.7	1.3	5.1	0.567
8	0.213	6	4	8.7	1.3	4.6	0.589
9	0.213	9	4	8.7	1.3	4	0.444
10	0.213	6	2	8.7	1.3	4.9	0.544
11	0.213	6	6	8.7	1.3	4.4	0.489
12	0.213	3	4	11.2	1.3	6.3	0.7
13	0.213	6	4	11.2	1.3	6	0.667
14	0.213	9	4	11.2	1.3	5.75	0.689
15	0.213	6	2	11.2	1.3	6.85	0.761
16	0.213	6	6	11.2	1.3	5	0.556
17	0.213	3	4	8.3	1.6	5.93	0.659
18	0.213	6	4	8.3	1.6	4.6	0.511
19	0.213	9	4	8.3	1.6	4.2	0.467
20	0.213	6	2	8.3	1.6	5.7	0.633
21	0.213	6	6	8.3	1.6	4.4	0.489
22	0.213	3	4	9.1	1	5.8	0.644
23	0.213	6	4	9.1	1	4.7	0.522
24	0.213	9	4	9.1	1	4	0.444
25	0.213	6	2	9.1	1	5.2	0.578
26	0.213	6	6	9.1	1	4.1	0.456

3. Study on Scour Characteristics around a Monopile under Wave–Current–Vibration Interaction

3.1. The Variation Law of Wave Height and Reference Point Velocity

The regular wave conditions used in the test are as follows: when the wave period is 1.3 s, the wave heights are 5.4 cm, 8.7 cm, and 11.2 cm, respectively; when the wave period is 1.6 s, the wave height is 8.3 cm; when the wave period is 1 s, the wave height is 9.1 cm. The undisturbed flow velocity at the distance of 1D (9 cm) from the bed surface is taken as the reference point velocity [13]. The horizontal velocity at this reference point represents the velocity of the water near the bottom. The measured velocity at the reference point of each working condition of the wave is: 0.074 m/s, 0.123 m/s, 0.155 m/s, 0.138 m/s, 0.087 m/s. The pure flow only uses one velocity, and the measured velocity at the reference point is 0.213 m/s. In the test, an ADV current meter is used to measure the flow velocity of the reference point, and the contact wave height meter is used to measure the wave height. The pre-experiment is carried out before the test, and it is determined that the wave-making system and the current-making system can stabilize the wave-making and current-making for a long time to meet the accuracy required for the test.

3.1.1. The Variation Law of Velocity and Wave Height at the Reference Point under Wave Action

- Water particle trajectory at the reference point;

The small amplitude wave theory holds that the trajectory of water particles at each position is an ellipse under finite water depth, and its long half-axis is defined as $a = A\cos \text{hk}(Z_0 + h)/\sin \text{hk}h$. The short half-axis is defined as $b = A\sin \text{hk}(Z_0 + h)/\sin \text{hk}h$, A is the amplitude of the water point, and the long and short half-axes of the water point trajectory gradually decrease from the water surface to the bed surface. Therefore, the parameters of the water particle trajectory at the reference point of each working condition of the wave in this experiment are shown in Table 5.

Table 5. Parameter table of water particle trajectory at the reference point.

Wave Period T/s	Wave Height H/cm	Long Half-Axis a/cm	Short Half-Axis b/cm	Moving Trajectory
1.3	5.4	1.530	0.367	ellipse
1.3	8.7	2.468	0.591	ellipse
1.3	11.2	3.174	0.761	ellipse
1.6	8.3	3.496	0.635	ellipse
1	9.1	1.241	0.443	ellipse

- Wave height curve and horizontal velocity change at the reference point;

The free surface curve of the pure wave under various working conditions is shown in Figure 2 below. It can be seen from the figure that the wave crest, wave period, wave trough, and other forms are relatively regular under the wave conditions used in the test. Because the size of the wave flume used in the test is large, the pile diameter is relatively small, and the reflection of the wave is less. Therefore, after a long time of wave loading, the waveform and parameters can still be maintained well, so that the test can carry out a long time of wave loading.

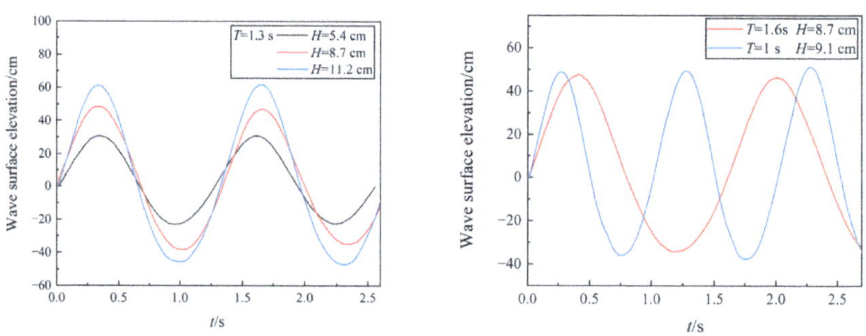

Figure 2. Free surface curve of pure wave condition.

The horizontal velocity at the reference point of each working condition under the action of waves is shown in Figure 3. It can be seen from the diagram that the phase of the horizontal velocity at the reference point is the same as that of the free surface curve, the velocity changes periodically, the peak velocity is the same as the trough velocity, and the direction is opposite. The trajectory of the water particle is elliptical, which conforms to the small amplitude wave theory. Figure 4 shows the comparison between the calculated value and the theoretical value of the wave reference point flow velocity. It can be seen from the figure that the theoretical value of the test flow velocity is relatively close to the calculated value, and the test conditions meet the requirements.

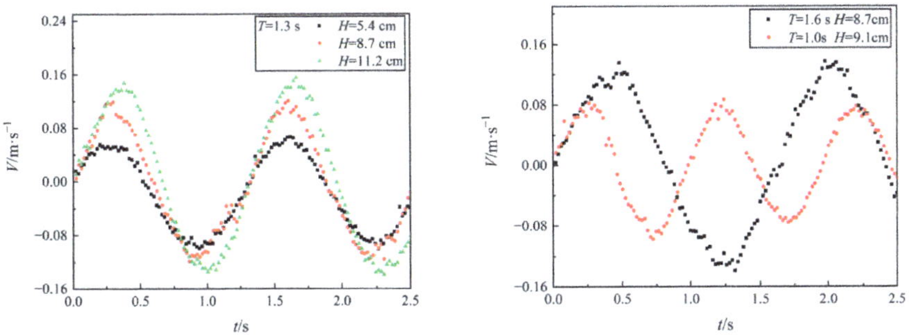

Figure 3. The horizontal velocity at the reference point of wave condition.

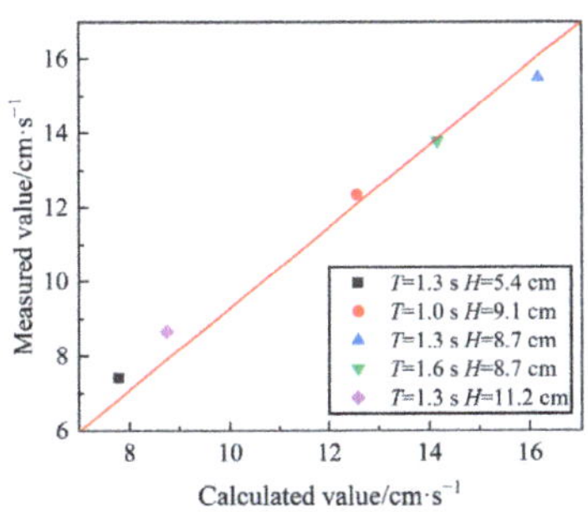

Figure 4. The calculated value of wave reference point velocity is compared with the theoretical value.

3.1.2. The Variation Law of Velocity and Wave Height at the Reference Point under the Action of Wave–Current

The free surface curve under various working conditions of the wave–current is shown in Figure 5 below. It can be seen from the diagram that after the wave superposition, for the free surface curve, compared with the pure wave under the same working condition, the peak and trough are obviously flatter, the wave height is reduced, and the wave period is increased, but the waveform can still be maintained well, so the wave–current test can be loaded for a long time.

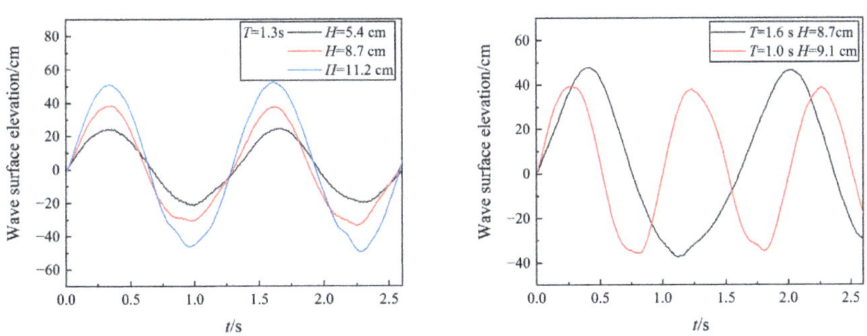

Figure 5. Free surface curve of wave–current condition.

The change of horizontal velocity at the reference point under various working conditions of wave–flow is shown in Figure 6. Comparing Figure 5 with Figure 6, it can be seen that the horizontal velocity at the reference point is the same as the phase of the free surface curve under the combined action of waves. The maximum horizontal velocity in the figure reaches 0.357 m/s, and the minimum velocity is 0.052 m/s. Because the flow velocity is 0.213 m/s, which is much larger than the wave horizontal velocity, there is no reverse velocity, and the peak velocity and the trough velocity are symmetrically distributed along the flow velocity of 0.213 m/s. By adding the flow velocity at the reference point to the flow velocity, it can be found that the wave–current velocity is not a simple linear addition of the flow velocity under the action of water flow and waves, which is consistent with the phenomenon found by Qi [13]. The superposition of water flow on the wave action will increase the size and strength of the horseshoe vortex around the pile, and reduce the critical KC number of the horseshoe vortex, so that the horseshoe vortex also exists at a smaller KC number, and is accompanied by upward pore water pressure generated by the trough, which makes the initial transport of sediment easier, and ultimately affects the scouring process around the pile.

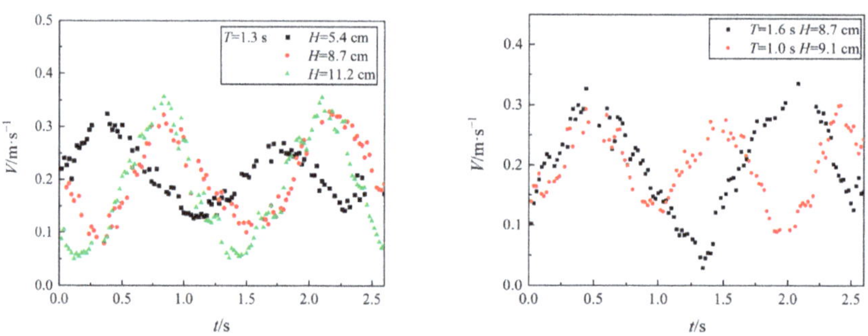

Figure 6. The horizontal flow velocity at each reference point in the wave–flow condition.

3.2. Study on the Development of the Local Scour Depth of a Monopile under the Combined Action of Wave–Current–Vibration

Due to the difference in wave height and wave period between the test conditions, the hydrodynamic characteristics around the monopile are different. The difference in vibration frequency and amplitude leads to a difference in sediment particle movement around the monopile, which leads to the great difference in the scouring process. The most obvious difference is in the development process of local scour depth. The change trend of local scour depth with time is a gradual process, which is one of the important indexes to reflect the scour development characteristics of a monopile under certain flow conditions and vibration loads. Before the test, a scale is pasted on the side wall of the monopile to facilitate the reading of the scour depth. During the test, the scale is read according to the plan, and the scale data are statistically processed after the test to obtain the erosion duration development curve. In this section, the influence of vibration parameters on local scour depth under different wave–current conditions is analyzed, and the difference between clear water scour and live-bed scour is compared. The development characteristics of local scour depth under wave–current–vibration are analyzed in detail.

3.2.1. The Influence of Vibration Frequency on the Duration of Scour Depth Development

Figure 7 is the local scour time curve of the monopile caused by different vibration frequencies under the action of wave–current–vibration. It can be clearly seen from the diagram that the local scour development model of the monopile under different vibration frequencies can also be divided into a rapid development stage, a stable development stage, and a balanced development stage. Under the same wave–current conditions, the equilibrium scour depth and scour rate of the monopile are the largest when the vibration frequency is small, reaching $0.66D$ at t = 210 min. The scour depth under the vibration load of f_6a_4 is $0.511D$, and the scour depth under the vibration load of f_9a_4 is $0.467D$, while the final scour depth of pure wave–current without vibration load is $0.833D$, which indicates that the vibration load will reduce the quasi-equilibrium scour depth, and with the increase in vibration frequency, the quasi-equilibrium scour depth will gradually decrease. This is because the increase in vibration frequency will increase the backfill efficiency of the sand bed around the monopile, resulting in a decrease in the quasi-balanced scour depth [23].

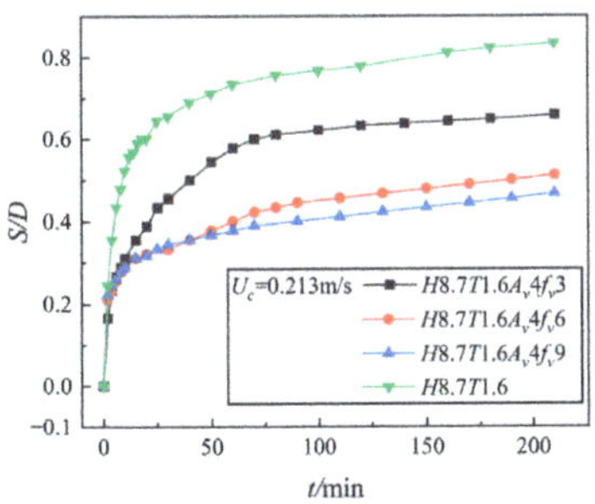

Figure 7. Effect of vibration frequency on scour duration.

3.2.2. The Influence of Amplitude on the Duration of Scour Depth Development

Figure 8 is the local scour time curve of the monopile caused by different amplitudes under the action of wave–current–vibration. It can be clearly seen from the figure that the influence of amplitude on the scour development curve is similar to that of vibration frequency. As the vibration frequency increases, the quasi-equilibrium scour depth decreases. This is because the increase in the amplitude will increase the influence of the sand bed around the monopile, which will increase the backfill efficiency of the scour hole and reduce the quasi-equilibrium scour depth [23].

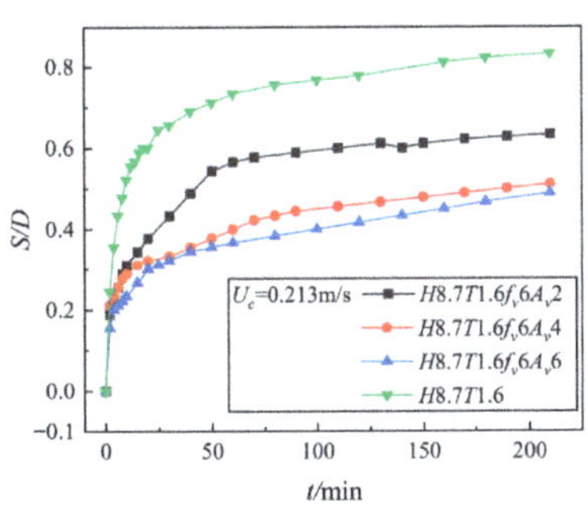

Figure 8. Effect of amplitude on scour duration.

3.3. Development Process and Morphological Characteristics of the Scour Hole

The exploration of the development law of local scour holes changing with time under the combined action of wave–current–vibration is an important part of studying the scour mechanism. In this paper, through the scour test under the action of convection–vibration and the scour test under the action of wave–current–vibration, t = 20 min, 60 min, 210 min, and 450 min are selected as typical moments. At typical moments, the topographic data are scanned by the underwater topographic scanner, and the topographic map is obtained after processing. The development process of the scour hole and the shape of the scour hole are qualitatively described through the topographic map, so as to obtain the characteristics of the local scour development process of the monopile under the action of current–vibration and wave–current–vibration. Taking U_c = 0.213 m/s, f_v = 6 Hz, A_v = 4 mm (working condition 1 in Table 3), and U_c = 0.213 m/s, H = 8.7 cm, T = 1.3 s, f_v = 6 Hz, A_v = 4 mm (working condition 8 in Table 4) as examples, the development process and morphological characteristics of local scour of the monopile under convection–vibration and wave–current–vibration are analyzed, respectively.

In the process of local scour of a monopile under wave–current load, horseshoe vortex, and descending flow are the main factors causing local scour around the pile. It can be seen from Figure 9 below that under the action of pure flow, the forward water flow moves to the front of the pile. Under the obstruction of the pile, the velocity potential energy of the water flow is transformed into pressure potential energy, and the flow velocity in the vertical direction of the water flow presents a logarithmic distribution. The resulting pressure gradient also shows a near-logarithmic distribution with a large pressure potential energy on the upper part of the monopile and a small pressure potential energy on the lower part. Under the action of this pressure difference, a downward flow is formed. When the downward flow passes through the monopile, the flow section decreases due to the existence of the monopile, and the flow will accelerate to bypass the monopile. The downward flow will touch the bottom in front of the pile, accompanied by the movement of the flow, forming a circular horseshoe vortex, rotating around the bottom of the monopile, carrying a large amount of sediment to the downstream movement. The flow field structure around the monopile under the action of wave and current is shown in Figure 10. Similar to the flow field under the action of pure current, it is also composed of a descending flow in front of the pile, an annular horseshoe vortex at the bottom of the ring pile and a trailing vortex behind the pile. However, due to the obvious periodicity of the wave, the intensity of the vortex formed is also cyclical, and the strength of the sediment around the monopile is also cyclical. In the process of local scour of the monopile under the combined action of current–vibration or wave–current–vibration in this experiment, there is not only scour under the action of the above-mentioned falling flow and vortex, but also densification of sediment and ratchet convection. The action of vibration has an effect on the formation and strength of the falling flow and vortex, so this section discusses the local scour characteristics of the monopile under this complex action.

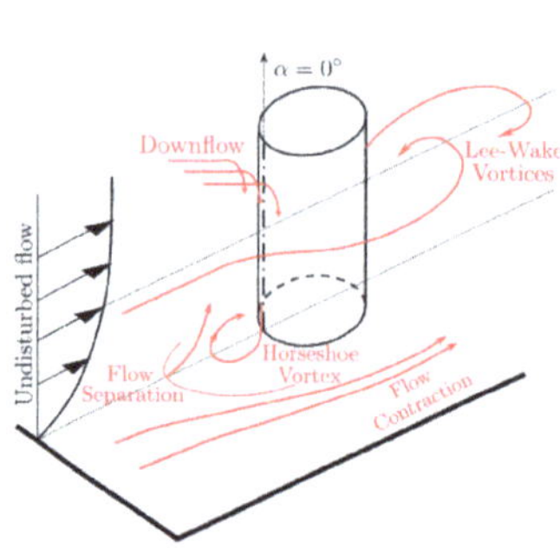

Figure 9. Flow field diagram around a monopile under pure flow [26].

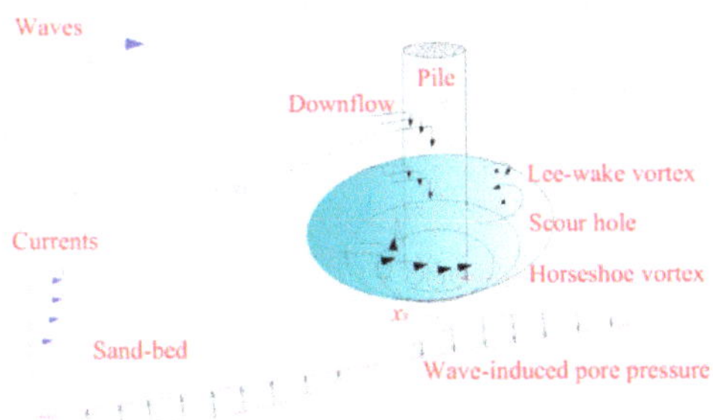

Figure 10. Flow field diagram around a monopile under wave–current interaction [13].

3.3.1. The Development Process of the Scour Hole under the Combined Action of Current and Vibration

Under the influence of a horseshoe vortex and falling water flow, the sediment around the monopile will move, resulting in some sediment being transported to the pile by the vortex. During the transportation process, the sediment will gradually settle with the decrease in the strength of the wake vortex, and gradually accumulate behind the pile to form a large dune. There is only a ring-like scour hole in front of the pile without deposition, and the area behind the pile is alternately changed with erosion and deposition. Under the combined action of current and vibration, in addition to the influence of the horseshoe vortex and falling water flow, it is also affected by the densification and ratchet convection motion under the action of vibration load, resulting in the sand bed sediment around the monopile sliding into the scour hole, which will affect the scouring process around the monopile under the action of water current.

Figure 11a–d is the topographic map at typical times of the sand bed around the monopile under the action of U_c = 0.213 m/s, f_v = 6 Hz, A_v = 4 mm (working condition 1 in Table 3). In the first 20 min of the scour, the scour around the monopile develops rapidly, and the sediment transport rate is high at this time. A large amount of sediment moves rapidly to the back of the pile under the action of the horseshoe vortex and tail vortex to settle, forming a sand ripple connected with erosion and deposition. The vibration load causes the sediment around the monopile to slide into the scour hole, increasing the width of the scour hole and slowing down the slope of the scour hole. Under the coupling effect of water current and vibration load, the scour hole topography shown in the figure is formed. When t = 60 min, the scour depth around the monopile increases slowly, while the width of the semi-annular-like scour hole increases rapidly, and the height of the dune behind the pile also increases further. The maximum scour depth is in front of the pile. When t = 210 min, the development of scour depth and width slows down, and the development of the scour hole is basically mature. At this stage of 210 min~450 min, the scour depth develops slowly, the width of the scour hole increases slowly, and the scour depth reaches a quasi-equilibrium state.

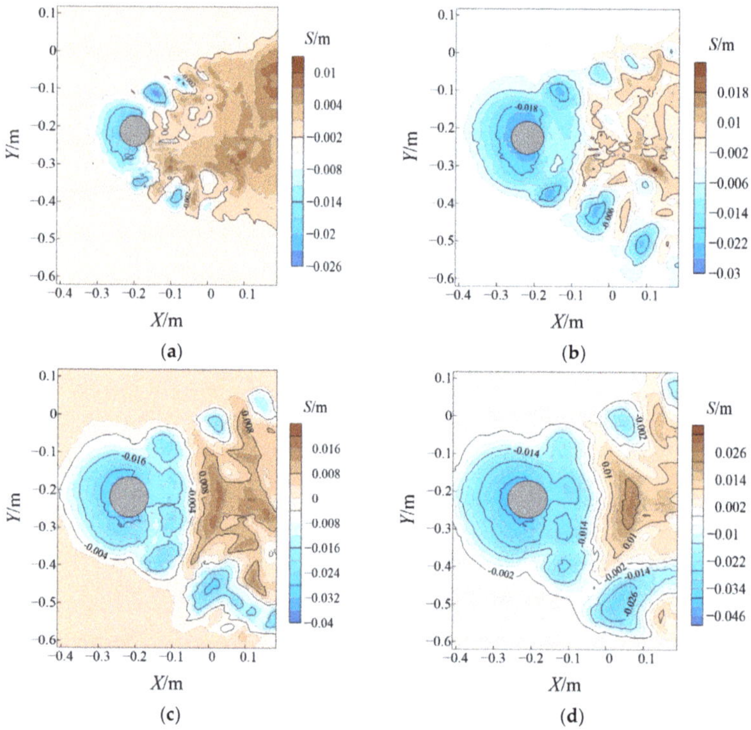

Figure 11. The development process of the scour hole under the action of current and vibration: (**a**) t = 20 min; (**b**) t = 60 min; (**c**) t = 210 min; (**d**) t = 450 min.

3.3.2. The Development Process of the Scour Hole under the Action of Current and Vibration

Figure 12a–d is the typical time topographic map of the sand bed around the monopile under the action of U_c = 0.213 m/s, H = 8.7 cm, T = 1.3 s, f_v = 6 Hz, A_v = 4 mm (working condition 8 in Table 4).Under the combined action of wave–current–vibration, the scour hole around the monopile develops rapidly under the action of periodic sand lifting of waves, sediment carrying of water current, densification of vibration load, and ratchet convection. In the first 20 min, the erosion around the monopile develops rapidly. It can be observed that the sediment around the monopile is rolled up under the action of the vortex, and continues to move backward under the action of the water flow and wake vortex. At t = 60 min, it can be clearly observed that the maximum scour depth is in front of the pile, the width of the scour hole develops rapidly, and the dunes behind the pile move obviously backward. When t = 210 min, the sand bed appears as a whole, and the scour hole also develops from a semi-ring-like shape in front of the pile to a ring-like shape around the pile. This is because $\theta/\theta_c = 1.02 > 1$ in this working condition, which belongs to the live-bed scour, but it is close to the critical Shields number, so the formation time of sand ripple is slow. At t = 450 min, the scour around the pile reaches a quasi-equilibrium state. From Figure 12d, it can be observed that the height of the dune decreases significantly. This is because under the action of wave–current, the dune behind the pile gradually moves backward, and the main peak position is outside the topographic map. The scour terrain in the quasi-equilibrium state is highly symmetrical, and the maximum scour hole is located on the side of the pile.

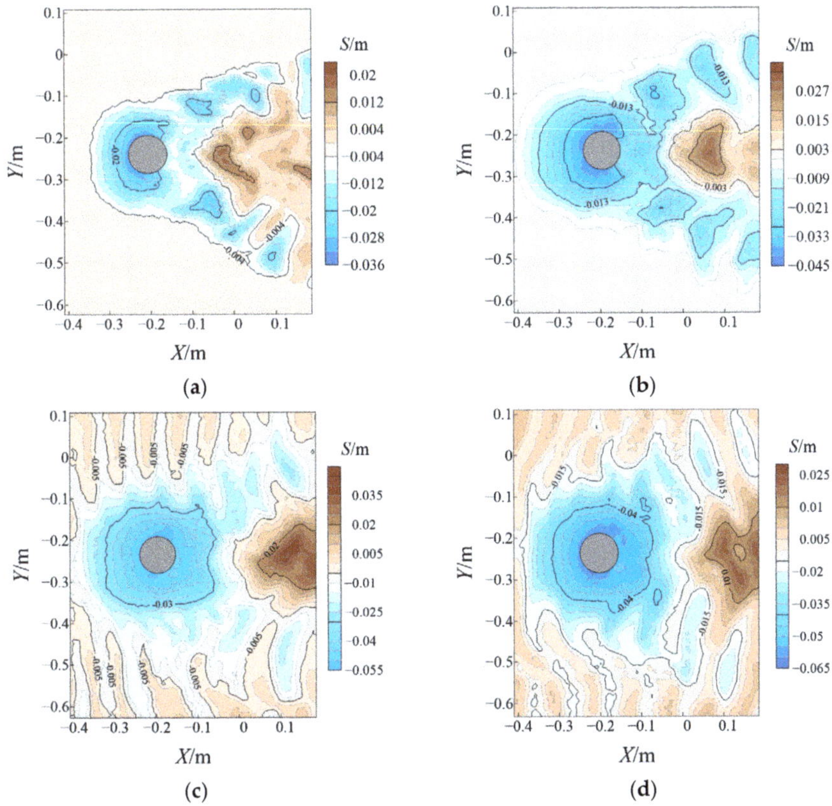

Figure 12. The development process of the scour hole under the combined action of wave–current–vibration: (**a**) t = 20 min; (**b**) t = 60 min; (**c**) t = 210 min; (**d**) t = 450 min.

3.3.3. The Development Process of the Scour Hole under the Action of Current and Vibration

Figures 13 and 14 are the longitudinal profiles of the scour hole parallel to the current direction and passing through the center of the monopile under the combined action of current–vibration and wave–current–vibration, respectively. It can be seen from Figure 13 that with the continuous loading of current–vibration, the depth and width of the scour hole around the monopile and the height of the dune behind the tail continue to increase, and the sand peak of the dune behind the tail continues to move downstream. With the continuous loading of current–vibration, although the depth of the scour hole continues to develop, the shape of the front wall at each time during the development process is relatively consistent with the slope. And the shape of the back wall is also relatively consistent, but the back wall will continue to steepen during the development process. This is due to the gradual backward movement and increase in the sand dunes behind the tail, so that the effect of ratcheting intensity on the sand dunes on the back wall is reduced. Therefore, with the development of loading time, the slope of the front wall of the pile remains relatively unchanged, and the slope of the back wall of the pile continues to increase. It can be seen from Figure 14 that under the continuous action of wave–current–vibration load, the development law of the front wall of the scour hole and the back wall of the scour hole is similar to that under the action of current and vibration. The shape of the front wall at each time is relatively consistent with the slope, and the shape of the back wall changes with the development of time. Also, the sand dunes behind the pile are sand ripples, and this is due to the live-bed scour under the combined action of wave–current–vibration. The back wall of the pile is relatively longer than the front wall, so there are sand ripples.

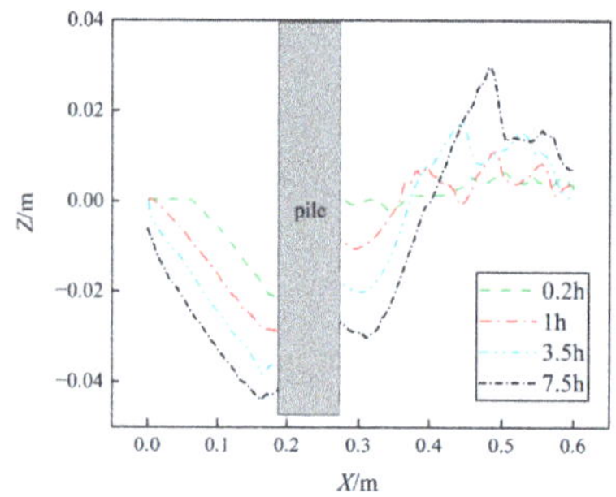

Figure 13. Longitudinal profile of scour hole development under current and vibration action.

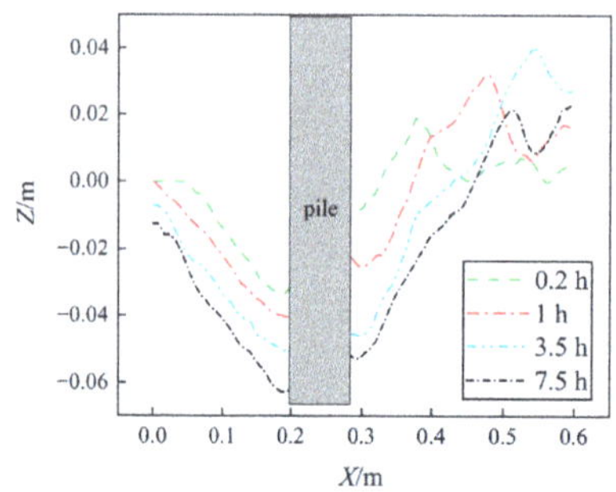

Figure 14. Longitudinal profile of scour hole development under wave–current–vibration.

3.4. The Influence of Vibration Load on the Profile of the Scour Hole

Figure 15 shows the longitudinal section of the scour hole under the influence of different vibration loads under the combined action of wave–current–vibration. It can be observed from Figure 15a that under the same wave–current and vibration frequency conditions, with the increase in vibration amplitude, the width of the scour hole in front of the pile increases significantly, the depth of the scour hole changes little, the slope of the front wall decreases, and the wall behind the pile gradually moves backward with the increase in the amplitude. It can be seen from the whole that the increase in amplitude will significantly increase the width of the scour hole, and with the increase in amplitude, the height of the dune behind the pile shows a step-like increase, and the dune as a whole shows a streamlined structure. It can be observed from Figure 15b that the influence of vibration frequency on the scour hole around the pile shows different characteristics from the amplitude. Under the same wave–current and amplitude conditions, with the increase in vibration frequency, the scour hole in front of the pile is obviously backfilled, the depth of the scour hole is obviously reduced, the width of the scour hole changes little, and the slope of the front wall is obviously slowed down. When the sand dunes behind the pile are acted on at a small frequency, the shape of the back wall remains highly consistent. With a continuous increase in the vibration frequency, the sand dunes behind the pile are affected by the vibration, and the sand dunes near the pile are obviously collapsed, and the slope of the back wall is obviously slowed down. It can be known as a whole that an increase in vibration frequency will backfill the scour hole and significantly reduce the depth of the scour hole.

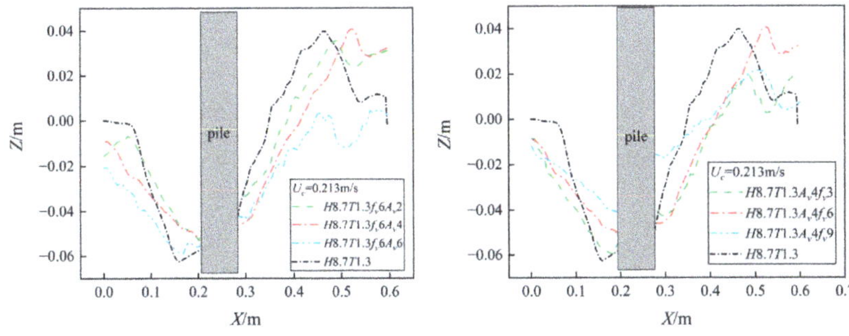

Figure 15. Effect of vibration load on scour hole profile: (**a**) Longitudinal profile of scour hole under different amplitudes; (**b**) Longitudinal profile of scour hole under different vibration frequencies.

4. Analysis of Influencing Factors on the Local Scour Depth of a Monopile under the Combined Action of Wave–Current–Vibration

4.1. The Influence of Vibration Intensity on the Maximum Scour Depth

Vibration frequency and amplitude have an effect on the local scour depth of a monopile. The vibration frequency and amplitude of offshore wind turbines also change during normal operation. Therefore, a comprehensive parameter for vibration intensity Λ [27,28] is introduced to measure the influence of vibration load on local scour. The calculation formula is as follows:

$$\Lambda = \frac{A_v \times (2\pi f_v)^2}{g}, \tag{5}$$

In this formula: A_v is the vibration frequency; f_v is the amplitude; the value of π is 3.14; g is the gravitational acceleration, which is 9.81 m/s^2.

Figure 16 shows the relationship between vibration intensity and relative scour depth under different wave conditions. The five curves in the diagram represent five different wave and current conditions. According to the Hilbert number, the scouring test can be divided into two types: live-bed scouring and clear water scouring. It can be seen from Figure 16 that under the condition of clear water scour, the relative scour depth shows a decreasing trend with the increase in vibration intensity. Under the condition of live-bed scour, when $\theta/\theta_c \leq 1.1$, the relationship between relative scour depth and vibration intensity is similar to that of clear water scour. With the increase in the relative Shields number, when $\theta/\theta_c > 1.1$ (red curve and green curve), the relative scour depth increases first, then decreases, and finally increases with the increase in vibration intensity.

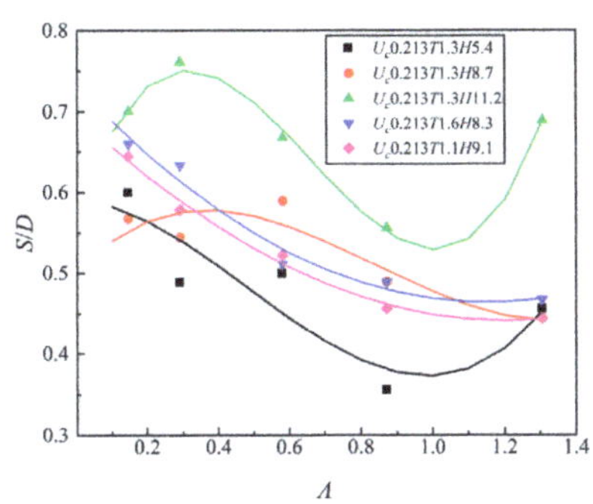

Figure 16. The influence of vibration intensity on the maximum scour depth.

4.2. The Influence of the Froude Number Fr on the Maximum Scour Depth

The Froude number Fr is one of the important parameters affecting the local scour of a monopile foundation. The Froude number under wave and current conditions is defined as $Fr = U_a / \sqrt{gD}$; this parameter is closely related to the flow field structure (horseshoe vortex, wake vortex) around the monopile under the combined action of wave–current–vibration. The local relative scour depth of a monopile under wave–current–vibration is fitted with the test data of Qi and Gao [13] and Sumer [29,30], and the curve shown in Figure 17 is obtained. After the dimensionless processing of the test data, it is found that the test data are located below the fitting curve, but the growth law is similar to the fitting curve. The relative scour depth S/D shows an increasing trend with the increase in the Fr number, and the backfill caused by vibration is the main reason for the test data to be located below the fitting curve.

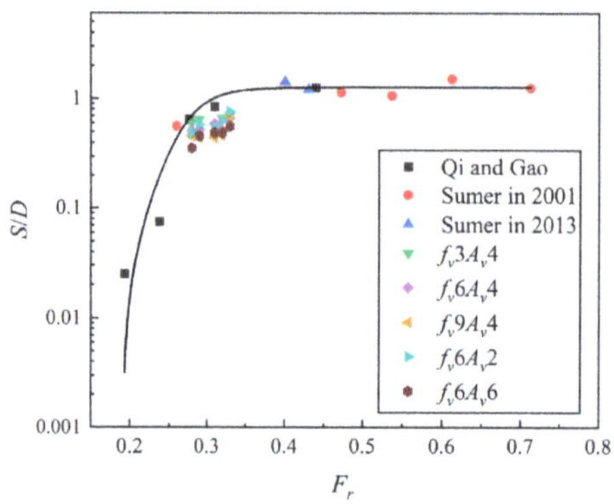

Figure 17. Relationship between Fr and relative scour depth [13,29,30].

4.3. The Influence of the KC Number on the Maximum Scour Depth

The Keulegan–Carpenter (KC) number is one of the important parameters to describe the local scour process of pile foundation under wave action, which shows the relative ratio between viscous force and inertial force. The change in KC reflects the change degree in wave height, so the local scour depth of the monopile in this test should be in a functional relationship with KC. Figure 18 is a fitting curve of relative scour depth S/D with KC number under different U_{cw} by fitting the data of Sumer [29,30], Qi and Gao [13], Rudolph and Bos [31]. The test data are classified according to three intervals: U_{cw} = 0.4–0.5, U_{cw} = 0.5–0.7, U_{cw} = 0.7–0.85. There are 25 groups of working conditions under the combined action of wave–current–vibration in this test. U_{cw} can be divided into two intervals, which are classified and represented in Figure 18. It can be seen from the figure that all the relevant data are located near the corresponding curve, but it can be clearly observed that as the vibration frequency and amplitude increase, the corresponding data points gradually shift downward and fall below the curve. This is because as the vibration frequency and amplitude increase, the densification of the sand bed around the pile and the ratcheting convection movement causes the backfill effect of the scour hole to increase [23], resulting in a decrease in the relative scour depth. The resulting data points deviate from the curve, but the overall data trend with KC. The growth trend is basically consistent with the fitting curve, and it shows an increasing trend as the KC number increases.

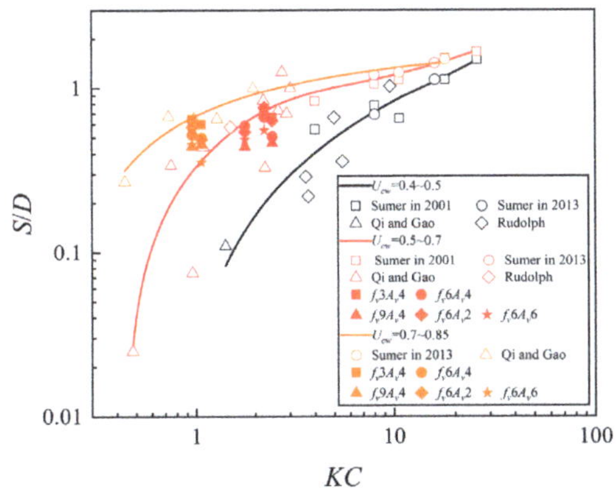

Figure 18. Relationship between KC number and relative scour depth [13,29–31].

4.4. Effect of U_{cw} on the Maximum Scour Depth

U_{cw} is a dimensionless number reflecting the relative strength of current velocity and wave–current velocity under the combined action of wave and current. It can be defined as $U_{cw} = U_c/(U_c + U_{wm})$, where U_c is the current velocity at the current reference point, and U_{wm} is the maximum horizontal velocity of the water particle at the wave reference point. Figure 19 shows the curves of relative scour depth S/D with U_{cw} under different vibration load parameters. The distribution of five curves with different colors in the figure represents the scour test under five groups of vibration load parameters. According to the Shields number, the scour test can be divided into two types: live-bed scour and clear water scour. It can be seen from the diagram that under the action of vibration load, with the increase in U_{cw}, the relative scour depth under clear water scour and live-bed scour shows a downward trend.

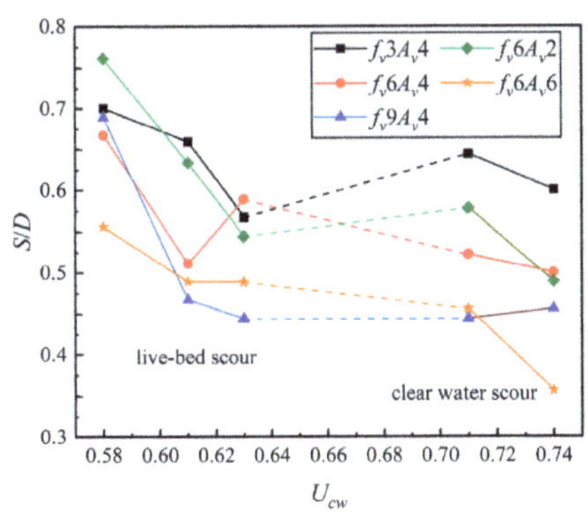

Figure 19. Relationship between U_{cw} and relative scour depth.

4.5. Empirical Formula of Local Scour Depth of a Monopile under Combined Action of Wave–Current–Vibration

There are many empirical equations for predicting the local scour depth of a monopile under the action of wave and current. The commonly used equation is the large KC number equation ($KC \geq 6$) (Equation (6)) proposed by Sumer [32]. Recently, Dogan [33] modified the local scour equations of fine piles and large-diameter piles under wave action. Compared with the estimation formula of local scour depth under pure wave action, the local scour depth of a pile foundation under wave–current action is also closely related to various dimensionless parameters such as the Fr number and U_{cw}. Qi and Gao [34] summarized the empirical equation for predicting the local scour depth of a monopile based on multiple sets of test data ($0.4 < KC < 26$) (Equation (7)). However, the above empirical formula only considers the wave and water current, and lacks the empirical formula of local scour depth of a pile foundation under vibration load.

$$\frac{S}{D} = 1.3\{1 - \exp[-0.03(KC - 6)]\}, \tag{6}$$

$$lg\left(\frac{S}{D}\right) = -0.8\exp\left(\frac{0.14}{Fr}\right) + 1.11, (0.1 < Fr < 1.1,\ 0.4 < KC < 26), \tag{7}$$

Whitehouse [24] believed that the local scour depth of a pile foundation is related to the current conditions, sediment characteristics, pile size, water depth, and other factors. In this research, the monopile foundation under vibration load is taken as the research object, the vibration strength Λ is introduced, and the dimensionless parameter equation of scour depth S is established by Equation (8).

$$S = f(H,\ T,\ U_c,\ U_{wm},\ d_{50},\ D,\ h,\ g,\ \rho_s,\ \rho_w,\ f_v,\ A_v), \tag{8}$$

where S is the maximum scour depth, ρ_s is the volume weight of sediment, ρ_w is the volume weight of water, g is the acceleration of gravity, and the other parameters have been described above.

Within the parameter range covered in this study, further dimensionless processing is performed based on the Pi theorem. The final empirical prediction model is shown in Equation (9). The four dimensionless parameters of the Keulegan–Carpenter number (KC), vibration intensity (Λ), velocity ratio (U_{cw}), and Froude number (Fr) are used as multivariate regression model variables, where k_1, k_2, k_3, k_4, k_5, and k_6 are the control coefficients of equations and variables. In this paper, a total of five scour depth data points without vibration, only wave–current action, and 25 scour depth data points under the combined wave–current–vibration action of this experiment are used as training data samples for fitting the formula coefficients. The final equation and coefficient values are shown in Equation (10).

$$\frac{S}{D} = \exp(k_1 \cdot KC + k_2 \cdot U_{cw} + k_3) \cdot Fr^{k_4} + k_5 \cdot \Lambda^{k_6}, \tag{9}$$

$$\frac{S}{D} = \exp(0.014 \cdot KC - 0.065 \cdot U_{cw} + 0.962) \cdot Fr^{0.949} - 0.182 \cdot \Lambda^{0.264}, \tag{10}$$

The results of Equation (10) in the case of no vibration are calculated and compared with the results of Equation (7) proposed by Qi and Gao [34] under the test conditions in this paper, and the obtained comparisons are shown in Table 6. From the table, it can be seen that the errors of the two formulas are below 15%, which can indicate that the results of the fitted formulae in the absence of vibration are also reliable.

Table 6. Comparison table of values calculated by Formulas (6) and (9).

Working Condition	KC	U_{wm}	Fr	Calculated by Formula (6) [34]	Calculated by Formula (9)	Data Comparison
1	1.07	0.74	0.28	0.618	0.756	−13.84%
2	1.78	0.63	0.31	0.713	0.847	−13.41%
3	2.23	0.58	0.33	0.771	0.908	−13.66%
4	2.45	0.61	0.32	0.743	0.883	−13.99%
5	0.96	0.71	0.29	0.651	0.782	−13.14%

The error between the measured value of the test and the calculated value of the Equation (10) is shown in Figure 20. It can be seen from the diagram that the formula can better predict the local scour depth of a monopile under the action of wave–current–

vibration. Only a preliminary exploration is carried out on the basis of existing test data. The empirical formula with wider application range and higher precision needs further research in the future.

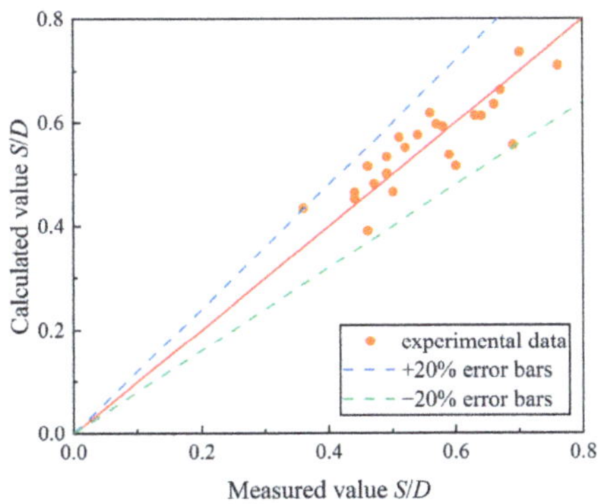

Figure 20. Comparison between calculated values and measured values.

5. Conclusions

In this paper, the experimental study on the local scour of a monopile under the combined action of wave–current–vibration is carried out by arranging the vibration load loading device in the wave–current flume. Firstly, the local scour characteristics of the monopile under the action of wave–current–vibration are tested, and the development of maximum scour depth, the development of the scour hole shape, and the shape of the scour hole profile are analyzed. Further, the influence of vibration intensity Λ and Froude number Fr, KC number, and U_{cw} on the local scour depth of a monopile is analyzed. Finally, the dimensionless influencing factors are analyzed, and an empirical formula for predicting the local scour depth of a monopile under the combined action of wave–current–vibration is fitted. The main conclusions are as follows:

(1) The maximum scour depth under the combined action of wave–current–vibration will be significantly smaller than the maximum scour depth under wave action. Vibration load will reduce the quasi-equilibrium scour depth, and with the increase in vibration frequency and amplitude, the quasi-equilibrium scour depth around the monopile will gradually decrease. The local scour topography of the monopile under the combined action of current and vibration is highly symmetrical, and the maximum scour depth is on the side of the pile, with the scour hole type presenting a spoon-like shape. As the amplitude increases, the width of the scour hole will increase significantly. As the vibration frequency increases, the backfill will be more obvious, and the maximum scour depth in front of the pile will be significantly reduced.

(2) Under the combined action of wave–current–vibration, the relative scour depth decreases with the increase in vibration intensity under the scouring of clear water. Under live-bed scour, when $\theta/\theta_c \leq 1.1$, the relationship between relative scour depth and vibration intensity is similar to that of clear water scour. When $\theta/\theta_c > 1.1$, the relative scour depth increases first, then decreases, and finally increases with the increase in vibration intensity.

(3) S/D is the function of vibration intensity Λ and Froude number Fr, KC number, U_{cw} under the wave–current–vibration interaction. The relative scour depth S/D increases with the increase in Fr and KC, and has the same trend as the vertical pile fitting curve. However, with the increase in U_{cw}, the relative scour depth under clear water scour and live-bed scour shows a downward trend.

(4) The four dimensionless parameters of the Keulegan–Carpenter number (KC), vibration intensity (Λ), velocity ratio (U_{cw}), and Froude number (Fr) are used as multivariate regression model variables. The prediction formula of the local scour depth of a

monopile under wave–current–vibration is obtained by fitting. A comparison is made between the measured and calculated values of the error in this formula. The results show that this formula can predict the local scour depth of a monopile under wave–current–vibration.

Author Contributions: Conceptualization, L.S. and Y.Z.; methodology, L.S., Y.Z. and B.X.; software, L.S.; validation, L.S. and Y.Z.; formal analysis, L.S. and Y.Z.; investigation, L.S. and Y.Z.; resources, Y.C.; data curation, L.S., Y.Z. and X.H.; writing—original draft preparation, L.S.; writing—review and editing, Y.C., B.X. and X.H.; visualization, L.S. and Y.Z.; supervision, Y.C.; project administration, L.S., Y.Z. and Y.C.; funding acquisition, Y.C. All authors have read and agreed to the published version of the manuscript.

Funding: This work was financially supported by the National Natural Science Foundation of China (Grant Nos. 52071031; 52371258).

Institutional Review Board Statement: Not applicable.

Informed Consent Statement: Not applicable.

Data Availability Statement: Data is contained within the article.

Conflicts of Interest: The authors declare that they have no known competing financial interests or personal relationships that could have appeared to influence the work reported in this paper.

References

1. He, K.; Ye, J. Seismic Dynamics of Offshore Wind Turbine-Seabed Foundation: Insights from a Numerical Study. *Renew. Energy* **2023**, *205*, 200–221. [CrossRef]
2. Yu, L.-Q.; Wang, L.-Z.; Guo, Z.; Bhattacharya, S.; Nikitas, G.; Li, L.-L.; Xing, Y.-L. Long-Term Dynamic Behavior of Monopile Supported Offshore Wind Turbines in Sand. *Theor. Appl. Mech. Lett.* **2015**, *5*, 80–84. [CrossRef]
3. Cuéllar, P.; Georgi, S.; Baeßler, M.; Rücker, W. On the Quasi-Static Granular Convective Flow and Sand Densification around Pile Foundations under Cyclic Lateral Loading. *Granul. Matter* **2012**, *14*, 11–25. [CrossRef]
4. Nanda Kishore, Y.; Narasimha Rao, S.; Mani, J.S. The Behavior of Laterally Loaded Piles Subjected to Scour in Marine Environment. *KSCE J. Civ. Eng.* **2009**, *13*, 403–408. [CrossRef]
5. Shi, S.; Zhai, E.; Xu, C.; Iqbal, K.; Sun, Y.; Wang, S. Influence of Pile-Soil Interaction on Dynamic Properties and Response of Offshore Wind Turbine with Monopile Foundation in Sand Site. *Appl. Ocean Res.* **2022**, *126*, 103279. [CrossRef]
6. Van Der Tempel, J.; Diepeveen, N.F.B.; Salzmann, D.J.C.; De Vries, W.E. Design of Support Structures for Offshore Wind Turbines. In *WIT Transactions on State of the Art in Science and Engineering*; WIT Press: Southampton, UK, 2010; Volume 1, pp. 559–591, ISBN 978-1-84564-205-1.
7. Wang, X.; Li, S.; Li, J. Load Bearing Mechanism and Simplified Design Method of Hybrid Monopile Foundation for Offshore Wind Turbines. *Appl. Ocean Res.* **2022**, *126*, 103286. [CrossRef]
8. Matutano, C.; Negro, V.; López-Gutiérrez, J.-S.; Esteban, M.D. Scour Prediction and Scour Protections in Offshore Wind Farms. *Renew. Energy* **2013**, *57*, 358–365. [CrossRef]
9. Negro, V.; López-Gutiérrez, J.-S.; Esteban, M.D.; Matutano, C. Uncertainties in the Design of Support Structures and Foundations for Offshore Wind Turbines. *Renew. Energy* **2014**, *63*, 125–132. [CrossRef]
10. Luengo, J.; Negro, V.; García-Barba, J.; López-Gutiérrez, J.-S.; Esteban, M.D. New Detected Uncertainties in the Design of Foundations for Offshore Wind Turbines. *Renew. Energy* **2019**, *131*, 667–677. [CrossRef]
11. Wu, M.; Stratigaki, V.; Fazeres-Ferradosa, T.; Rosa-Santos, P.; Taveira-Pinto, F.; Troch, P. Experimental Uncertainty Analysis of Monopile Scour Protection Stability Tests. *Renew. Energy* **2023**, *210*, 174–187. [CrossRef]
12. Sumer, B.M.; Hatipoglu, F.; Fredsøe, J. Wave Scour around a Pile in Sand, Medium Dense, and Dense Silt. *J. Waterw. Port Coast. Ocean Eng.* **2007**, *133*, 14–27. [CrossRef]
13. Qi, W.-G.; Gao, F.-P. Physical Modeling of Local Scour Development around a Large-Diameter Monopile in Combined Waves and Current. *Coast. Eng.* **2014**, *83*, 72–81. [CrossRef]
14. Zhang, B.; Li, J.; Liu, W.; Zhang, H.; Shi, P.; Fu, X. Experimental Study on Dynamic Characteristics of a Monopile Foundation Based on Local Scour in Combined Waves and Current. *Ocean Eng.* **2022**, *266*, 113003. [CrossRef]
15. Gautam, S.; Dutta, D.; Bihs, H.; Afzal, M.S. Three-Dimensional Computational Fluid Dynamics Modelling of Scour around a Single Pile Due to Combined Action of the Waves and Current Using Level-Set Method. *Coast. Eng.* **2021**, *170*, 104002. [CrossRef]
16. Shi, Y.; Cheng, Y.; Huang, X.; Li, Y. Numerical Simulation of Unsteady Flow Field around Monopile under the Combined Influence of Wave and Vibration. *Ocean Eng.* **2023**, *276*, 114238. [CrossRef]
17. Yu, Z.; Cheng, Y.; Cheng, H. Experimental Investigation on Seabed Response Characteristics Considering Vibration Effect of Offshore Monopile Foundation. *Ocean Eng.* **2023**, *288*, 116000. [CrossRef]

18. Yang, W.; Tan, S.K.; Wang, H.; Yu, G. Rheological Properties of Bed Sediments Subjected to Shear and Vibration Loads. *J. Waterw. Port Coast. Ocean Eng.* **2014**, *140*, 109–113. [CrossRef]
19. Awad-Allah, M.F.; Yasufuku, N.; Abdel-Rahman, A.H. Cyclic Response of Wind Turbine on Piles in Unsaturated Sand. *Int. J. Phys. Model. Geotech.* **2017**, *17*, 161–176. [CrossRef]
20. Al-Hammadi, M.; Simons, R.R. Local Scour Mechanism around Dynamically Active Marine Structures in Noncohesive Sediments and Unidirectional Current. *J. Waterw. Port Coast. Ocean Eng.* **2020**, *146*, 04019026. [CrossRef]
21. Guan, D.; Chiew, Y.-M.; Melville, B.W.; Zheng, J. Current-Induced Scour at Monopile Foundations Subjected to Lateral Vibrations. *Coast. Eng.* **2019**, *144*, 15–21. [CrossRef]
22. Qin, B.; Xie, Y.; Yang, W.; Qu, R.; Geng, F. A Further Study on the Scour around the Monopile Foundation of Offshore Wind Turbines. *Sustain. Energy Technol. Assess.* **2023**, *57*, 103198. [CrossRef]
23. Cheng, Y.; Xie, H.; Zheng, Y.; Shi, L. Experimental Study on the Sediment Movement around Monopile under Vibration Action. *Ocean Eng.* **2023**, *288*, 116206. [CrossRef]
24. Whitehouse, R. *Scour at Marine Structures*; Thomas Telford Publishing: London, UK, 1998.
25. Cuéllar, P.; Baeßler, M.; Rücker, W. Ratcheting Convective Cells of Sand Grains around Offshore Piles under Cyclic Lateral Loads. *Granul. Matter* **2009**, *11*, 379. [CrossRef]
26. Yang, Y.; Melville, B.W.; Macky, G.H.; Shamseldin, A.Y. Experimental Study on Local Scour at Complex Bridge Pier under Combined Waves and Current. *Coast. Eng.* **2020**, *160*, 103730. [CrossRef]
27. Mawatari, Y.; Koide, T.; Tatemoto, Y.; Uchida, S.; Noda, K. Effect of Particle Diameter on Fluidization under Vibration. *Powder Technol.* **2002**, *123*, 69–74. [CrossRef]
28. Xu, C.; Zhu, J. Parametric Study of Fine Particle Fluidization under Mechanical Vibration. *Powder Technol.* **2006**, *161*, 135–144. [CrossRef]
29. Sumer, B.M.; Fredsøe, J. Scour around Pile in Combined Waves and Current. *J. Hydraul. Eng.* **2001**, *127*, 403–411. [CrossRef]
30. Sumer, B.M.; Petersen, T.U.; Locatelli, L.; Fredsøe, J.; Musumeci, R.E.; Foti, E. Backfilling of a Scour Hole around a Pile in Waves and Current. *J. Waterw. Port Coast. Ocean Eng.* **2013**, *139*, 9–23. [CrossRef]
31. Rudolph, D.; Bos, K.J. Scour around a Monopile under Combined Wavecurrent Conditions and Low KC-Numbers. In Proceedings of the 3rd International Conference on Scour and Erosion (ICSE-3), Amsterdam, The Netherlands, 1–3 November 2006; pp. 582–588.
32. Sumer, B.M.; Fredsøe, J.; Christiansen, N. Scour Around Vertical Pile in Waves. *J. Waterw. Port Coast. Ocean Eng.* **1992**, *118*, 15–31. [CrossRef]
33. Dogan, M. The Equilibrium Depth of Wave Scour around Both Slender and Large Piles. *Ocean Eng.* **2021**, *236*, 109474. [CrossRef]
34. Qi, W.; Gao, F. Equilibrium Scour Depth at Offshore Monopile Foundation in Combined Waves and Current. *Sci. China Technol. Sci.* **2014**, *57*, 1030–1039. [CrossRef]

Journal of
Marine Science and Engineering

MDPI

Article

Investigations on Free Span Rectification for a Submarine Pipeline on Sand Wave Seabeds

Xing Zou [1], Botao Xie [1], Xuechun Bai [2], Yuanyuan Guo [3], Qingqing Yuan [3] and Zhipeng Zang [4,5,*]

1 CNOOC Research Institute Co., Ltd., Beijing 100028, China; zouxing@cnooc.com.cn (X.Z.); xiebt@cnooc.com.cn (B.X.)
2 School of Civil Engineering, Hohai University, Nanjing 213022, China; baixuechun@hhu.edu.cn
3 CRRC SMD (Shanghai) Ltd., Shanghai 201306, China; guoyy@crrcsmd.com.cn (Y.G.); yuanqq@csrzic.com.cn (Q.Y.)
4 State Key Laboratory of Hydraulic Engineering Simulation and Safety, Tianjin University, Tianjin 300350, China
5 Key Laboratory of Earthquake Engineering Simulation and Seismic Resilience of China Earthquake Administration, Tianjin University, Tianjin 300350, China
* Correspondence: zhipeng.zang@tju.edu.cn

Abstract: Sand waves are large-scale bed forms commonly occurring on the continental shelf seabed and can result in free spans of submarine pipelines, which may have an influence on the stability of the pipelines. Existing span rectification procedures have primarily focused on local rectification methods for free spans caused by local scour or individual spans resulting from seabed unevenness. This paper aims to present a span rectification design applicable to the pipeline crossing sand wave region, and to offer practical guidance on sand wave intervention strategies. A large-scale approach is necessary for the rectification of multiple spans across the field, which may involve the use of either a mass flow excavator (MFE) or a remotely operated vehicle (ROV) jetting tool. A comparative analysis of the estimated durations for post-lay trenching using the MFE and ROV jetting tools is also provided. In instances where the large-scale method fails to achieve span lengths suitable for long-term operation, a localized approach is necessary to address individual spans. The desired trench depth can be attained through a combination of pre-lay and/or post-lay trenching techniques. The analysis of on-bottom roughness and free span has demonstrated that, given the natural seabed profile without trenching, there are no spans surpassing the ultimate limit state (ULS) or fatigue limit state (FLS) criteria for the temporary installation scenario. Consequently, pre-lay rectification is not necessary. However, the analysis indicates that post-lay rectification is essential to meet ULS and FLS criteria under operating conditions. All spans that exceed the ULS and FLS criteria can be effectively rectified by trenching to a depth of 1 m.

Keywords: free span rectification; submarine pipeline; sand waves; trench depth

1. Introduction

Sand waves are large-scale undulating bedforms predominantly found in shallow waters with a dominant tidal current [1,2]. However, they have also been identified at depths exceeding 300 m on continental shelves, such as those in the Lufeng oil fields of the South China Sea [3]. These sand waves are capable of migrating due to the effects of water currents, wave action, and other external environmental factors [4,5]. The irregularity and dynamic nature of the seabed, attributed to the presence of sand waves, pose significant challenges for the construction of offshore oil and gas infrastructure. These challenges include the suspension of submarine pipelines and the exposure of platform foundations. Addressing these issues has become a pressing priority in the development of oil and gas fields in regions characterized by sand wave-affected seabeds [6].

Subsea pipelines are critical infrastructures that play a vital role in the oil and gas sector. These structures are surrounded by an extreme environment, in which free spanning is a major issue that occurs normally in pipeline at an uneven seabed, dynamic seabed, or pipeline crossing [7]. Additionally, subsea pipelines are particularly vulnerable to damage from both anthropogenic and natural factors, making them difficult to access for maintenance and inspection. This vulnerability highlights the need for effective and cost-efficient inspection and monitoring technologies to extend their lifespan. To address this need, we propose employing techniques based on transient testing [8,9]. Our approach involves a numerical framework that utilizes a one-dimensional model to simulate transient flow conditions, combined with analytical methods to evaluate the transient behavior of common anomalies, such as leaks and wall degradation [9]. The results obtained from this methodology are validated through diver surveys, reinforcing the accuracy of our findings.

These challenges pose significant risks to the safety standards of subsea pipelines, necessitating a thorough probabilistic analysis of pipelines affected by free spanning and corrosion degradation [10]. Research indicates the importance of early-inspection data for planning the suitable operating regime of the pipeline in terms of internal pressure, which can be used to estimate the allowable critical spanning lengths and corrosion defect geometries, to ensure the appropriate actions to avoid failures [11].

Corrosion is a leading cause of failure in metallic transmission pipelines [12].To safeguard metallic pipelines from corrosion, especially in challenging environments and corrosive soils, appropriate corrosion protection methods can be employed. External corrosion protection strategies can be categorized into three main groups: passive, active, and hybrid techniques. Passive methods encompass various approaches such as coatings, linings, barriers, material selection, electrical isolation, use of inhibitors, and multi-passive techniques [13], whereas active corrosion protection techniques include sacrificial anodes and impressed current cathodic protections [13,14]. This detailed review aims to support researchers, practitioners, and industry stakeholders in establishing priorities for their policies. It will help in selecting suitable techniques for both external and internal corrosion protection while also addressing existing research voids and highlighting future avenues for study [15,16].

For the submarine pipeline traversing the sand wave region, seabed treatment is conducted in accordance with the maximum allowable suspension span value determined through an unevenness analysis of the seabed. This treatment involves addressing the seabed in areas where the maximum allowable suspension span is exceeded by utilizing trenching equipment to remove protrusions. This process ensures that the pipeline maintains full contact with the seabed, thereby minimizing the free suspension span of the submarine pipeline. Such measures are critical to ensuring the safety of the submarine pipeline during installation [17].

During the design process, it is imperative to thoroughly consider and propose effective treatment measures. Certain segments of the suspension span, once laid, do not comply with the temporary operational conditions, necessitating pretreatment prior to pipe installation. The operational phase imposes more stringent requirements on the suspension span of the pipeline, often necessitating post-installation processing, referred to as post-treatment. In the case of large and extensive sand slopes, both pretreatment and post-treatment are required to ensure structural integrity and compliance with operational standards [18]. In general, span mitigation strategies can be classified into two primary categories: (1) subtractive methods, which involve the removal of material, such as ploughing, jetting, mechanical cutting, mass flow excavation, and mechanical grabbing; and (2) additive methods, which entail the addition of material, including rock dumping, gravel dumping, concrete mattresses, fabric formwork grout bags, sandbags, steel supports, and strakes [19].

The analysis of the free suspension span was primarily conducted in accordance with DNV GL-RP-F105 [20], adhering to the principles of fatigue screening (SFC) and the ultimate limit state (ULS) for load-bearing capacity. These principles are aligned with those outlined in DNV GL-RP-F105 and are further supported by the guidelines in DNV GL-ST-F101 [21]. The analysis of free suspension spans is performed using the combined loading principle. Prior to the laying of submarine pipelines, an assessment of seabed unevenness is conducted based on the design routing, the pre-survey results of the seabed, the maximum allowable suspension span, and the geological conditions of the seabed. This process simulates the interaction between the pipeline and the seabed soil post-laying, generating data on the positions and lengths of free suspension spans. By comparing these results with the maximum allowable span length, it is determined whether the requirements are met.

To date, existing span rectification procedures have primarily focused on local rectification methods for free spans caused by local scour or individual spans resulting from seabed unevenness. However, large-scale rectification of the seabed for multiple spans in sand wave regions is infrequently addressed. This study investigates a pipeline traversing an extensive sand wave field in the Lufeng oil field. The rectification of the free spans of the pipeline can be accomplished using either pre-lay or post-lay techniques, both of which are analyzed in this study. The results provide practical guidance on span rectification strategies for both pre-lay and post-lay methods applicable to pipelines situated on sand wave seabeds. This paper is organized as follows: Section 1 provides an overview of the research background and motivation. Section 2 presents a review of span mitigation methods along with an assessment of their feasibility. Section 3 offers an evaluation of the plough as a technique for pre-lay span rectification. Section 4 examines methods appropriate for post-lay span rectification. Section 5 delineates the requirements for span rectification. Finally, Section 6 offers concluding remarks.

2. Span Mitigation Methods and Feasibility Assessment

It is mentioned in the Introduction that span mitigation strategies can be classified into subtractive methods and additive methods. A brief description of each method is provided below.

2.1. Subtractive Methods for Span Mitigation

(1) Ploughing: Ploughing tools are suitable for displacing large quantities of seabed material and are commonly used for span mitigation.

(2) Jetting: Jetting tools are effective for trenching pipelines in uncemented sands and certain clay substrates, and they are particularly suitable for high-volume span mitigation. These tools operate by utilizing high-pressure water to fluidize the seabed, thereby facilitating the lowering of the pipeline into the seabed.

(3) Mechanical cutter: Mechanical cutting tools are well-suited for the displacement of substantial volumes of seabed material and are frequently employed in span mitigation processes. These tools are particularly effective in harder soils and clays where alternative methods such as ploughing or jetting may prove impractical [22].

(4) Mass flow excavator: An MFE employs propelled water to fluidize and displace the seabed. These methods are appropriate for high-volume span mitigation; however, trenching rates are slower compared to ploughing and jetting techniques. Their effec-

tiveness is heavily dependent on the suitability of the seabed, as they are ineffective against hard or cemented soils.

(5) Mechanical grab: The removal of substantial volumes of seabed material at a specific location can be accomplished through mechanical excavation.

2.2. *Additive Methods for Span Mitigation*

(1) Rock dump: Rock dumping operations are conducted using either a fall pipe vessel or a side dump vessel, contingent upon the water depth and the precision required. This technique is applicable for both pre-lay and post-lay span mitigation. However, pre-lay operations necessitate larger volumes of rock to adequately prepare the pipelay corridor. The operational water depth ranges from a minimum of 10–12 m to a maximum of up to 2000 m.

(2) Gravel dump and sand bags: Gravel or sand-filled bags can be strategically positioned beneath pipelines to function as span supports. These bags are applicable for both pre-lay and post-lay span mitigation; however, larger quantities are necessary during the pre-lay phase to adequately cover the pipelay corridor. It should be noted that ensuring contact between the pipe and the bags is not always feasible when they are installed beneath an existing pipeline.

(3) Concrete mattresses: Concrete mattresses are predominantly utilized for pre-lay span mitigation. However, in certain situations, they can be employed as post-lay interventions by positioning the mattress over the pipeline to either secure it to the seabed or protect the span from hydrodynamic forces. Their effectiveness is generally limited when installed beneath an existing pipeline.

(4) Fabric formwork grout bags: Fabric formwork grout bags are strategically placed on the seabed and subsequently inflated with grout delivered from a surface vessel. These bags are effective for rectifying spans with a minimum clearance of approximately 0.2 m and a maximum clearance not exceeding 5 m. Grout bags may not be suitable for pre-lay correction methods due to the substantial grout volumes required. Furthermore, the potential loss of control over the grout within the downline imposes a constraint on the maximum water depth at which grout bags can be effectively employed.

(5) Steel supports: Purpose-built steel supports represent a viable solution for pre-lay and post-lay span correction. However, these supports are occasionally subject to foundation failures, necessitating a comprehensive design process to mitigate this risk. Considering the documented mobility of the seabed, scour has been recognized as a potential threat to structural integrity, warranting further evaluation prior to the acceptance and implementation of steel supports.

(6) Strakes: Strakes may be installed on a pipeline during the pipelaying process or retrofitted post-installation to mitigate vortex-induced vibration (VIV). However, they are ineffective for spans that exceed the ultimate limit state (ULS) thresholds.

2.3. *Criteria and Assessment*

For the present studied sea area, each method mentioned above was evaluated for feasibility based on the following criteria: (1) appropriateness for use in water depths ranging from 130 m to 330 m; (2) suitability for application in environments characterized by sand waves, such as seabeds with mobile sediments, scour, and steep slopes; (3) the speed at which mitigation can be executed, taking into account the extensive requirements for span mitigation; (4) viability as a pre-lay span mitigation technique; and (5) viability as a post-lay span mitigation technique. The feasibility assessment of methods for both pre-lay and post-lay span mitigation is detailed in Table 1. In the table, a value of 3 indicates that the method is feasible and associated with low risk, whereas a value of 1 signifies that the method is not feasible.

Table 1. Method feasibility assessment.

	Rectification Method	Suitable for Water Depth (130–330 m)	Suitable for Sand Waves	Speed of Mitigation	Pre-Lay	Post-Lay
Subtractive Methods	Ploughing	3	2	3	2	1
	Jetting	3	2	3	1	2
	Mechanical Cutter	3	2	3	2	1
	Mass Flow Excavator	3	3	2	3	3
	Mechanical Grab	3	3	1	3	1
Additive Methods	Rock Dump	3	3	3	3	3
	Gravel Dump	3	3	Not Suitable for Large Quantities of Span Rectification	3	3
	Mattresses	3	3		3	1
	Fabric Formwork Grout Bags	2	2		1	2
	Sand Bags	3	2		2	2
	Steel Supports	3	1		1	1
	Strakes	3	3		3	2

Key

3	Feasible, Low Risk	2	Carries Risk	1	Not Feasible, High Risk

3. Pre-Lay Sand Waves Rectification

Pre-lay rectification is conducted before the commencement of pipelaying operations. The primary objective of this procedure is to mitigate the occurrence of pipeline spanning conditions, which may adversely affect the structural integrity of the pipeline during its temporary state. The criteria for span acceptance in pre-lay rectification are derived from the fatigue limit state (FLS) and ultimate limit state (ULS) standards, specifically considering the pipeline's temporary empty and flooded conditions.

This section examines the application of a subsea plough, specifically the VMP 500 model, for the pre-lay rectification of sand waves. The analysis of the pre-lay rectification encompasses the following procedures: (1) trench backfill rate, (2) control of plough positioning and depth, (3) trench slope stability, (4) achievable trench depth and trench/pipelay acceptance criteria, (5) transition length of pipeline, (6) plough tow forces on flat seabed, and (7) impact of sand waves on plough tow forces. Sections 3.1–3.7 provide a detailed elaboration of the aforementioned analytical procedures.

3.1. Trench Backfill Rate

The calculation for trench backfill is conducted according to the methodology proposed by Fredsøe [23]. Two transport mechanisms are considered: (1) bed load, which comprises larger sediments transported along the seabed, and (2) suspended load, which consists of smaller sediments suspended within the water column [24]. It is acknowledged that backfill resulting from bed load transport is the predominant mechanism. The contribution of suspended load to backfilling is not considered in this analysis, as the gravitational infill from bed load is the predominant mechanism of re-sedimentation. Despite the suspended load being significantly greater than the bed load, the narrow width of the trench prevents suspended particles from settling effectively, resulting in their transport across the trench by the current. Consequently, sedimentation from suspended load accounts for only 2.5% to 6% of the sedimentation attributed to bed load transport.

The calculations are conducted with the trench only considered, without the presence of a pipeline. The primary direction of the current velocities under consideration ranges from 0.40 to 0.80 m/s. Given the alignment of the pipeline route, the angle between the trench and the current direction is determined to be 44 degrees. Certain locations may encounter a more acute angle, potentially resulting in slower backfill rates.

The sedimentation of bed load within the trench is described by [23]:

$$A(t) = 2D_{\text{tm}}\frac{\sqrt{\delta}}{\sqrt{\pi}}\left(\sqrt{t+t_0}-\sqrt{t_0}\right) \tag{1}$$

where D_{tm} is the depth of trench and t is time and δ and t_0 can be calculated as:

$$\delta = \left(q_{\text{b}}\frac{\cos^2\beta}{\tan\theta} + K\frac{dq_{\text{b}}}{d\tau_{\text{d}}}\sin^2\beta\right)\frac{1}{(1-\phi)} \tag{2}$$

$$t_0 = \frac{\pi}{64}\frac{a^2}{\delta(\tan\alpha_0)^2} \tag{3}$$

where q_{b} is the bedload transport rate; β is the angle between current direction and trench; θ is the sand dynamic friction angle; K is a constant = 0.1; $dq_{\text{b}}/d\tau_{\text{d}}$ is rate of change of bed load transport with change in dimensionless bed shear stress; and φ is the porosity of soil.

Figure 1a illustrates the backfill rates of trenches with slopes of 21 and 37 degrees. While the data suggest a smooth progression, it is important to note that backfilling occurs in discrete steps, characterized by rapid increases during periods of high current and minimal advancement during periods of lower current. The curves depicted represent the average anticipated backfill over time. To guarantee a trench depth of at least 1 m, the backfill criterion is set at a maximum of 25% of the trench cross-sectional area. Figure 1b illustrates an idealized scenario in which the trench retains its shape and fills from the bottom upwards. Based on this scenario and the previously presented backfill results, it can be inferred that if pipelaying occurs within 60 days post-trenching, the requisite trench depth is likely to be maintained. In cases where backfill exceeds 25%, trench rectification using a mass flow excavator may be conducted prior to pipelaying to restore the trench to its minimum target depth. As per standard practice, the suitability of the trench depth for pipelaying can be verified during the pre-lay survey.

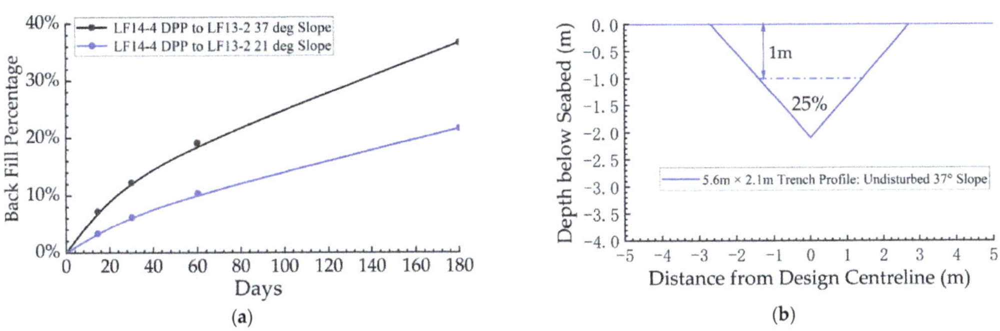

Figure 1. (**a**) Trench back fill rate and (**b**) sketch of backfill of 2 m deep trench (1 m depth is achieved).

3.2. Control of Ploughing Position and Depth

(1) Position control: Ploughs have been effectively employed in previous subsea projects to address sand wave formations. Nonetheless, a significant challenge associated with this rectification method involves the precise control of the plough's positioning.

The Lufeng pipeline routes intersect sand waves at various angles relative to the sand wave crests. Navigating the plough along these routes may present challenges due to the resultant side slope when crossing at such angles. Despite the plough's direct steering capability, maintaining the stringent trench tolerances is anticipated to be difficult in all scenarios. The result is anticipated to be a trench that, in certain areas, deviates from the specified tolerances, thereby jeopardizing the assurance that pipeline installation will achieve the requisite minimum depth at these points. Consequently, additional rectification measures, either pre-lay or post-lay if permissible, are likely to be required. Regarding the ploughing direction, it is projected that controlling the plough, alongside managing tow forces, will be most effective when the plough ascends the gentle slope of the sand wave and descends the steeper slope.

(2) Depth control: The VMP 500 subsea plough, in its standard configuration, is capable of achieving a minimum trench depth of approximately 1.3 m. In regions characterized by sand waves, the span lengths are determined by the elevation differential between the crests and troughs of the sand waves. Consequently, with a trench depth of 2 m at the crest and 1.3 m at the trough, the effective gain from trenching is estimated to be approximately 0.7 m.

3.3. Trench Slope Stability

Figure 2 illustrates the prospective trench configuration resulting from a single trenching pass by VMP 500, contingent upon the expected plough geometry. The trench exhibits a side slope angle of approximately 37°, which may be attainable if the sand remains undisturbed. However, this angle exceeds the angle of repose for disturbed sand, which averages between 21° and 25° along the pipeline routes. Consequently, there exists a risk of trench sidewall slippage following the trenching process.

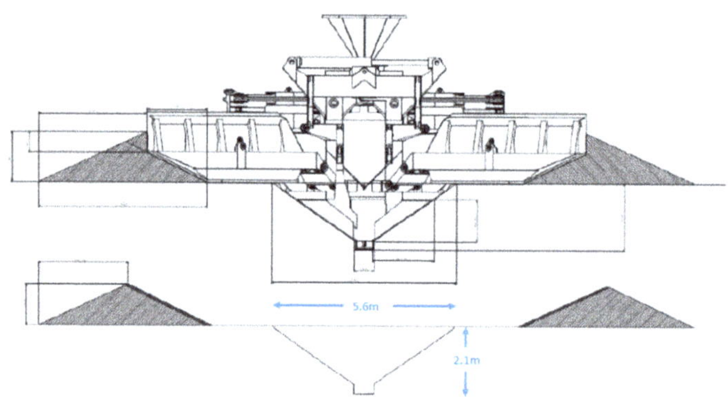

Figure 2. Plough trench profile for one trenching pass; undisturbed 37° slope.

Figure 3a illustrates a modified trench profile characterized by a 21° side slope angle, an 8 m top width, and a maximum depth of 1.5 m. These dimensions were determined based on the principle of mass conservation, ensuring that the cross-sectional area remains equivalent to that of a trench with a 37° slope. The figure facilitates a comparison between the two trench configurations with both disturbed and undisturbed soil profiles. In most instances, soil is likely to remain 'undisturbed'; however, in certain areas, 'disturbed' soil may be encountered. The causes of soil disturbance can include slope instability, alterations in the depth of excavation, adjustments made by the mass flow excavator to align the trench with the designated centerline, and the decision to conduct multiple passes with the plough, which increases the probability of a disturbed trench profile.

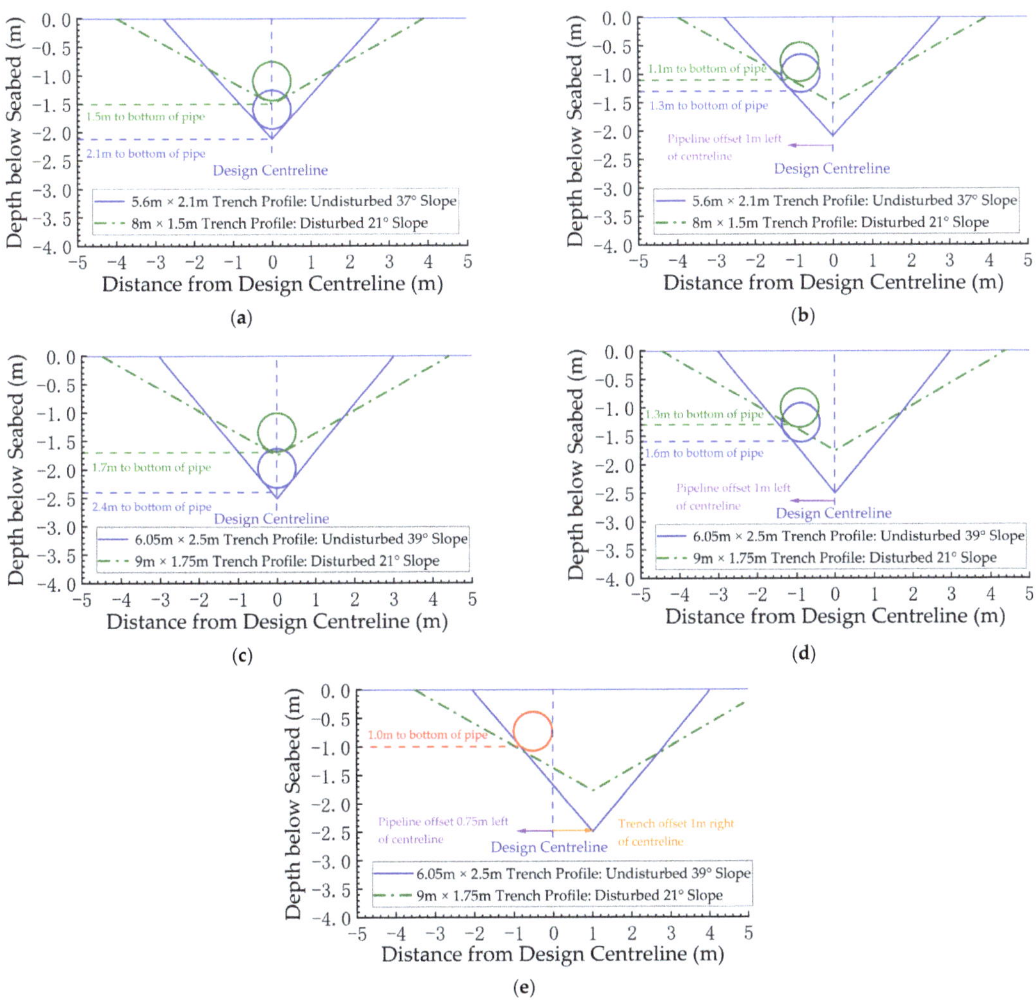

Figure 3. (**a**) Pipeline on design centerline, with both undisturbed profiles for 37° and 21°, one trenching pass; (**b**) pipeline offset 1 m to left of design centerline, one trenching pass; (**c**) pipeline on design centerline, with both undisturbed profiles for 37° and 21°, two trenching passes; (**d**) pipeline offset 1 m to left of design centerline, two trenching passes; (**e**) pipeline offset 0.75 m to left of design centerline with trench offset 1 m to right of centerline, two trenching passes.

3.4. Achievable Trench Depth and Trench/Pipelay Acceptance Criteria

The initial assumption during the ploughing process is that the trench will consistently be constructed along the design centerline. In instances where the plough deviates from the specified tolerance, trench rectification is performed using mass flow excavation. Consequently, the attainable pipeline depth is contingent solely upon the trench slope angle and a pipelay tolerance of ±1 m. As illustrated in Figure 3b, a depth ranging from 1.1 m to 1.3 m (measured at the bottom of the pipe) can be achieved, contingent upon the trench side slope. Nevertheless, this depth assumption is unrealistic, as it presumes the trench profile will invariably align with the designated route. The trench must be constructed in accordance with a specified straightness tolerance criterion, which may necessitate deviations from the route centerline in certain areas. The minimum straightness tolerance permissible for trench construction is ±0.5 m, representing a notably stringent requirement.

The trench depths previously discussed are based on a single trenching pass with the plough. Opting for two passes with the plough is expected to enhance the pipeline depth, as illustrated in Figure 3c,d. When the pipeline is positioned at the center of the trench, achieving a depth of 1.7 m or greater is feasible. However, this depth cannot be assured due to the pipelay tolerances previously mentioned. Conversely, if the pipeline is offset by 1 m from the trench center, a minimum depth of 1.3 m is a more realistic expectation that can be guaranteed.

Figure 3e illustrates an alternative pipeline alignment, indicated by the red circle. This scenario considers two plough passes, with a trench tolerance of ±1 m and a pipelay tolerance of ±0.75 m. This configuration provides enhanced construction flexibility during installation compared to those previously discussed and may substantially decrease the scope of MFE rectification. Nevertheless, the second plough pass is anticipated to encounter control-related challenges, as elaborated in the subsequent paragraph.

It is noteworthy that monitoring the touchdown point will generally enable the pipeline to be installed at or near the center of the trench. Nonetheless, in instances where trench straightness is not maintained, there exists a risk that, as pipelaying progresses, the lay tension may draw the pipeline into a straight alignment. Consequently, the final position of the pipeline may be situated higher along the slope of the trench than initially observed during touchdown monitoring, thereby reducing the achieved as-laid depth.

In summary, a trench depth of 1 m (measured to the bottom of the pipe) represents the minimum design depth that should be reliably ensured. If higher precision is typically attained, or if two passes of the plough are employed, this depth may be increased to 1.3 m. However, achieving guaranteed depths exceeding 1.3 m may necessitate a second pass and/or a substantial amount of rectification work using the mass flow excavation (MFE) tool. Alternatively, to minimize MFE rectification, a depth of 1 m is likely attainable by employing two passes of the plough and adopting a combined trench and pipe tolerance of ±1.75 m.

3.5. Transition Length of Pipeline

Generally, a necessary pipeline length for a seamless transition from a flat seabed to the bottom of the trench is required. Figure 4 illustrates the analytical trench transition ensuring full contact with the seabed for a trench depth of 1 m and 2 m. The estimation of transition length incorporates considerations of the pipeline's bending stiffness and its self-weight. The findings indicate that a trench with a depth of 1 m necessitates a transition length of 45 m, whereas a trench with a depth of 2 m requires a transition length of 53 m. Consequently, the transition into and out of a trench necessitates an approximate total length of 100 m. This target should be pursued wherever feasible. However, it is important to note that this length exceeds the wavelengths of many smaller sand waves.

Therefore, more aggressive transitions are typically required. Specifically, for sand wave profiles characterized by very steep slopes, a shorter transition can reduce spanning. To optimize the trenched profile, varying transition lengths are necessary, contingent upon the properties of the sand waves being traversed. Although this approach may increase management tasks during trenching operations, it will enhance the final trenching and spanning outcomes.

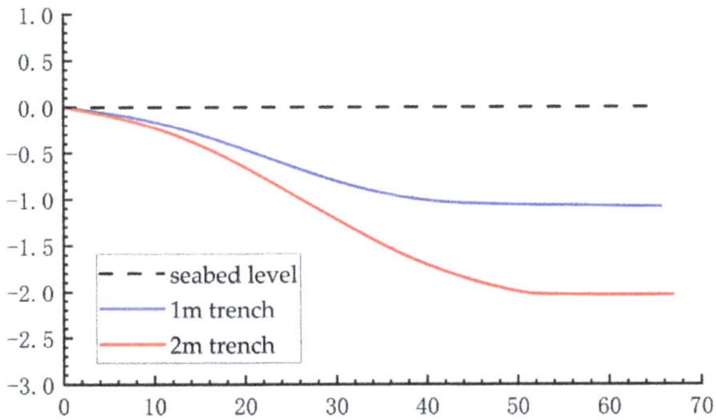

Figure 4. Analytical trench transition (applies to a flat seabed).

3.6. Plough Tow Forces on Flat Seabed

Conventional approaches for estimating plough tow forces have predominantly concentrated on scenarios involving ploughing across a flat seabed. Consequently, the initial determination of plough tow forces relevant to the Lufeng area is conducted under this assumption. The estimation of plough tow forces on a flat seabed is performed utilizing the methodology outlined in Thusyanthan [25], specifically for cohesionless soil, such as sand. This model calculates ploughing resistance by incorporating both static and dynamic components:

$$F = \underbrace{C_w w + C_s \gamma' D^3}_{\text{Static}} + \underbrace{C_d V D^2}_{\text{Dynamic}} \tag{4}$$

where F is the tow force; C_w is the friction coefficient between plough and sand; W is the submerged weight of the plough; C_s is a coefficient related to passive pressure; γ' is effective unit weight of soil; D is the depth; C_d is dynamic force coefficient (increases slightly with soil density and strongly as soil permeability reduces); and V is velocity. The static component characterizes the ploughing resistance at extremely low velocities, allowing for complete drainage. In contrast, the dynamic component addresses soil dilation during shearing, which varies with plough speed. This methodology has been developed utilizing scale models. Plough towing forces are influenced by various parameters, such as trench depths, plough velocities, sand types, etc.

Figure 5a–c illustrate the predicted plough tow forces on a flat seabed for trench depths of 1 m, 2 m, and 2.5 m, respectively. Specifically, loose and medium dense are both considered, which are typically found in the sand wave region. It is highlighted that the dynamic force coefficient, C_d, is extremely sensitive to the soil D_{10} (mm). A review of soil D_{10} values across Lufeng was undertaken, which found values ranging between 0.1 mm to 0.5 mm. Calculations are based on the smaller D_{10} = 0.1 mm value, which gives a higher tow force estimate. Tow forces within the field could vary quite significantly based on this parameter. The tow forces exhibit an increase concomitant with both the escalation of tow velocity and the augmentation of trench depth. Notably, the tow force required for medium dense sand is approximately 50% greater than that necessary for loose sand. The results pertaining to loose sand warrant careful consideration, as physical modeling conducted in Robinson et al. [26] has substantiated anecdotal concerns that the loose sand model tends to underestimate tow forces. Tow forces derived from medium dense sand are deemed more reliable.

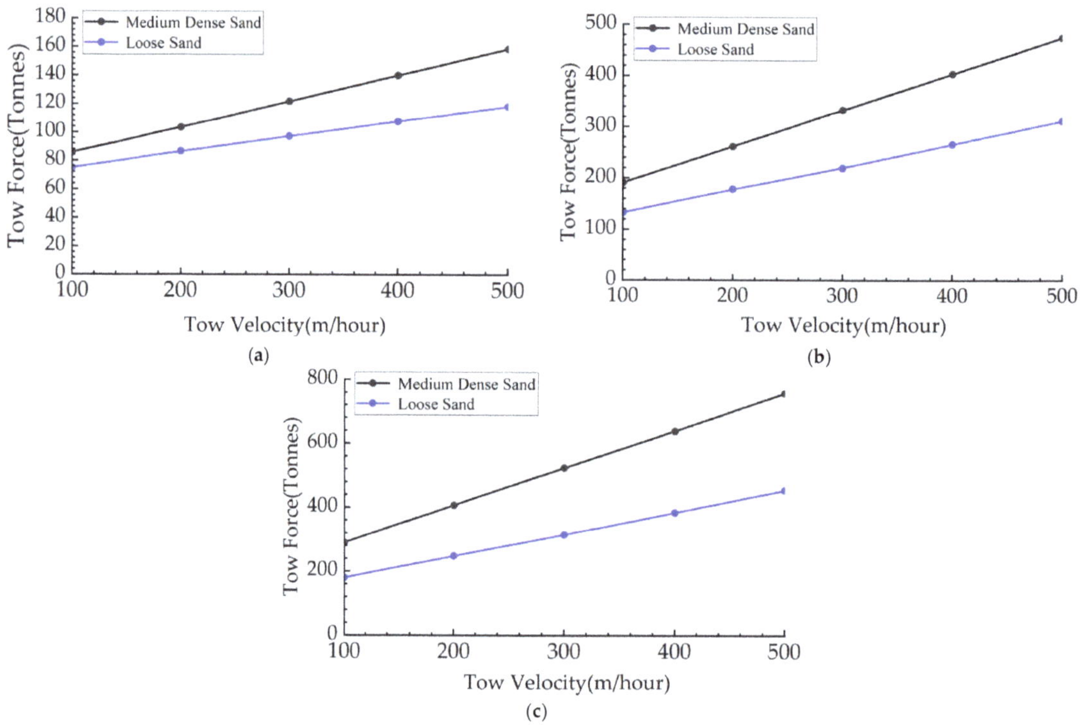

Figure 5. Predicted plough tow force on flat seabed, D_{10} = 0.1 mm, (**a**) 1 m trench depth, (**b**) 2 m trench depth, (**c**) 2.5 m trench depth.

3.7. Impact of Sand Waves on Plough Tow Forces

The study presented in Fredsøe [23] conducted small-scale model testing to examine ploughing in regions characterized by sand waves. The findings indicated that predicting ploughing loads with high confidence is challenging when navigating multiple sand waves. The research evaluated various sand wave geometries, including a flat bed and sand waves with inclinations of 9°, 17°, and 32° and the corresponding of sand wave steepness of h/L = 0.05, 0.10, and 0.20, respectively. The results, illustrated in Figure 6, demonstrated that tow forces increased with both the amplitude and inclination of the sand waves. Under identical plough settings, the towing force required to traverse the sand wave exhibited a 50% increase at a 9° inclination, doubled at a 17° inclination, and quintupled at a 32° inclination, in comparison to the flat bed scenario.

Despite potential variations in the modeling of the Lufeng plough and trench, the findings provide an indication of approximate increases in tow force. The gentle sides of the sand waves along the pipeline route generally exhibit slope angles of up to 5°, with a maximum observed angle of 8°. These observations imply that steeper sand wave slopes should be avoided. This can be effectively managed by strategically selecting the ploughing direction to ascend the gentler slopes and descend the steeper ones.

Several additional factors may contribute to a reduction in tow force: (1) the VMP500 plough is equipped with an active share track, which diminishes the tow force requirements; (2) the model testing cited employed a constant ploughing depth of cut throughout the entire sand wave. This approach contrasts with the Lufeng ploughing process, which involves transitioning in and out of the trench for each sand wave, potentially leading to lower peak tow forces; (3) a reduction in tow force can also be achieved by decreasing the speed of the vessel or plough.

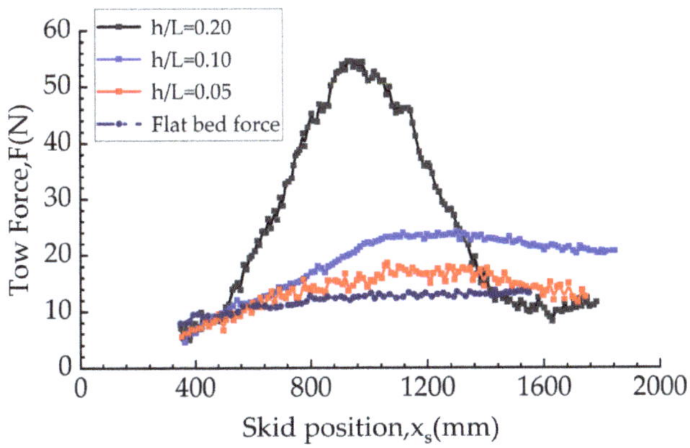

Figure 6. Comparison of tow force for various sand wave heights, for maximum seabed inclinations of 0° (flatbed), 9° (h/L = 0.05), 17° (h/L = 0.10), and 32° (h/L = 0.20).

4. Post-Lay Span Rectification

Post-lay span rectification is performed following pipelay and is required for spans which are not suitable for long term operation. This can occur under the following scenarios: (1) if pre-lay trenching is performed: trenching does not achieve an acceptable span; (2) if no pre-lay trenching is performed: the as-laid span length is not acceptable.

The acceptance criteria for span rectification following pipeline installation are determined by the fatigue limit state (FLS) and ultimate limit state (ULS) under operational conditions. The most effective solution for post-lay span rectification will involve multiple intervention techniques. These techniques can be classified into large-scale and localized methods. A large-scale method is appropriate for the rectification of multiple spans across the field. The chosen method must efficiently and swiftly level the seabed. In instances where the large-scale method fails to achieve span lengths conducive to long-term operation, one or more localized methods will be necessary for the rectification of individual spans.

4.1. Large-Scale Span Rectification Method

The feasibility assessment of the methods has identified the mass flow excavator (MFE) and ROV jetting as suitable techniques for large-scale post-lay span rectification. The primary functional requirements for these large-scale intervention methods are as follows: (1) the method must not pose a threat to the integrity of the pipeline; (2) the method must be appropriate for pipelines experiencing severe free spanning; and (3) the method must be rapid and efficient in leveling the seabed. Detailed analyses of these two methods are provided below.

4.1.1. Mass Flow Excavator

The suitability of MFE tools for post-lay span rectification is evaluated in Table 2. The criteria include pipeline integrity, suitability for severe free spanning, trench rates, and depth and geotechnical conditions.

Table 2. Evaluation of MFE for post-lay span rectification.

Criteria	Evaluation
Pipeline integrity	(1) Non-contact method, therefore minimal integrity threat. (2) Normal construction risk assessments will consider the threat posed by the MFE tool impacting the pipeline in uneven seabed conditions (sand waves).
Suitable for severe free spanning	Non-contact method, therefore suitable for pipelines subject to severe free spanning.
Trench rate and depth	(1) A trench rate of 30 m/h is taken to be an appropriate and robust estimate for the MFE tool for estimating purposes. Detailed evaluation should be performed by equipment vendor. (2) A maximum practical trench depth of 3 m is taken as a basis. The tool does not have a hard limit; however, deeper depths become less practical due to the required trench side slopes and volumes required to be excavated.
Geotechnical conditions	(1) CPT and BH data within the sand wave region have found sands ranging from loose to medium dense. (2) As an example, the T8000 MFE (James Fisher and Sons Plc., Cumbria, UK) tool is suitable for sands, as well as gravel and rock dump. This MFE tool is therefore deemed suitable for the Lufeng sand waves. (3) In some cases, the CPT and BH data do not extend down to 3 m (maximum practical trench depth). However, sand waves are expected to consist of sand down to at least the natural seabed level. (4) Whilst no clays are expected, for information, the T8000 is specified as being suitable for clays up to 50 kPa shear strength, or up to 300 kPa with auxiliary jets.

4.1.2. ROV Jetting

The suitability of ROV Jetting tools for post-lay span rectification is evaluated in Table 3. The criteria for the evaluation are same as those for MFE.

Table 3. Evaluation of ROV jetting for post-lay span rectification.

Criteria	Evaluation
Pipeline integrity	(1) The ROV does not contact the pipeline; however, contact between the jetting nozzles (swords) and pipeline is to be expected. (2) Whilst in free flying mode, the ROV is operated in a near neutrally buoyant configuration, thereby minimizing pipeline interaction forces. Contact between the swords and pipeline is unlikely to pose an integrity threat to the pipeline; however, should be assessed as part of normal construction risk assessments.
Suitable for severe free spanning	At least one vendor has indicated a preference for 'free flying' mode to minimize interaction forces in a pipeline subject to severe free spanning.

Table 3. *Cont.*

Criteria	Evaluation
Trench rate and depth	(1) Trenching rates are dependent on machine specifications, soil conditions and pipeline diameter. Typical manufacturer provided rates range from around 250 m/h to 1000 m/h. (2) 300 m/h is taken to be an appropriate and conservative basis for estimating purposes for Lufeng, noting that full rates may not be achieved due to the start-stop nature of trenching through sand waves. (3) Vendor datasheets typically specify trench depth capabilities of up to 3 m. However, if the pipeline is pre-lay trenched, the achievable depth could be limited to less than 3 m due to potential clearance constraints between the ROV and trench spoil pile. (4) The typical sword length is 3 m. One vendor has advised the sword length can be extended to 4 m, limited by clearance constraints with the ROV chassis.
Geotechnical conditions	(1) CPT and BH data within the sand wave region have found sands ranging from loose to medium dense sand. (2) As an example, the QT1600 jetting tool by SMD (Newcastle, UK), is capable of trenching in soils up to 100 kPa. This ROV jetting tool is therefore suitable for the Lufeng sand waves. (3) The burial suitability shows that sands with density up to and including 'medium dense', are within the operational envelope for jetting. (4) In some cases, the CPT and BH data do not extend down to 3 m (maximum practical trench depth). However, sand waves are expected to consist of sand down to at least the natural seabed level.

4.2. Local Span Mitigation Method

The feasibility assessment of the methods has identified those suitable for local post-lay span rectification, including gravel bags, fabric formwork grout bags, sandbags, steel supports, and strakes.

The primary functional requirements for the local intervention method are as follows: the method must provide adequate long-term vertical support (excluding strakes) and must be suitable for use in regions with a mobile seabed. The shortlisted local rectification methods are evaluated for their suitability against these functional requirements, as detailed in Table 4.

4.3. Recommended Solution

4.3.1. Estimated Durations for MFE vs. ROV Jetting

The potential to decrease the duration of the post-lay campaign through the utilization of an ROV jetting tool, which has an estimated trenching rate of 300 m per hour compared to the 30 m per hour estimated for the MFE tool, has been assessed. It is determined that no pre-lay intervention will be conducted on this pipeline.

The durations for post-lay trenching are estimated differentially for the MFE and ROV jetting options. To assess the sensitivity to various potential intervention outcomes, both Lower Bound (LB) and Upper Bound (UB) estimates are provided. The underlying methodology for this assessment is detailed in Table 5.

Table 4. Suitability of local rectification methods against functional requirements.

Methods	Suitability Against Functional Requirements	
	Vertical Support	**Seabed Mobility**
Gravel	-	Less susceptible to scour due to pliant nature of support
Fabric Formwork Grout Bags	-	High potential for scour leading to loss of contact with pipe.
Sand Bags	Contact with pipe not always achieved when installed post-lay. High potential for long term settlement leading to loss of support.	High potential for scour leading to loss of contact with pipe.
Steel Supports	-	High potential for scour leading to loss of contact with pipe.
Strakes	N/A	-

Table 5. Basis for estimating differential durations.

Step No	Description	Methodology
1	Post-Lay Intervention Requirements	(1) Estimated post-lay rectification requirements are approximately 11 sand wave crests with a total length of around 1.4 km. (2) However, final post-lay requirements cannot be confirmed until following pipelay. Future span rectification requirements are therefore bounded based on ±25% of the estimated 1.4 km length.
2	Trenching Rates	Estimated trenching rates: - Mass flow excavator = 15 m/h to 30 m/h. - ROV Jetting = 150 m/h to 300 m/h.
3	Differential Estimate	(1) Durations are estimated on a differential basis. Activities such as mob/Dembo, span surveys and transits between spans have the same duration for the MFE and ROV jetting options, are not a differentiator and are therefore excluded. (2) Durations are estimated based on total length of seabed requiring intervention and estimated trenching rates. (3) UB duration includes a combined 25% allowance for mechanical downtime and waiting on weather and LB estimate a combined 10% allowance.

4.3.2. Differential Estimate

The differential estimate of post-lay trenching durations is presented in Table 6.

Table 6. Differential estimate of post-lay trenching durations.

Parameter	Values	
	LB	UB
Total seabed length requiring rectification	1050 m	1750 m
Estimated MFE trench rate	30 m/h	15 m/h
Estimated ROV Jetting trench rate	30 m/h	15 m/h
Allowances—mechanical downtime and waiting on weather	10%	25%
Rectification duration excludes trans-its/survey and	38 h	145 h
includes allowances (A/trench rate) × D	4 h	14 h
Time saved using ROV jetting tool	34 h 1.4 days	131 h 5.5 days

The differential analysis indicates that the duration of post-lay trenching could potentially be decreased by approximately 1 to 6 days through the utilization of ROV jetting. Nonetheless, the advantages of this reduced duration must be weighed against the additional effort necessary to mobilize the new equipment.

Trenching durations are estimated using a differential approach. Other campaign activities, including span surveys and transits between spans—activities that remain consistent for both MFE and ROV jetting options—do not serve as differentiators and are thus excluded from consideration. The proposed solution for post-lay span rectification is detailed in Table 7.

Table 7. Recommended post-lay span rectification solution.

Method	Recommendation	Justification
Large Scale	Mass Flow Excavator or ROV jetting	(1) Suitable methods are the MFE and ROV Jetting tools. (2) The differential estimate of post-lay trenching durations has shown that the ROV jetting tool could reduce the overall campaign duration by up to approximately 6 days. (3) Both tools comply with the key functional requirements; they do not pose an integrity threat to the pipeline (subject to risk assessment, minimal threat expected) and are suitable for a pipeline subject to severe spanning. (4) Both tools are capable of achieving the maximum required trench depth of 1 m.
Local	None (gravel contingency)	(1) Based on the recommended rectification solution, there is no requirement for a local method. A contingency method should be planned for. (2) Gravel based supports are less susceptible to scour due to the pliant nature of the support arrangement. Other more rigid forms of support, i.e., sand bags, fabric formwork grout bags and steel supports are generally more susceptible. (3) Discrete gravel-based supports could be constructed using individual gravel bags or a fall pipe vessel. The optimum installation method is dependent on the number of spans requiring rectification and ultimately required volumes. (4) The gravel solution complies with the functional requirement to provide adequate long-term vertical support.

5. Span Rectification Requirement

5.1. Trench Depth Requirements

The trench depth is characterized as the vertical distance from the crest of the undisturbed sand wave to the Bottom of Pipeline (BOP), as depicted in Figure 7. Rectification requirements are specified in terms of the minimum trench depth necessary to meet span acceptance criteria. Figure 7 demonstrates that the target trench depth can be attained through various combinations of pre-lay and/or post-lay trenching techniques. Consequently, the requirement for a minimum trench depth of 1 m, in order to fulfill the operational acceptance criteria, can be achieved through either pre-lay or post-lay trenching methods, or a combination thereof. The minimum trench depth profile serves as the basis for establishing the acceptance criteria for trenching construction.

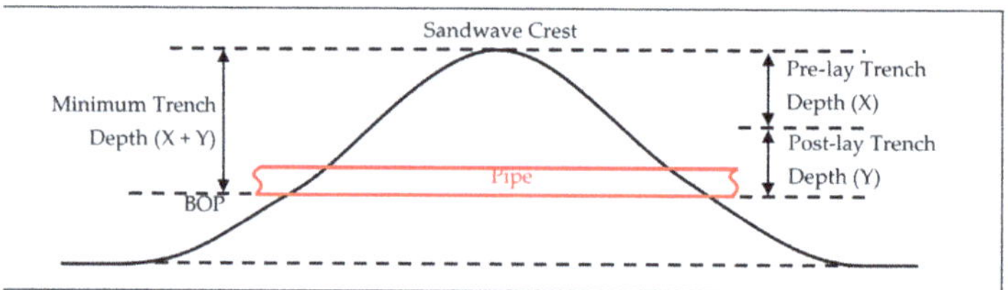

Figure 7. Definition of trench depth.

5.2. Post-Lay Requirements (Operating Condition)

The analysis of on-bottom roughness and free span has indicated that certain spans necessitate rectification to meet the ultimate limit state (ULS) and fatigue limit state (FLS) criteria under operational conditions. It is anticipated that local rectification will not be required for the present pipeline, given the maximum trench depth requirement of 1 m. All spans that exceed the ULS and FLS acceptance criteria can be addressed through a global rectification approach, specifically trenching. Table 8 provides a summary of the trenching requirements.

Table 8. Post-lay (operating condition) rectification requirements summary.

Trench Depth (m)	Quantity of Sand Waves to Be Rectified (-)	Total Length (m)	Trench Angle (°)	Trench CSA (m^2)	Estimated Earthworks Volume (m^3)
1.0	11	1380	20	2.75	3792

The volumes of trench earthworks for the MFE solution have been estimated based on an idealized V-trench cross-section. It is anticipated that the actual volumes displaced by the MFE tool will exceed these estimates and will be influenced by the operator's proficiency. The trench slope, denoted as θ, is calculated using a worst-case soil friction angle of 20°, which is suitable for the sand wave region, as illustrated in Figure 8. It is important to note that this estimation is not applicable to the ROV jetting solution, as this method nominally does not involve soil displacement.

The post-lay rectification requirements outlined in Table 8 encompass the correction of the three crests located between KP 9.2 and KP 9.6, which are associated with the potential upheaval buckle. It is advisable to undertake the post-lay rectification promptly following the pipelay process to minimize uncertainties arising from scour effects, as well as to reduce the risk of vortex-induced vibrations (VIV) and the accumulation of fatigue damage [27]. Comprehensive results pertaining to the post-lay rectification solution under operating conditions are depicted in Figure 9. A summary of the post-lay rectification requirements under operating conditions is provided in Table 8.

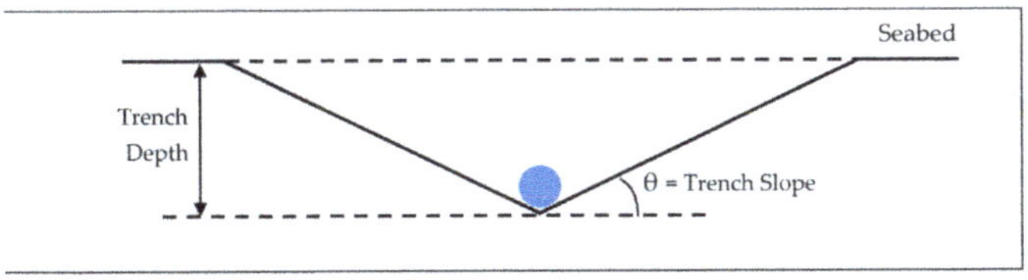

Figure 8. Idealized post-lay trench cross-section (MFE solution).

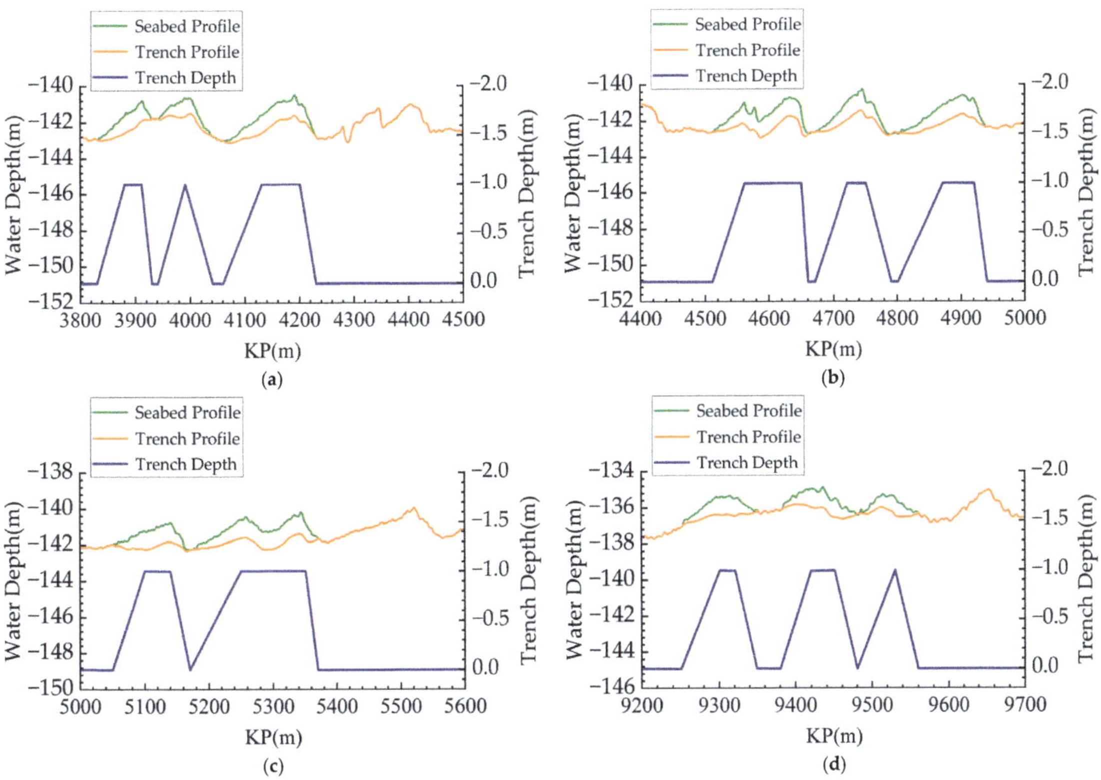

Figure 9. Details of the post-lay (operating condition) rectification solution, (**a**) KP 3.8–KP 4.5, (**b**) KP 4.4–KP 5.0, (**c**), KP 5.0–KP 5.6, (**d**) KP 9.2–KP 9.7.

6. Conclusions

This paper aims to present a design for span rectification specifically tailored for pipeline installations traversing sand wave regions, alongside offering practical guidance on sand wave intervention techniques. The pre-lay rectification process, conducted before the pipeline is laid, is intended to mitigate pipeline spanning conditions that may compromise the integrity of the pipeline during temporary phases. Post-lay span rectification is conducted subsequent to the pipelay process and is necessitated for spans deemed unsuitable for long-term operation.

For the pre-lay span rectification, the analysis generally encompasses the following steps: (1) estimating the trench backfill rate, (2) controlling the plough position and depth,

(3) analyzing the trench slope stability, (4) determining the trench depth and trench/pipelay acceptance criteria, (5) estimating the transition length of pipeline, (6) calculating the plough tow forces on flat seabed, and (7) accounting for the impact of sand waves on plough tow forces.

In the absence of pre-lay trenching, this procedure may be required under the following circumstances: (1) the as-laid span length is deemed unacceptable; or (2) the as-laid span length increases in situ due to seabed mobility or scour. The optimal solution for post-lay span rectification necessitates the utilization of multiple intervention methods. For the rectification of multiple spans across the field, a large-scale approach is essential. A comparative analysis of post-lay trenching durations employing the MFE and ROV jetting tools is provided. In instances where the large-scale method fails to achieve span lengths adequate for long-term operation, a localized approach is necessary to address individual spans.

The desired trench depth may be attained through any combination of pre-lay and/or post-lay trenching techniques. The minimum trench depth profile serves as the basis for establishing the trenching construction acceptance criteria. Analysis of on-bottom roughness and free span indicates that, given the natural seabed profile (without trenching), there are no spans surpassing the ultimate limit state (ULS) or fatigue limit state (FLS) criteria for the temporary installation scenario. Consequently, pre-lay rectification is not necessary. The analysis of on-bottom roughness and free span has indicated that post-lay rectification is essential to meet the ultimate limit state (ULS) and fatigue limit state (FLS) criteria under operating conditions. All spans surpassing the ULS and FLS criteria can be rectified by trenching to a depth of 1 m.

The analysis of the suspension span of submarine pipelines in sand wave regions necessitates a detailed examination based on the specific seabed characteristics of the project location. The exposure of submarine pipelines on the seabed can lead to variations in the suspension span due to the dynamic movement of sand waves, which may either increase or decrease the span. Given that this is a prolonged and gradual process, it is essential to account for the impact of sand wave movement on alterations in seabed topography and pipeline span during seabed treatment considerations. Based on the adaptive analysis of trenching in sand wave regions, along with the results from mechanism analysis and the feasibility evaluation criteria, the most appropriate method for addressing overspan issues can be proposed to ensure the successful implementation of the project.

Author Contributions: Conceptualization, X.Z. and Z.Z.; data curation, X.Z., B.X. and X.B.; methodology, X.Z., Y.G., Q.Y. and Z.Z.; investigation, X.Z., B.X. and X.B.; resources, X.Z. and Z.Z.; writing—original draft preparation, X.Z. and B.X.; writing—review and editing, X.Z., B.X., Y.G., Q.Y. and Z.Z.; visualization, X.B., Y.G. and Q.Y.; supervision, B.X. and Z.Z.; funding acquisition, X.Z. and Z.Z. All authors have read and agreed to the published version of the manuscript.

Funding: This research was funded by the National Natural Science Foundation of China (52371289) and the China National Offshore Oil Corporation (Major science and technology project during the 14th Five-year plan).

Institutional Review Board Statement: Not applicable.

Informed Consent Statement: Not applicable.

Data Availability Statement: Data from the present experiment appear in the submitted manuscript.

Conflicts of Interest: Xing Zou and Botao Xie were employed by CNOOC Research Institute Co., Ltd.; Yuanyuan Guo and Qingqing Yuan were employed by CRRC SMD (Shanghai) Ltd. The remaining authors declare that the research was conducted in the absence of any commercial or financial relationships that could be construed as a potential conflict of interest.

References

1. Luan, X.; Peng, X.; Wang, Y.; Qiu, Y. Characteristics of Sand Waves on the Northern South China Sea Shelf and Its Formation. *Acta Geol. Sin.* **2010**, *84*, 233–245.
2. Guo, L.; Ma, X.; Yan, J. Distribution pattern and control factors of sand waves in southeast Beibu Gul. *Mar. Geol. Quateranry Geol.* **2017**, *37*, 67–76.
3. *Mobility Monitoring of Seabed Sand Waves in Lufeng Area, Engineering Geophysical Survey Report*; COOEC Offshore Division: Tianjin, China, 2020.
4. Németh, A.A.; Hulscher, S.J.M.H.; de Vriend, H.J. Modelling sand wave migration in shallow shelf seas. *Cont. Shelf Res.* **2002**, *22*, 2795–2806. [CrossRef]
5. Zang, Z.; Xie, B.; Cheng, L.; He, F.; Zou, X. Numerical Investigations on the Transient Behavior of Sand waves in Beibu Gulf Under Normal and Extreme Sea Conditions. *China Ocean Eng.* **2023**, *37*, 1–15. [CrossRef]
6. Sun, Y.; Wang, C.; Zhou, Q.; Liu, Z. A review on sand waves evolution and its influnece on pipeline engineering. *Adv. Mar. Sci.* **2018**, *36*, 490–498.
7. Rezazadeh, K.; Zhu, L.; Bai, Y.; Zhang, L. Fatigue Analysis of Multi-Spanning Subsea Pipelines. In Proceedings of the International Conference on Ocean, Offshore and Arctic Engineering, Shanghai, China, 6–11 June 2010.
8. Meniconi, S.; Brunone, B.; Tirello, L.; Rubin, A.; Cifrodelli, M.; Capponi, C. Transient Tests for Checking the Trieste Subsea Pipeline: Toward Field Tests. *J. Mar. Sci. Eng.* **2024**, *12*, 374. [CrossRef]
9. Meniconi, S.; Brunone, B.; Tirello, L.; Rubin, A.; Cifrodelli, M.; Capponi, C. Transient Tests for Checking the Trieste Subsea Pipeline: Diving into Fault Detection. *J. Mar. Sci. Eng.* **2024**, *12*, 391. [CrossRef]
10. Xiao, R.; Zayed, T.; Meguid, M.; Sushama, L. Rapid failure risk analysis of corroded gas pipelines using machine learning. *Ocean Eng.* **2024**, *313*, 119433. [CrossRef]
11. Seghier, M.E.A.B.; Mustaffa, Z.; Zayed, T. Reliability assessment of subsea pipelines under the effect of spanning load and corrosion degradation. *J. Nat. Gas Sci. Eng.* **2022**, *102*, 104569. [CrossRef]
12. Liang, X.; Huang, Y.; Galedari, S.A.; Azarmi, F. Pipeline corrosion assessment using embedded Fiber Bragg grating sensors. *Proc. SPIE Int. Soc. Opt. Eng.* **2015**, *9435*, 94351Z. [CrossRef]
13. Farh, H.M.H.; Seghier, M.E.A.B.; Zayed, T. A Comprehensive Review of Corrosion Protection and Control Techniques for Metallic Pipelines. *Eng. Fail. Anal.* **2022**, *143*, 106885. [CrossRef]
14. Xu, H.; Yin, S. Surface Corrosion Rate of Gas Pipeline Based on Dynamic Stray Current and Carbon Dioxide Interference. *Surf. Technol.* **2017**, *46*, 206–210.
15. Melo, C.; Dann, M.R.; Hugo, R.; Janeta, A. Extreme value modeling of localized internal corrosion in unpiggable pipelines. *Int. J. Press. Vessel. Pip.* **2020**, *182*, 104055. [CrossRef]
16. Zhao, S.; Lan, W. Present Status and Research Progress of Anti-corrosion Technology in Pipeline. *Surf. Technol.* **2015**, *44*, 112–118.
17. Han, P.; Gao, J.; Wang, F.; Su, Z.; Yang, Z.; Huang, S.; Jing, S. Bottom roughness analysis and span treatment method of offshore pipeline laid on sand wave seabed. *Pet. Eng. Constr.* **2021**, *47* (Suppl. S1), 25–31.
18. Huang, M.; Xu, J.; Luan, Z.; Liu, M.; Li, X.; Liu, B. Analysis of DF1-1 subsea pipeline free-span distribution characteristics and rectification effects. *Mar. Sci.* **2021**, *45*, 77–87.
19. *Lufeng Sandwave Treatment Proposal*; COOEC Offshore Division: Tianjin, China, 2020.
20. *DNV GL-RP-F105*; Free Spanning Pipelines. Det Norske Veritas: Oslo, Norway, 2017.
21. *DNV GL-ST-F101*; Submarine Pipeline Systems. Det Norske Veritas: Oslo, Norway, 2017.
22. Lauder, K. The Performance of Pipeline Ploughs. Ph.D. Thesis, University of Dundee, Dundee, UK, 2010.
23. Fredsøe, J. Natural backfilling of pipeline trenches. *J. Pet. Technol.* **1979**, *31*, 1223–1230. [CrossRef]
24. Lowe, S. Using natural sedimentation to backfill a dredged pipeline trench. *J. Mar. Eng. Technol.* **2010**, *9*, 11–16. [CrossRef]
25. Thusyanthan, I. State of the art ploughability assessment. In Proceedings of the Twenty-Sixth (2016) International Ocean and Polar Engineering Conference, Rhodes, Greece, 26 June–2 July 2016; pp. 327–334.
26. Robinson, S.; Brown, M.; Matsui, H.; Brennan, A.; Augarde, C.; Coombs, W.; Cortis, M. Centrifuge Testing to Verify Scaling of Offshore Pipeline Ploughs. *Int. J. Phys. Model. Geotech.* **2019**, *19*, 305–317. [CrossRef]
27. Zou, X.; Xie, B.; Zang, Z.; Chen, E.; Hou, J. Vortex-induced vibration and fatigue damage assessment for a submarine pipeline on a sand wave seabed. *J. Mar. Sci. Eng.* **2023**, *11*, 2031. [CrossRef]

Journal of
Marine Science and Engineering

MDPI

Article

Wave-Induced Instantaneous Liquefaction of a Non-Cohesive Seabed around Buried Pipelines: A Liquefaction-Associated Non-Darcy Flow Model Approach

Shichong Han [1], Mozhen Zhou [1,*], Dingli Zhang [1], Wengang Qi [2,3], Chaodong Xue [1] and Qian Fang [1]

1 Key Laboratory for Urban Underground Engineering of the Ministry of Education, Beijing Jiaotong University, Beijing 100044, China; 22121071@bjtu.edu.cn (S.H.); dlzhang@bjtu.edu.cn (D.Z.); 20121130@bjtu.edu.cn (C.X.); qfang@bjtu.edu.cn (Q.F.)

2 Key Laboratory for Mechanics in Fluid Solid Coupling Systems, Institute of Mechanics, Chinese Academy of Sciences, Beijing 100190, China; qiwengang@imech.ac.cn

3 School of Engineering Science, University of Chinese Academy of Sciences, Beijing 100049, China

* Correspondence: mzzhou@bjtu.edu.cn

Abstract: In complex marine environments, the wave-induced instantaneous liquefaction of the seabed is a key issue for the long-term safety control of marine structures. Existing computational frameworks for instantaneous liquefaction result in unreasonable tensile stresses in a non-cohesive seabed. To address this issue, a liquefaction-associated non-Darcy flow model has been proposed, but it has only been applied to the scenario of a pure seabed without a structure. In this study, we applied the previously proposed non-Darcy flow model to investigate the mechanism of wave–seabed–structure interactions under extreme wave loading considering a pipeline fully buried in a non-cohesive seabed. By comparing the liquefaction depths in the presence and absence of structures, it was found that the existence of structures weakens the attenuation of the pore pressure amplitude and influences the overall pore pressure distribution. Parametric studies were conducted. It was found that the liquefaction depth from the non-Darcy model is approximately 0.73 times that from the traditional Darcy model, regardless of whether or not a pipeline is involved. A quantitative relationship between the wave loading and structural size was established. The liquefied zone above the buried pipeline was found to be smaller than that in a pure seabed without a structure. A tentative explanation is provided for this phenomenon.

Keywords: extreme wave loading; wave-induced seabed liquefaction; non-Darcy flow model; wave–seabed–structure interactions; non-cohesive seabed

1. Introduction

In the wave-dominated coastal zone, the wave-induced liquefaction of the seabed often occurs due to the propagation of waves inducing excess pore pressure within the seabed, leading to a decrease in effective stress. Seabed liquefaction has a close relation with the erosion or resuspension of marine sediment [1–5] and can affect the stability of coastal structures [6,7]. Therefore, it has become a key concern within the scope of marine engineering. Based on laboratory experiments and field measurements [8,9], wave-induced seabed liquefaction can be categorized into instantaneous liquefaction and residual liquefaction, induced by instantaneous pore pressure and residual pore pressure, respectively. This study focuses on the instantaneous liquefaction occurring in a non-cohesive seabed.

In this scenario, the behavior of the non-cohesive seabed can be described using the pore-elasticity assumption for the following reasons. After the occurrence of instantaneous liquefaction, the soil becomes loose, and then it rearranges into a dense state through wave-induced compaction or consolidation. This results in subsequent wave-induced seabed responses following a "reloading–unloading" pattern [7,10,11]. For the case of a pure seabed without a structure, several analytical solutions have been developed under various conditions [12–15]. However, for a seabed including structures such as submarine pipelines [16–26], breakwaters [27–29], coastal slopes [30], offshore wind turbine foundations [31–37], gravity-based structure (GBS) offshore platforms [38], dumbbell-shaped cofferdams [39], and immersed tunnels [40,41], the presence of the structures complicates the boundary conditions, generally requiring numerical methods for analyses. Recently, Li et al. [42] developed an open-source numerical toolbox for simulating the interaction between a porous seabed, waves, and structures. This toolbox greatly enhances the efficiency of research on wave–seabed–structure interactions. When marine structures undergo rocking motion under external forces, additional excess pore pressure can be induced [43–46]. It has also been found that marine structures can exhibit the phenomenon of principal stress rotation [47,48]. Furthermore, nonlinear contact behavior at the interface between the structure and seabed should also be considered [49]. These scenarios have been extensively analyzed in previous studies.

However, the existing computational methods and analytical models associated with instantaneous liquefaction can result in unreasonable tensile stresses in a non-cohesive seabed [50,51]. This nonphysical phenomenon further influences the overall distribution of pore pressure [50,51]. In fact, when instantaneous liquefaction occurs in a non-cohesive seabed, the physical and mechanical parameters of the seabed should change accordingly. However, most existing models employ an invariant poro-elastic assumption [14,35].

In order to eliminate unreasonable tensile behavior, Zhou et al. [52] was inspired by the literature [53,54] and introduced a dynamic permeability model. This model uses soil permeability k_s that is determined by pore pressure p instead of the constant permeability assumed in conventional Darcy's law. This treatment mitigates the occurrence of unrealistic tensile stresses in the liquefaction zone. Compared with the previous dynamic permeability models [53,54], the new model built by Zhou et al. [52] agrees with the concept of an increase in permeability during soil liquefaction [55–60]. However, this model [52] faces challenges in nonlinear convergence. When using larger model parameters or finer computational grids, numerical divergence may even occur, limiting its applications. Therefore, Zhou et al. [61] further converted the instantaneous liquefaction problem into a mathematical problem, i.e., a nonlinear complementarity problem (NCP). They employed specific Karush–Kuhn–Tucker (KKT) conditions for instantaneous liquefaction and a Lagrange multiplier method to solve the problem, while they optimized it using the primal-dual active set strategy (PDASS) [62]. Additionally, they improved computational efficiency by employing direct delta function interpolation to condense the multipliers. This method addresses the challenge of nonlinear convergence caused by the previous dynamic permeability model [52]. However, solving the NCP problem falls within the scope of a constrained variational theory framework and is not a standard module in most finite element codes (such as PORO-FSSI-FOAM [20,21]). Applying the NCP approach to broader scenarios, such as pipelines that are partially buried in the seabed [16,23], gravity foundations for offshore wind turbines [32], and monopile foundations [33,35], is also not convenient.

Recently, Zhou et al. [63] proposed a new KKT condition for wave-induced instantaneous liquefaction through the variational equivalence principle; thus, a non-Darcy flow model was proposed for instantaneous liquefaction. This model not only eliminates the phenomenon of tensile stress in the liquefied zone but also provides significant convenience for numerical implementations, because only the formulation of a seepage constitutive equation is required to be incorporated into finite element codes. Furthermore, it provides a unified explanation for various existing phenomena, such as non-cohesive granular soils with an upward seepage flow [64]; experimental results on piping in sandy gravels [65]; field trials [3]; and DEM-PFV simulations [66]. However, these works were still limited to the scenario of wave–seabed interactions without a structure.

In this work, we further took into account a pipeline fully buried in a non-cohesive seabed under wave loadings in order to investigate wave–seabed–structure interactions with the use of our non-Darcy flow model [63]. First, we extended our previous finite element code to a scenario that can consider a structure fully buried in the seabed. This extension was validated by comparing our numerical results with the experimental results of Turcotte [67]. Then, a computational example under extreme wave loading was conducted to compare the liquefaction depths in the presence and absence of structures. Parametric studies were further accomplished considering wave, seabed, and structural parameters. Non-dimensional parametric studies were also carried out.

The remainder of this paper is as follows. Section 2 provides a brief overview of the fundamental theory and governing equations of the non-Darcy flow model considering a fully buried structure. In Section 3, the wave–seabed–structure model is validated based on classical experimental results without liquefaction. Section 4 presents the numerical results under extreme wave loading conditions, which induce seabed liquefaction. A quantitative relationship between the wave loading and structural response is exploited. Section 5 tries to provide an explanation for the liquefaction phenomenon in a seabed including a buried structure. Finally, Section 6 draws several conclusions.

2. The Liquefaction-Associated Non-Darcy Flow Model

2.1. Wave–Seabed–Structure Model and Liquefaction Criteria

Figure 1 depicts the wave–seabed–structure model and the process of seabed liquefaction, where d_t represents the thickness of the seabed, h represents the water depth, and H represents the wave height. The coordinate value on the seabed surface is 0, and positive values exist below the seabed. Ω_L denotes the zones in the seabed where instantaneous liquefaction occurs. During the process of wave loading, when the soil is located at or near a wave trough, the wave-induced seepage force is upward. When the upward seepage force equals the initial vertical effective stress of the seabed soil, the effective stress of the soil becomes zero, leading to an instantaneous liquefaction state. It should be noted that the wave in Figure 1 is linear. If considering nonlinear waves, Stokes waves [17] or cnoidal waves [68] should be employed. To simplify the model, this study applies linear wave theory [69]:

$$P_b = p_0 \cos\left(\frac{2\pi}{L}x - \frac{2\pi}{L}t\right), \tag{1}$$

where P_b, whose amplitude is represented by p_0, represents the wave pressure acting on the seabed surface (z = 0). L denotes the wavelength, while T denotes the wave period. The values of p_0 and L can be calculated by the following equations:

$$p_0 = \frac{\gamma_w H}{2\cosh(2\pi h/L)}, \ L = \frac{gT^2}{2\pi}\tanh\left(\frac{2\pi h}{L}\right), \tag{2}$$

where γ_w denotes the unit weight of water, while g represents the acceleration of gravity. The governing equations of the wave–seabed–structure model are as follows:

$$\nabla \cdot (\sigma' - p\mathbf{I}_{2\times 2}) + \mathbf{b} = 0 \text{ in } \Omega\,, \tag{3a}$$

$$\frac{\partial \varepsilon_v}{\partial t} + n\beta\frac{\partial p}{\partial t} - \nabla \cdot \left(\frac{k_s}{\gamma_w}\nabla p\right) = 0 \text{ in } \Omega\,, \tag{3b}$$

$$\mathbf{u} = \hat{\mathbf{u}} \text{ on } \Gamma_u\,, \tag{3c}$$

$$\sigma \cdot \mathbf{n}_\sigma = \hat{\mathbf{t}} \text{ on } \Gamma_\sigma\,, \tag{3d}$$

$$p = \hat{p} \text{ on } \Gamma_p\,, \tag{3e}$$

$$\mathbf{v}_w \mathbf{n}_v = \hat{v}_w^n \text{ on } \Gamma_v\,, \tag{3f}$$

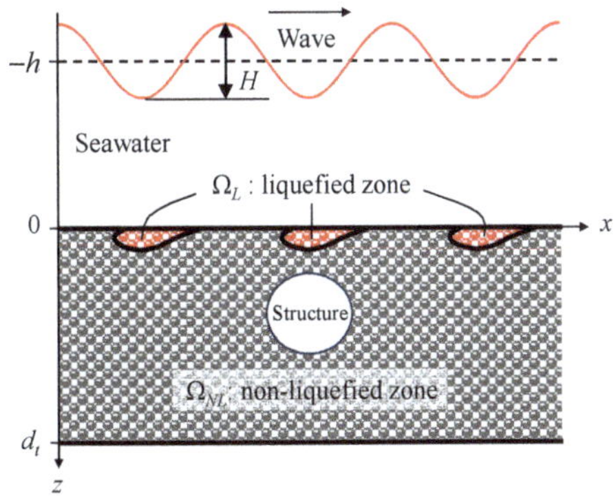

Figure 1. Simplified wave–seabed–structure model with instantaneous liquefaction.

The first two equations are derived from the theory of poro-elasticity theory [70]. Equation (3a) denotes the equilibrium of the solid–fluid mixture, in which σ' represents the effective stress, $\mathbf{I}_{2\times2}$ denotes the second-order unit tensor, and $\mathbf{b}$ denotes the body force per unit volume. Equation (3b) denotes the conservation of mass, in which ε_v represents the volumetric strain, t represents time, k_s is Darcy's coefficient of permeability, and β represents the pore-fluid compressibility, which is determined by [71]

$$\beta = \frac{1}{K_{w0}} + \frac{1 - S_r}{P_{abs}}, \tag{4}$$

where K_{w0} represents the true bulk modulus of pore water, which can be taken as 2.0×10^9 Pa [14]. S_r denotes the degree of saturation, while P_{abs} represents the hydrodynamic pressure of water. The remaining expressions in Equation (3c–f) denote boundary conditions. Γ_u and Γ_σ denote two boundaries of the solid phase, where Γ_u represents the Dirichlet boundary and Γ_σ represents the Neumann boundary, while Γ_p and Γ_v denote two boundaries of the fluid phase, where Γ_p represents the Dirichlet boundary and Γ_v represents the Neumann boundary. In Equation (3c), $\mathbf{u}$ represents the vector of displacement, while $\hat{\mathbf{u}}$ denotes the displacement that is constrained. In Equation (3d), σ denotes the tensor of total stress, $\mathbf{n}_\sigma$ denotes the unit normal vector pointing outward from Γ_σ, and $\hat{\mathbf{t}}$ denotes the surface force per unit area acting on the boundary. In Equation (3e), $\hat{p}$ represents the pore pressure. In Equation (3f), $\mathbf{v}_w$ denotes the vector of pore-fluid velocity, $\mathbf{n}_v$ denotes the unit normal vector pointing outward from Γ_v, and $\hat{v}_w^n$ represents the boundary Darcy velocity.

It should be noted that the research presented in this paper aims to extend the non-Darcy flow model from pure seabed scenarios without structures to scenarios with structures. The presence of a structure in the seabed implies increased boundary conditions compared to the pure seabed. In order to simplify the computational model, the boundary condition around the structure is considered an impermeable boundary, corresponding to Equation (3), where $\hat{v}_w^n$ is taken as zero.

When the seabed experiences instantaneous liquefaction, there are two widely applied liquefaction criteria for determining the liquefaction zone (Ω_L), which use the tensile region of the sandy soil as the liquefaction zone:

$$p - P_b \geq \gamma' z, \tag{5}$$

$$j_z \geq \gamma', \tag{6}$$

where p is the excessive pore pressure caused by waves. Its gradient along the vertical direction is expressed as $j_z = \partial p / \partial z$. γ' denotes the buoyant unit weight of the seabed and obeys the following expression: $\gamma' = (G_s - 1)(1 - n)\gamma_w$, where G_s represents the specific gravity of sand particles, and n represents the porosity of the sand, which is related to the void ratio e: $n = e/(1+e)$.

2.2. Numerical Implementation of the Non-Darcy Flow Model

In general, the liquefaction zone may vary based on the aforementioned two liquefaction criteria. Qi and Gao's [51] research revealed that, in Equations (5) and (6), the parts where $p - P_b > \gamma' z$ and $j_z > \gamma'$ represent the presence of tensile stress in the non-cohesive seabed. This is obviously nonphysical and leads to the inconsistency in liquefaction zones predicted by the two criteria. Therefore, when liquefaction occurs, even if the wave height continues to increase, the inequality signs in Equations (5) and (6) should always be equal signs. Thus, we revised Equations (5) and (6) as follows, respectively:

$$p - P_b \equiv \gamma' z \tag{7}$$

$$j_z \equiv \gamma' \tag{8}$$

By introducing Equation (8) as an additional constraint to Equation (3), Equation (3) can be rewritten as follows:

$$\nabla \cdot \left(\sigma' - p\mathbf{I}_{2\times 2}\right) + \mathbf{b} = 0 \text{ in } \Omega\,, \tag{9a}$$

$$\frac{\partial \varepsilon_v}{\partial t} + n\beta \frac{\partial p}{\partial t} - \nabla \cdot \mathbf{v}_w = 0 \text{ in } \Omega\,, \tag{9b}$$

$$\mathbf{u} = \hat{\mathbf{u}} \text{ on } \Gamma_u\,, \tag{9c}$$

$$\sigma \cdot \mathbf{n}_\sigma = \hat{\mathbf{t}} \text{ on } \Gamma_\sigma\,, \tag{9d}$$

$$p = \hat{p} \text{ on } \Gamma_p\,, \tag{9e}$$

$$\mathbf{v}_w \mathbf{n}_v = \hat{v}_w^n \text{ on } \Gamma_v\,, \tag{9f}$$

where the pore-fluid velocity $\mathbf{v}_w$ is determined by

$$v_{wx} = -\frac{k_s}{\gamma_w}\frac{\partial p}{\partial x},\ v_{wz} = -\frac{k_s}{\gamma_w}\frac{\partial p}{\partial z} - \kappa\left(\frac{\partial p}{\partial z} - \gamma'\right), \tag{10}$$

where v_{wx} and v_{wz} represent the horizontal and vertical components of $\mathbf{v}_w$, respectively. κ is the penalty factor and makes Darcy's law hold in the horizontal direction, while the significant nonlinearity caused by seabed liquefaction appears in the vertical direction.

Because the pressure gradient is related to the hydraulic gradient $\mathbf{i}$ ($\Delta p = \gamma_w \mathbf{i}$), v_{wz} can be written as

$$v_{wz} = -k_s i_z - \kappa(i_z - i_{cr}) = \begin{cases} -k_s i_z, & if\ i_z < i_{cr} \\ -k_s i_{cr} - (k_s + \kappa_\infty \gamma_w)(i_z - i_{cr}), & if\ i_z \geq i_{cr} \end{cases}, \tag{11}$$

where i_{cr} equals γ'/γ_w and represents the critical value that determines the occurrence of liquefaction. i_z represents the vertical component of the hydraulic gradient. κ_∞ represents the penalty value and is generally taken as a large value, e.g., 1000.

The equations described above are shown in Figure 2 in the $v_{wz} - i_z$ space, which appears as the non-Darcy flow model.

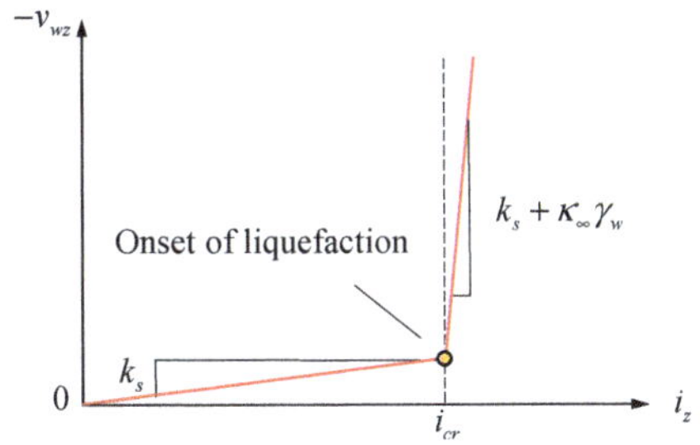

Figure 2. Schematic diagram of non-Darcy flow model.

After discretization, linearization, and other derivation steps (the specific derivation process has been provided by Zhou [63] and is omitted here), the finite element formulation using the non-Darcy flow model is obtained as follows:

$$\begin{bmatrix} \mathbf{K} & \mathbf{G} \\ \mathbf{G}^{\mathrm{T}} & \mathbf{H}+\mathbf{L} \end{bmatrix} \begin{Bmatrix} \mathbf{d}_u^{t,k} \\ \mathbf{d}_p^{t,k} \end{Bmatrix} = \begin{Bmatrix} \mathbf{F}_u \\ \mathbf{F}_p \end{Bmatrix}, \tag{12}$$

with

$$\mathbf{K} = \int_\Omega \mathbf{B}_u^{\mathrm{T}} \mathbf{D} \mathbf{B}_u \mathrm{d}\Omega, \tag{13a}$$

$$\mathbf{G} = -\int_\Omega \mathbf{B}_u^{\mathrm{T}} \mathbf{m} \mathbf{N}_p \mathrm{d}\Omega, \tag{13b}$$

$$\mathbf{H} = -\int_\Omega n\beta \mathbf{N}_p^{\mathrm{T}} \mathbf{N}_p \mathrm{d}\Omega, \tag{13c}$$

$$\mathbf{L} = \frac{\tau}{\gamma_w} \int_\Omega \mathbf{B}_p^{\mathrm{T}} \mathbf{v}'_w \mathbf{B}_p \mathrm{d}\Omega, \tag{13d}$$

$$\mathbf{F}_u = -\int_\Omega \mathbf{N}_u^{\mathrm{T}} \mathbf{b} \mathrm{d}\Omega - \int_{\Gamma_\sigma} \mathbf{N}_u^{\mathrm{T}} \hat{\mathbf{t}} \mathrm{d}\Gamma, \tag{13e}$$

$$\mathbf{F}_p = \mathbf{G}^{\mathrm{T}} \mathbf{d}_u^{t-\tau} + \mathbf{H}\mathbf{d}_p^{t-\tau} + \mathbf{L}\mathbf{d}_p^{t,k-1} + \tau \int_{\Gamma_v} \mathbf{N}_p^{\mathrm{T}} v_w^n \mathrm{d}\Gamma - \tau \int_\Omega \mathbf{B}_p^{\mathrm{T}} \mathbf{v}_{w,h}^{t,k-1} \mathrm{d}\Omega, \tag{13f}$$

with

$$\mathbf{B}_u = \nabla \mathbf{N}_u,\ \mathbf{m} = \{1,1,0,0\}^{\mathrm{T}},\ \mathbf{B}_p = \nabla \mathbf{N}_p \tag{14}$$

where $\mathbf{d}_u$ and $\mathbf{d}_p$ denote the discrete unknown vectors corresponding to the displacement (u) and the excessive pore pressure (p), respectively. The shape function matrices of $\mathbf{d}_u$ and $\mathbf{d}_p$ are $\mathbf{N}_u$ and $\mathbf{N}_p$. $\mathbf{v}'_w$ denotes the partial derivative of $\mathbf{v}_w$ with respect to $\mathbf{i}$, and $\mathbf{v}_{w,h}^t$ denotes the current discrete velocity, while the subscripts k and $k-1$ are used to indicate the nonlinear iterations.

Based on the algebraic representations mentioned above, the iterative process for the solution (d_u^t, d_p^t) can be derived and conveniently applied in conventional finite element codes.

3. Validation of the Wave–Seabed–Structure Model without Seabed Liquefaction

3.1. Computational Model

The model without a structure and our finite element code have been validated by Zhou et al. [52,61,63], including cylinder tests under 1-D wave loading and two-dimensional (2-D) wave–seabed interactions. These validations are therefore not repeated here in this work.

This section focuses on the validation of our finite element code applied to the scenario involving a structure, which is taken as a pipeline fully buried in the non-cohesive seabed. In this scenario, a series of flume tests were conducted by Turcotte [67], wherein seabed liquefaction was not induced. The model of these tests is shown in Figure 3. In the tests, the wave period T was 0.9–2.3 s and the wave height H was 3.02–14.3 cm, while L was 3.53 m, h was 0.533 m, S_r was 0.997, n was 0.42, k_s was 0.001 m·s^{-1}, E was 3 MPa, and ν was 0.33. The computational model was then built according to the given conditions.

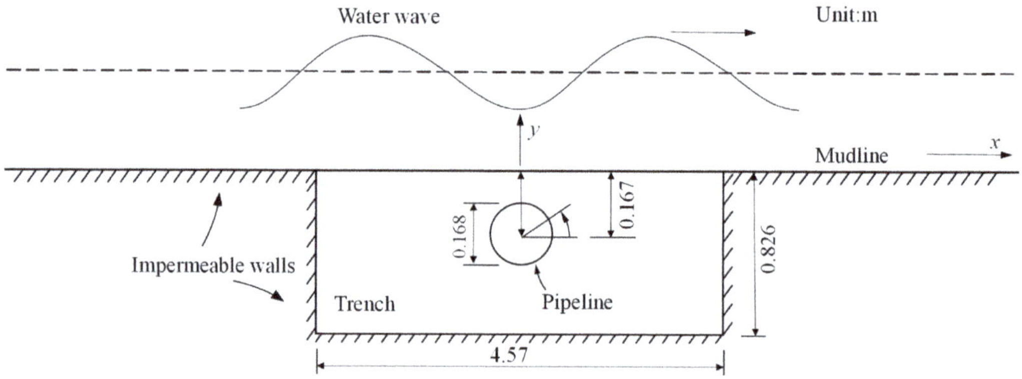

Figure 3. A schematic of the experiment by Turcotte [67].

3.2. Pore Pressure Response

The distribution of pore pressure around the pipeline is shown in Figure 4. The vertical axis represents the normalized pore pressure p/P_0. Cheng and Liu [72] also conducted numerical studies based on Turcotte's flume experiments. For a better visual comparison, the numerical results of this study were compared with the flume experimental results of Turcotte [67] and the numerical results of Cheng and Liu [72]. It can be seen that our numerical results are in basic agreement with the experimental results of Turcotte [67] and the numerical results of Cheng and Liu [72].

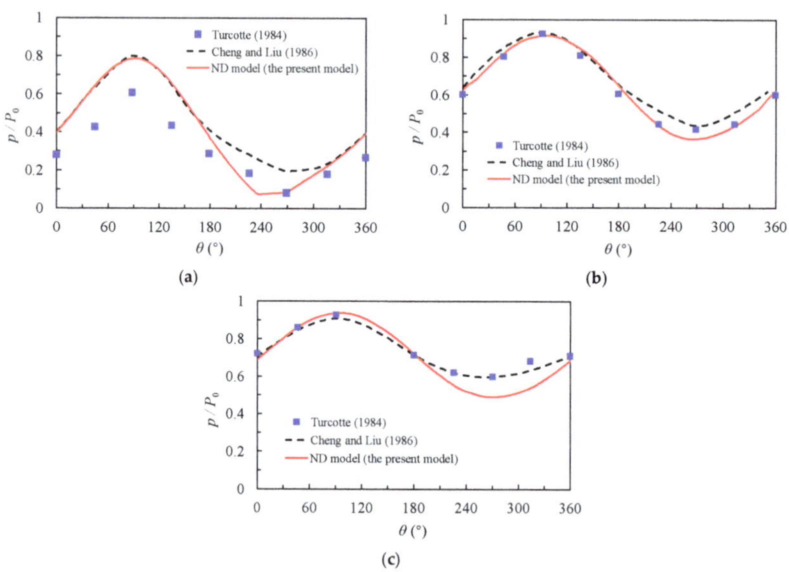

Figure 4. A comparison of the wave-induced pore pressure amplitude along the periphery of the fully buried pipeline between laboratory data by Turcotte [67] and numerical simulations by Cheng and Liu [72]: (**a**) T = 0.9 s, H = 5.24 cm; (**b**) T = 1.75 s, H = 14.3 cm; (**c**) T = 2.3 s, H = 3.02 cm.

Figure 5 shows the distribution of p/P_0 along the depth direction of the centerline using the constant-permeability model (which is hereafter called the CP model) and the liquefaction-associated non-Darcy model (which is hereafter called the ND model) both with and without a pipeline. Due to the absence of liquefaction in the seabed, the results of the CP and ND models are completely consistent. A comparative analysis reveals that the presence of the pipeline significantly affects the transmission of pore pressure in the seabed. Above the pipeline, the pore pressure attenuation is slower in the seabed with a pipeline, but below the pipeline, the pore pressure decreases significantly. This can also be clearly observed in the pore pressure variations along the centerline of the seabed model (e.g., Figure 6, in which the variation in pore pressure versus time step at different depths is displayed). The specific reasons for this phenomenon will be discussed in Section 5.

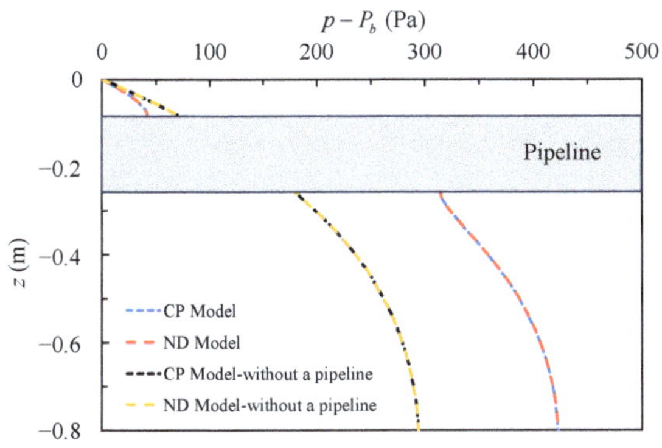

Figure 5. $p - P_b$ versus depth along the x-directional centerline.

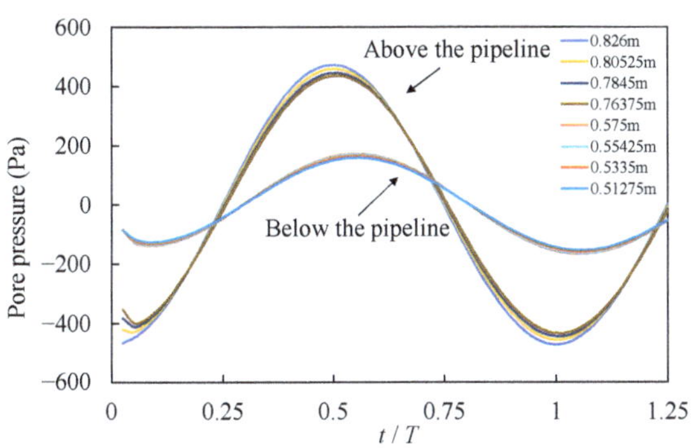

Figure 6. Pore pressure versus time step along the x-directional centerline at different depths.

4. Numerical Results of the Wave–Seabed–Structure Model with Seabed Liquefaction

4.1. Computational Model

In this subsection, our liquefaction-associated non-Darcy flow model is utilized to analyze the wave–seabed–pipeline interaction under extreme wave loadings, which induce seabed liquefaction. The computational model adopts a two-dimensional approach. The width of the seabed in the computational model is assumed to be 50 m, with a thickness of 20 m. The computational mesh consisting of quadrilateral elements is shown in Figure 7. The seabed mesh surrounding the pipeline is refined in the meshing process, with a minimum mesh size of 0.01 m. The computational parameters are listed in Table 1. The burial depth and diameter are denoted by d and D, respectively, while d is defined as the distance from the top position of the pipeline to the seabed surface.

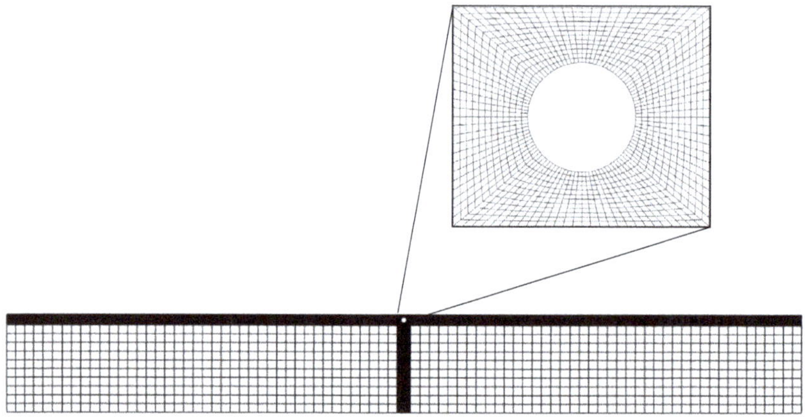

Figure 7. Schematic diagram of model grid.

Table 1. Computational parameters for the model with seabed liquefaction.

Parameter	T (s)	L (m)	H (m)	h (m)	S_r (-)	n (-)	γ' (kN/m^3)	k_s (m·s^{-1})	E (MPa)	ν (-)
Value	2	5.21	0.6	1	0.95	0.425	5.3	0.001	3	0.33

4.2. Pore Pressure Response

To investigate the influence of a fully buried pipeline on the pore pressure response in a non-cohesive seabed, three values of burial depth d are considered, as presented in Table 2, wherein the pipeline diameters are fixed at the same value of burial depth, i.e., $d/D = 1$. Figures 8–13 give the temporal distributions of pore pressure at the top position of the pipeline and the vertical distributions of pore pressure above the top of the pipeline under three different schemes. In these figures, P_b represents the wave pressure applied to the seabed surface.

Table 2. Computational schemes with various buried depths and diameters.

Scheme No.	Burial Depth d (m)	Diameter D (m)
I	0.042	0.042
II	0.084	0.084
III	0.126	0.126

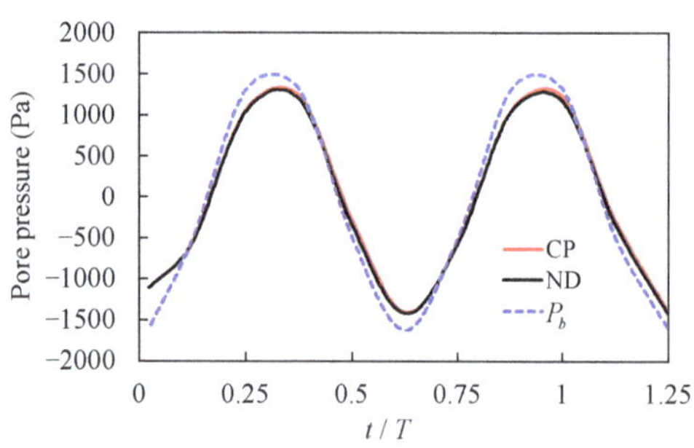

Figure 8. Pore pressure versus time step at the top position of the pipeline under Scheme I.

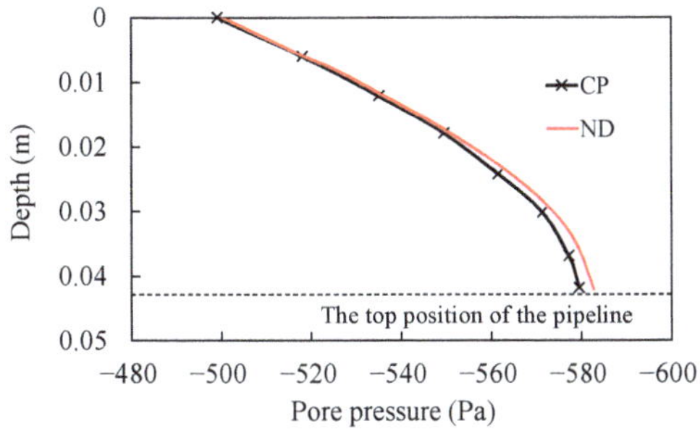

Figure 9. Pore pressure versus depth above the top position of the pipeline under Scheme I.

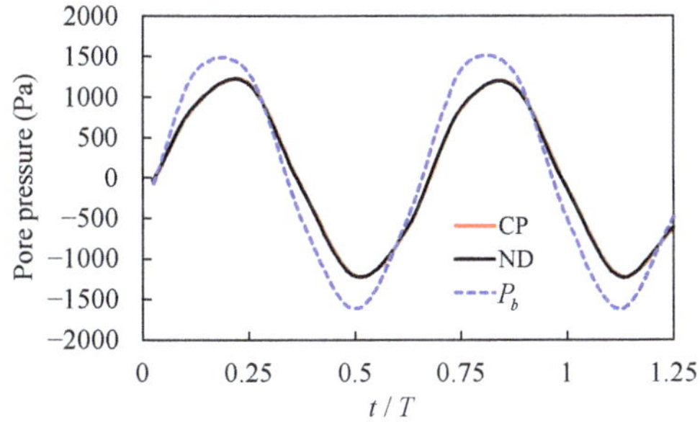

Figure 10. Pore pressure versus time step at the top position of the pipeline under Scheme II.

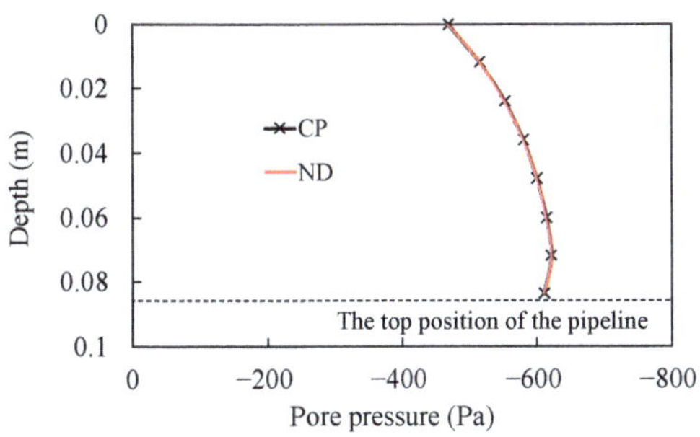

Figure 11. Pore pressure versus depth above the top position of the pipeline under Scheme II.

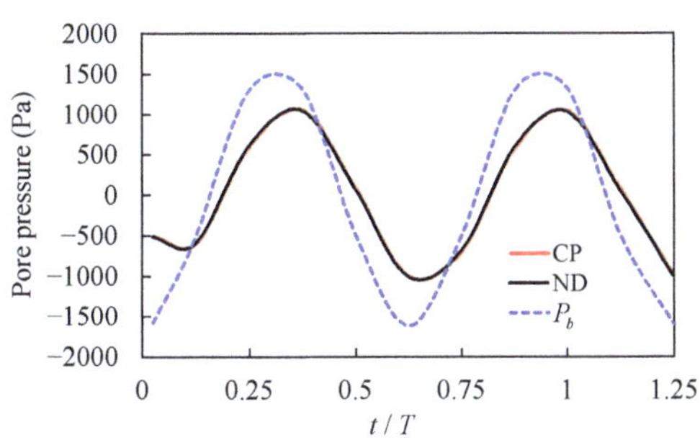

Figure 12. Pore pressure versus time step at the top position of the pipeline under Scheme III.

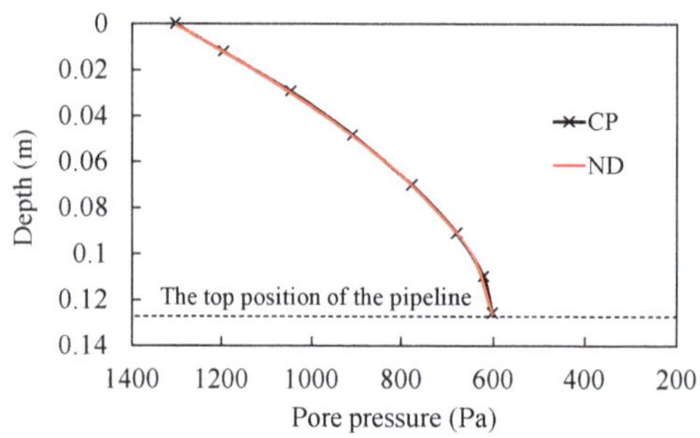

Figure 13. Pore pressure versus depth above the top position of the pipeline under Scheme III.

The wave pressure is instantaneously transmitted from the sea surface to the seabed surface. However, due to the unsaturated nature of the seabed and the compressibility of pore water flow, the wave pressure cannot be promptly transmitted within the seabed, resulting in phase lag over time (e.g., Figures 8, 10 and 12) and amplitude attenuation along the depth direction (e.g., Figures 9, 11 and 13). The amplitude attenuation and phase lag are inherent characteristics of the wave-induced instantaneous pore pressure response.

According to Figures 8, 10 and 12, the results obtained from the CP model and ND models are close to each other. The variation in pore pressure at the top position of the pipeline follows the waveform of the wave motion. Under Scheme I, with a pipeline burial depth of 0.042 m, the wave pressure amplitude p_0 at the seabed surface is 1500 Pa, while the pore pressure amplitude at the top position of the pipeline is 1300 Pa, resulting in an amplitude attenuation of 13.3%. Compared to the amplitude of the wave pressure P_b applied to the seabed surface, the pore pressure amplitude at the top position of the pipeline is delayed by about two computational steps, indicating a phase lag time of about 0.05T. Under Scheme II, with a pipeline burial depth of 0.084 m, p_0 at the seabed surface is 1500 Pa, while the pore pressure amplitude at the top position of the pipeline is 1200 Pa, resulting in an amplitude attenuation of 20.0%. Compared to the amplitude of P_b applied to the seabed surface, the pore pressure amplitude at the top of the pipeline is delayed by about three computational steps, indicating a phase lag time of about 0.075T. Under Scheme III, with a pipeline burial depth of 0.126 m, P_b at the seabed surface is 1500 Pa, while p_0 at the top position of the pipeline is 1100 Pa, resulting in an amplitude attenuation of 26.7%. The pore pressure amplitude at the top position of the pipeline appears at the same time as in Scheme II and is delayed by about three computational steps relative to the amplitude of P_b applied to the seabed surface, indicating a phase lag time of about 0.075T. According to the above analysis, it can be concluded that as the burial depth of the pipeline increases, the degree of amplitude attenuation and the phase lag also increase.

According to Figures 9, 11 and 13, the pore pressure attenuation at a depth z = 0.04 m can be analyzed. The attenuation under Schemes I, II, and III is 13%, 25%, and 26.9%, respectively. With an increase in the structural burial depth, the degree of pore pressure attenuation also increases. In other words, a fully buried pipeline can mitigate the amplitude attenuation of pore pressure transmission within the seabed.

It can be seen that pipelines of different sizes and at different burial depths have different effects on the transfer of pore pressure in the seabed. The weakening of the seabed by the fully buried pipeline in the seabed is mainly due to the significant influence of its diameter and burial depth on the pore pressure transfer in the seabed, which is mainly manifested in the pore pressure amplitude attenuation and phase lag. The amplitude attenuation and phase lag are also the main reasons for the instantaneous liquefaction of the seabed.

Figure 14 shows the liquefaction zones from the CP and ND models in the above three schemes. It can be seen that the liquefaction depth above the pipeline is nearly close to zero due to the liquefaction shielding effect, which is discussed in Section 5. As a result, the pore pressure distributions above the pipeline from the CP and ND models are close to each other, as shown in Figures 8–13. To better illustrate the difference between the CP and ND models, another computational scheme ($d = D = 0.168$ m) is considered, with the numerical results shown in Figure 15. Figure 15a,b are the liquefaction zones from the CP and ND models, respectively. It can be seen that the liquefaction zone from the ND model is apparently smaller than that from the CP model. Figure 15c is the vertical effective stress above the pipeline centerline, showing that the ND model effectively mitigates the presence of tensile stresses caused by the CP model.

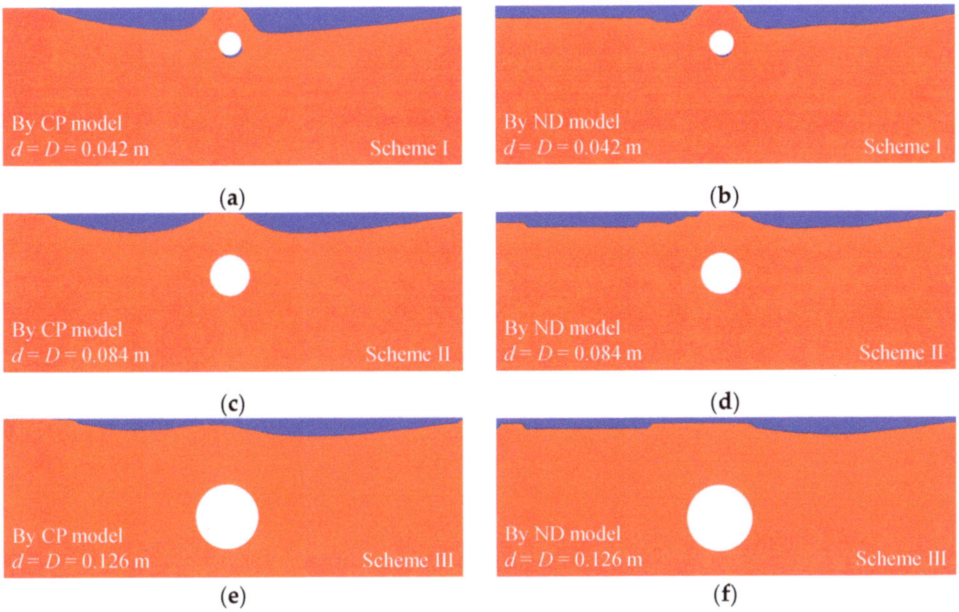

Figure 14. Liquefaction zones from CP and ND models in Schemes I~III: (**a**) from CP model in Scheme I; (**b**) from ND model in Scheme I; (**c**) from CP model in Scheme II; (**d**) from ND model in Scheme II; (**e**) from CP model in Scheme III; (**f**) from ND model in Scheme III.

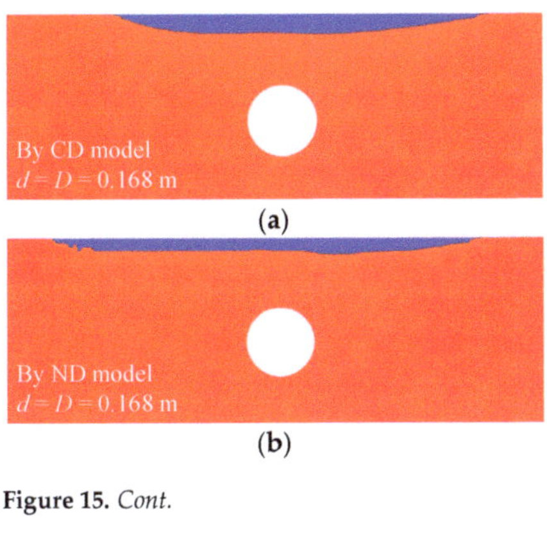

Figure 15. *Cont.*

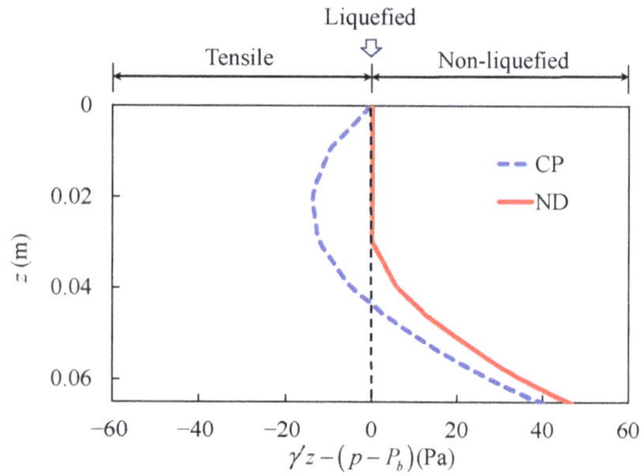

(c)

Figure 15. Liquefaction zones obtained by CP and ND models in the scheme of $d = D = 0.168$ m: (**a**) CP model; (**b**) ND model; (**c**) vertical distribution of $\gamma' z - (p - P_b)$.

4.3. Parametric Study on the Liquefaction Depth

According to the research conducted by Zhou et al. [52,61,63], when there is no liquefied zone in the non-cohesive seabed, the results obtained by the ND model are consistent with those by the CP model. Once liquefaction occurs in the non-cohesive seabed, the CP model results in unreasonable tensile stresses within the liquefied zone, whereas the ND model does not produce such results. Therefore, Zhou et al. [52,61,63] suggest using the CP model for calculations in practical engineering scenarios first and employing the ND model to eliminate tensile stresses if liquefaction is observed. Through the parametric study by Zhou et al. [52,61,63], it was found that the instantaneous liquefaction depth from the ND model is approximately 0.73 times that from the CP model. This relationship is obtained in the scenario of a pure seabed without structures. Here, in this subsection, we further describe a parametric study conducted for the scenario involving a pipeline fully buried in the seabed.

During the parametric study, a benchmark test was set with the computational parameters listed in Table 1, and we considered seven different parameters (Young's modulus E, permeability coefficient k_s, saturation degree S_r, soil porosity n, wave period T, water depth h, and wave height H). Building upon the model presented in Section 3.1, d and D of the pipeline are taken as 0.083 m and 0.168 m, respectively.

In Figure 16, CP represents the results of the CP model, while ND represents the results of the ND model, and the solid line represents the model with a pipeline, while the dashed line represents the model without a pipeline. Figure 16a shows that the liquefaction depth increases rapidly when the elastic modulus is between 1 and 3 MPa. As the elastic modulus exceeds 3 MPa, the rate of increase in liquefaction depth slows down and gradually approaches a constant value. In Figure 16b, it can be seen that the liquefaction depth decreases with the increase in the permeability coefficient. Figure 16c demonstrates the sensitivity of the liquefaction depth versus soil saturation. The presence of air in unsaturated soils is a key factor causing the amplitude attenuation and phase lag of pore pressure propagation in the seabed, thus directly affecting the liquefaction depth. Figure 16d shows that the liquefaction depth has a nearly linear relationship with the void ratio. In Figure 16e, it is evident that the liquefaction depth decreases linearly with the increase in water depth. Figure 16f reveals that the liquefaction depth increases and gradually approaches a constant value as the wave period increases. In Figure 16g, the change in liquefaction depth exhibits a roughly linear relationship with the wave height. The liquefaction depth increases with the increase in wave height and suddenly increases when the wave height reaches 0.8 m. This is because, during the process of increasing the wave height from 0.6 m to 0.8 m, a small liquefied zone also appears at the bottom of the pipeline, as shown in Figure 17. It can be therefore concluded that E, n, T, and H positively influence the liquefaction depth, while k_s, S_r, and h have a negative impact on it. Additionally, the liquefaction depth in the scenario with a pipeline is generally smaller than that in the scenario without a pipeline.

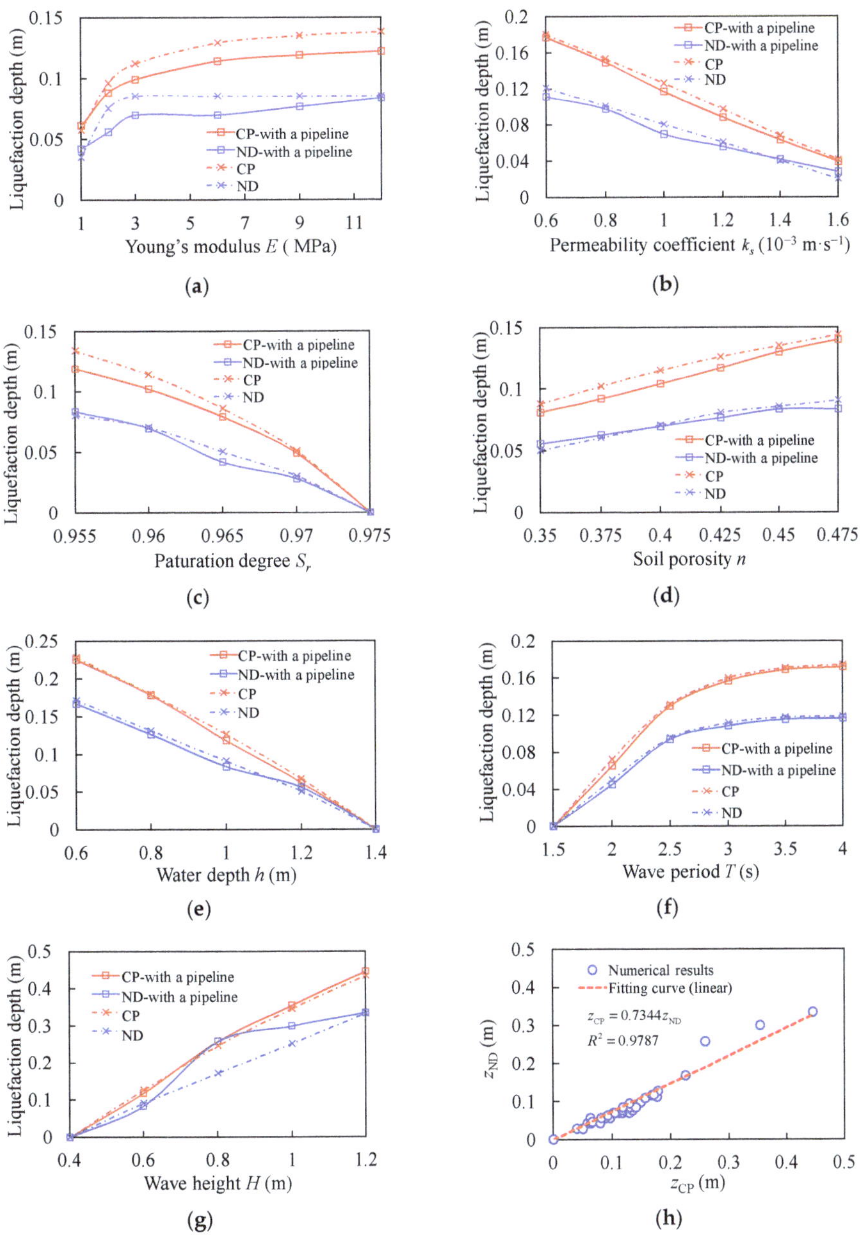

Figure 16. Liquefaction depth versus (**a**) Young's modulus E; (**b**) permeability coefficient k_s; (**c**) saturation degree S_r; (**d**) soil porosity n; (**e**) water depth h; (**f**) wave period T; (**g**) wave height H. (**h**) z_{ND} versus z_{CP}.

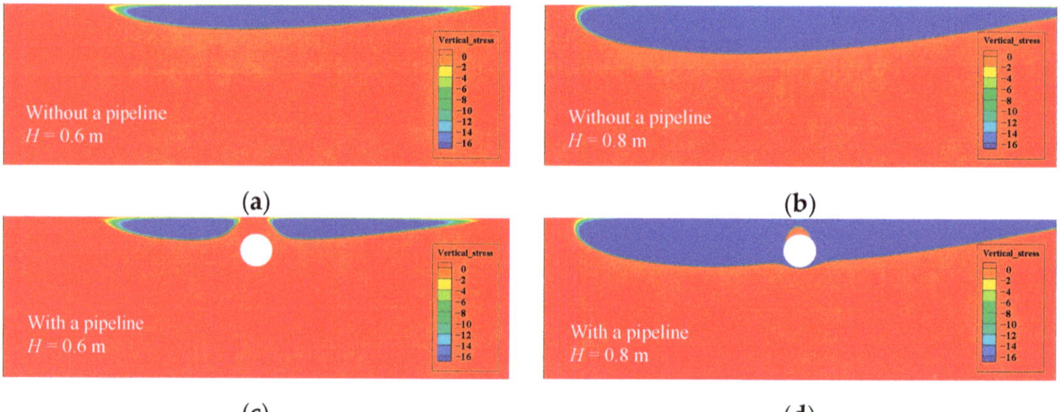

Figure 17. Liquefaction zones: (**a**) H = 0.6 m without a pipeline; (**b**) H = 0.8 m without a pipeline; (**c**) H = 0.6 m with a pipeline; (**d**) H = 0.8 m with a pipeline.

According to Figure 16h (where z_{CP} represents the liquefaction depth from the CP model and z_{ND} represents that from the ND model, which are all from Figure 16a–g), it can be observed that there is a significant linear correlation between z_{CP} and z_{ND} (with $R^2 = 0.9787$) in the scenario involving a pipeline. The ratio between the instantaneous liquefaction depth from the ND model and the CP model is approximately 0.73, which is close to that in the scenario of a pure seabed without structures.

4.4. Study on Non-Dimensional Parameters

To establish a quantifiable relationship between wave loads and structural responses with good universality, it is necessary to develop an expression based on non-dimensional parameters. To achieve this, adopting schemes with diameters ranging from 0.042 to 0.168 m and burial depths of 1–5 times the diameters (the specific values of burial depths and diameters are given in Table 3, while S_r is 0.96, ν is 0.3, and the other parameters are the same as those in Table 1), we conducted a large number of numerical calculations and attempted various combinations of horizontal and vertical parameters. Eventually, we found that when the horizontal coordinate is L/D (wavelength/diameter) and the vertical coordinate is z_{ND}/d (liquefaction depth/burial depth), there is a significant linear relationship between them, as shown in Figure 18. The correlation coefficient (R^2) was as high as 0.9223, indicating a good fitting effect.

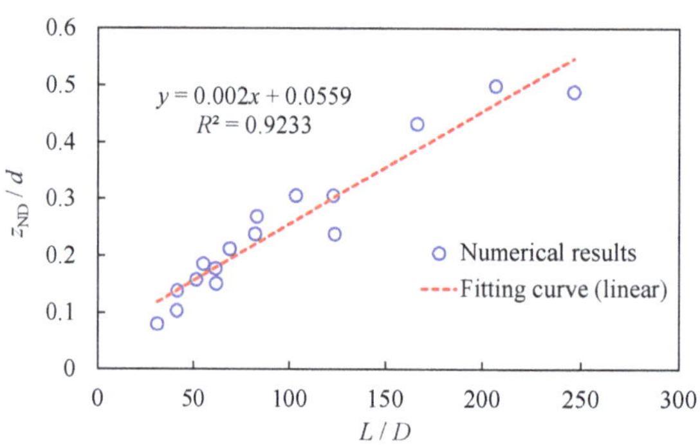

Figure 18. z_{ND}/d versus L/D in ND model.

Table 3. The scheme of the burial depth and diameter.

D (m)	$d = D$ (m)	$d = 2D$ (m)	$d = 3D$ (m)	$d = 4D$ (m)	$d = 5D$ (m)
0.042	0.042	0.084	0.126	0.168	0.21
0.084	0.084	0.168	0.252	0.336	0.42
0.126	0.126	0.252	0.378	0.504	0.63
0.168	0.168	0.336	0.504	0.672	0.84

5. The Shielding Phenomenon of a Structure in the Seabed under a Liquefaction Condition

5.1. The Shielding Effect of the Pipeline on the Liquefied Zone

In Section 4.3, it is shown that the presence of a pipeline in the seabed leads to a smaller liquefaction depth compared to the scenario without a pipeline. When a pipeline exists in the seabed, the liquefied zones above the pipeline exhibit a spatial distribution that "bypasses" the pipeline, resulting in a "liquefaction shielding phenomenon". Even if the liquefaction depth on both sides of the pipeline is below the top position of the pipeline, there will be a layer of non-liquefied soil above the pipeline, e.g., Figure 19a. As shown in Figure 19b–d, as the burial depth of the pipeline increases, keeping the diameter constant, the liquefaction depth also increases. However, when the burial depth is large enough, its further increase will not induce a further increase in the liquefaction depth. The reason is that instantaneous liquefaction generally occurs in a thin layer near the seabed surface. When the pipeline is far away from the seabed surface, the presence of the pipeline will no longer influence the liquefied zone. In contrast, when the buried depth is not large enough, the presence of the pipeline can have a significant influence on the liquefied zone. Below, a tentative explanation is provided for this phenomenon.

During the numerical simulation, the presence of the pipeline acts as an impermeable boundary in the seabed, causing a diversion of the flow field in the surrounding area, as shown in Figure 20. The flow does not reach the soil above the pipeline immediately, resulting in an impact on the phase lag and amplitude attenuation of the wave-induced seabed pore pressure response, and this phenomenon is referred to as the "liquefaction shielding phenomenon". In other words, when the structure is shallow, the structure will repel the liquefaction zone above it, and the area above the structure where liquefaction does not occur is referred to as the "shielding ring". It is therefore straightforward to imagine that when the pipeline is shallow enough, the liquefied zone above the pipeline can completely disappear. This prediction is studied in the following subsection.

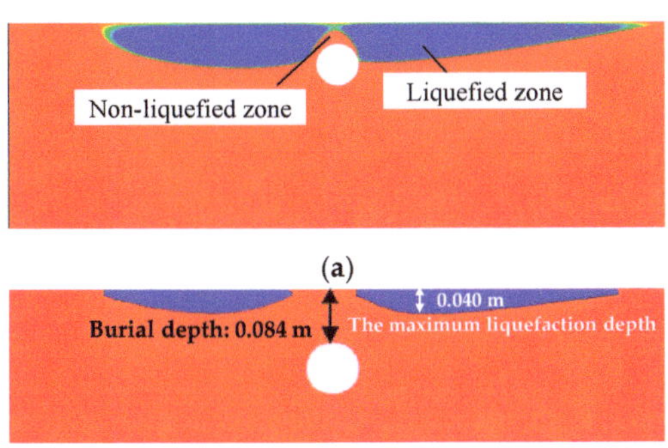

Figure 19. *Cont.*

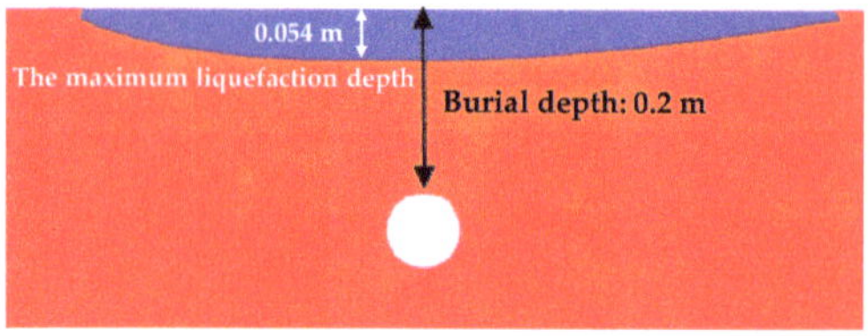

(**c**)

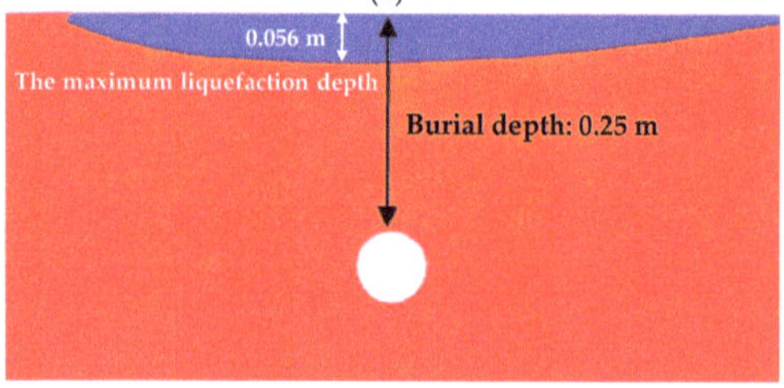

(**d**)

Figure 19. (**a**) Liquefaction shielding phenomenon above the pipeline; liquefied zones when (**b**) d = 0.084 m; (**c**) d = 0.2 m; (**d**) d = 0.25 m.

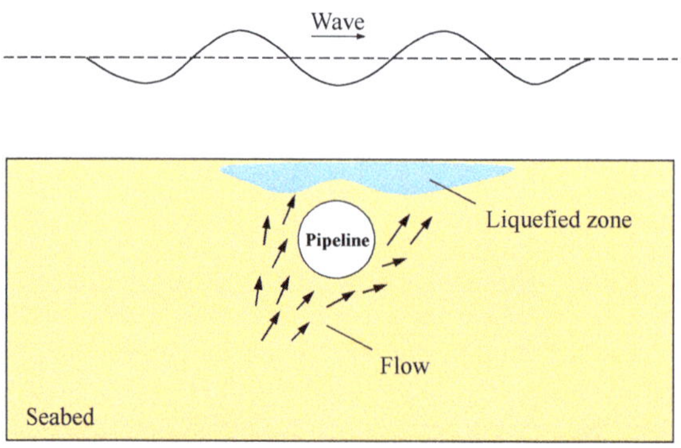

Figure 20. A schematic diagram of flow around the pipeline.

5.2. Onset Conditions for Liquefaction Occurrence above the Pipeline

According to Section 5.1, the shielding phenomenon of the pipeline on the liquefied zone weakens as the burial depth of the pipeline increases. This subsection will present a computational analysis of various combinations of burial depth d and diameter D to study the onset conditions for liquefaction occurrence above the pipeline.

Based on the numerical results in Section 4.4, we continued to calculate the liquefaction depths in the scenario involving a pipeline considering various schemes with diameters of 0.084 m, 0.126 m, and 0.168 m and burial depths ranging from 0.084 m to 0.4 m (please refer to Table 4 for specific parameters and results). The results in Section 4.4 and Table 4 are jointly plotted in Figure 21. When $d/D < 2$, although several schemes have liquefaction depths as high as 0.06 m, nearly half of the schemes did not show liquefied zones; i.e., the pipeline is near the seabed surface, and therefore, the aforementioned shielding effect is apparent. However, when $d/D \geq 2$, liquefaction zones occur in all the numerical results, with a minimum liquefaction depth of 0.015 m; i.e., the pipeline is far away from the seabed surface and therefore has a slight influence on the liquefied zone.

Table 4. Liquefaction depth under different working schemes.

D (m)	*d* (m)	*d*/*D*	Liquefaction Depths (m)
0.084	0.3	3.6	0.06
	0.25	3	0.06
	0.2	2.4	0.055
	0.15	1.8	0
	0.084	1	0
0.126	0.35	2.8	0.06
	0.3	2.4	0.06
	0.25	2	0.055
	0.2	1.6	0.05
	0.15	1.190476	0
	0.084	0.666667	0
0.168	0.4	2.380952	0.06
	0.3	1.785714	0.06
	0.25	1.488095	0.05
	0.2	1.190476	0.05
	0.15	0.892857	0
	0.084	0.5	0

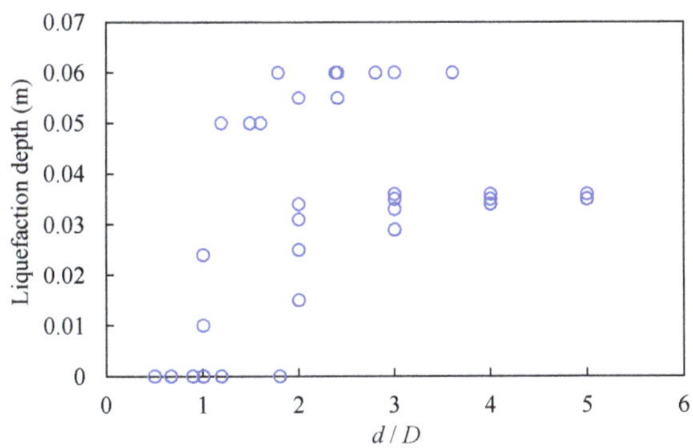

Figure 21. Liquefaction depth versus d/D.

It seems that a critical value of d/D can be quantitatively determined to distinguish whether liquefaction occurs above the pipeline. From Figure 21, this value can be taken as 2; i.e., when $d/D \geq 2$, liquefaction always occurs above the pipeline. It should be noted that this critical value can change if other computational parameters are applied. It is also notable that even if there is no liquefied zone above the pipeline, in the horizontal direction, if a position is far from the pipeline, the shielding effect of the pipeline can also disappear, which means that liquefied zones can also occur in a shallow layer of the seabed within these areas, e.g., Figure 19b.

6. Conclusions

This paper extends the application of our liquefaction-associated non-Darcy flow model from wave–seabed interactions to wave–seabed–pipeline interactions in order to investigate the influence of a fully buried pipeline on the pore pressure response and the instantaneous liquefaction of the non-cohesive seabed. Based on the numerical examples presented, the following conclusions can be drawn.

(1) Numerical simulations were conducted based on Turcotte's flume experiment, validating the reliability of our in-house code in modeling wave–seabed–structure interactions.
(2) A numerical study was conducted by varying the burial depth and diameter of the pipeline, revealing that the existence of a pipeline weakens the degree of amplitude

attenuation and the phase lag. Therefore, when the pipeline is shallow, the liquefied zone of the seabed with a pipeline is smaller than that in the pure seabed, which is called the "liquefaction shielding effect" in this work.

(3) Under some conditions, liquefaction can even completely disappear above the pipeline, while horizontally distant areas still have liquefied zones. The onset conditions for liquefaction occurrence above the pipeline are then discussed.

(4) As the burial depth of the pipeline increases, the liquefaction shielding effect weakens, resulting in an increase in the liquefaction depth above the pipeline. Once the pipeline is sufficiently far from the seabed surface, it no longer influences the liquefied zone.

(5) Based on the parametric study, it was observed that the liquefaction depth predicted by the non-Darcy model is approximately 0.73 times the value estimated by the conventional Darcy model, regardless of whether or not a pipeline is involved.

(6) The quantitative relationship between wave loadings and structural sizes is studied. A highly linear relationship between two sets of non-dimensional parameters, i.e., "liquefaction depth/burial depth" and "wavelength/diameter", is discovered.

Author Contributions: Conceptualization, M.Z., D.Z., W.Q. and Q.F.; methodology, S.H. and M.Z.; software, M.Z.; validation, S.H. and C.X.; writing—original draft preparation, S.H.; writing—review and editing, M.Z., D.Z., W.Q. and Q.F.; visualization, S.H. and C.X.; supervision, D.Z., W.Q. and Q.F. All authors have read and agreed to the published version of the manuscript.

Funding: This study was financially supported by the Fundamental Research Funds for the Central Universities (Grant No. 2023JBZY027); the National Natural Science Foundation of China (Grant Nos. 52271255, 11972036); and the Youth Innovation Promotion Association CAS (Grant No. 2021018) for financial support.

Data Availability Statement: Data is contained within the article.

Conflicts of Interest: The authors declare no conflict of interest.

References

1. Du, X.; Sun, Y.; Song, Y.; Zhu, C. In-Situ Observation of Wave-Induced Pore Water Pressure in Seabed Silt in the Yellow River Estuary of China. *J. Mar. Environ. Eng.* **2021**, *10*, 305–317.
2. Jia, Y.; Zhang, L.; Zheng, J.; Liu, X.; Jeng, D.-S.; Shan, H. Effects of wave-induced seabed liquefaction on sediment re-suspension in the Yellow River Delta. *Ocean Eng.* **2014**, *89*, 146–156. [CrossRef]
3. Mory, M.; Michallet, H.; Bonjean, D.; Piedra-Cueva, I.; Barnoud, J.M.; Foray, P.; Abadie, S.; Breul, P. A Field Study of Momentary Liquefaction Caused by Waves around a Coastal Structure. *J. Waterw. Port Coast. Ocean Eng.* **2007**, *133*, 28–38. [CrossRef]
4. Tian, Z.; Chen, T.; Yu, L.; Guo, X.; Jia, Y. Penetration depth of the dynamic response of seabed induced by internal solitary waves. *Appl. Ocean Res.* **2019**, *90*, 101867. [CrossRef]
5. Zhang, S.; Jia, Y.; Zhang, Y.; Liu, X.; Shan, H. In situ observations of wave pumping of sediments in the Yellow River Delta with a newly developed benthic chamber. *Mar. Geophys. Res.* **2018**, *39*, 463–474. [CrossRef]
6. Jeng, D.-S. *Mechanics of Wave-Seabed-Structure Interactions: Modelling, Processes and Applications*; Cambridge University Press: Cambridge, UK, 2018.
7. Sumer, T.B.M. *Liquefaction around Marine Structures*; World Scientific: Singapore, 2014.
8. Jeng, D.-S. Wave-induced sea floor dynamics. *Appl. Mech. Rev.* **2003**, *56*, 407–429. [CrossRef]
9. Zen, K.; Yamazaki, H. Mechanism of Wave-Induced Liquefaction and Densification in Seabed. *Soils Found.* **1990**, *30*, 90–104. [CrossRef] [PubMed]
10. Miyamoto, J.; Sassa, S.; Sekiguchi, H. Progressive solidification of a liquefied sand layer during continued wave loading. *Géotechnique* **2004**, *54*, 617–629. [CrossRef]
11. Mutlu Sumer, B.; Hatipoglu, F.; Fredsøe, J.; Kaan Sumer, S. The sequence of sediment behaviour during wave-induced liquefaction. *Sedimentology* **2006**, *53*, 611–629. [CrossRef]
12. Hsu, J.R.C.; Jeng, D.S. Wave-induced soil response in an unsaturated anisotropic seabed of finite thickness. *Int. J. Numer. Anal. Methods Geomech.* **1994**, *18*, 785–807. [CrossRef]
13. Madsen, O.S. Wave-Induced Pore Pressures and Effective Stresses in A Porous Bed. *Geotechnique* **1978**, *28*, 377–393. [CrossRef]
14. Yamamoto, T.; Koning, H.L.; Sellmeijer, H.; Vanhijum, E.P. Response of A Poro-Elastic Bed to Water-Waves. *J. Fluid Mech.* **1978**, *87*, 193–206. [CrossRef]
15. Zhang, Y.; Jeng, D.S.; Gao, F.P.; Zhang, J.S. An analytical solution for response of a porous seabed to combined wave and current loading. *Ocean Eng.* **2013**, *57*, 240–247. [CrossRef]

16. Duan, L.; Liao, C.; Jeng, D.; Chen, L. 2D numerical study of wave and current-induced oscillatory non-cohesive soil liquefaction around a partially buried pipeline in a trench. *Ocean Eng.* **2017**, *135*, 39–51. [CrossRef]
17. Gao, F.P.; Jeng, D.S.; Sekiguchi, H. Numerical study on the interaction between non-linear wave, buried pipeline and non-homogenous porous seabed. *Comput. Geotech.* **2003**, *30*, 535–547. [CrossRef]
18. Jeng, D.S.; Lin, Y.S. Wave–induced pore pressure around a buried pipeline in Gibson soil: Finite element analysis. *Int. J. Numer. Anal. Methods Geomech.* **1999**, *23*, 1559–1578. [CrossRef]
19. Li, K.; Guo, Z.; Wang, L.; Jiang, H. Effect of seepage flow on shields number around a fixed and sagging pipeline. *Ocean Eng.* **2019**, *172*, 487–500. [CrossRef]
20. Liang, Z.; Jeng, D.-S. PORO-FSSI-FOAM model for seafloor liquefaction around a pipeline under combined random wave and current loading. *Appl. Ocean Res.* **2021**, *107*, 102497. [CrossRef]
21. Liang, Z.; Jeng, D.-S.; Liu, J. Combined wave–current induced seabed liquefaction around buried pipelines: Design of a trench layer. *Ocean Eng.* **2020**, *212*, 107764. [CrossRef]
22. Lin, Z.; Guo, Y.; Jeng, D.S.; Liao, C.; Rey, N. An integrated numerical model for wave–soil–pipeline interactions. *Coast. Eng.* **2016**, *108*, 25–35. [CrossRef]
23. Zhao, H.Y.; Jeng, D.-S. Accumulated Pore Pressures around Submarine Pipeline Buried in Trench Layer with Partial Backfills. *J. Eng. Mech.* **2016**, *142*, 04016042. [CrossRef]
24. Zhao, H.Y.; Jeng, D.S.; Guo, Z.; Zhang, J.S. Two-Dimensional Model for Pore Pressure Accumulations in the Vicinity of a Buried Pipeline. *J. Offshore Mech. Arct. Eng.* **2014**, *136*, 042001. [CrossRef]
25. Chen, H.; Zhang, J.; Tong, L.; Sun, K.; Guo, Y.; Wei, C. Experimental study of soil responses around a pipeline in a sandy seabed under wave-current load. *Appl. Ocean Res.* **2023**, *130*, 103409. [CrossRef]
26. Duan, L.; Zhan, B.; Shen, L.; Fan, M.; Wang, D. Nonlinear Wave-Induced Uplift Force onto Pipelines Buried in Sloping Seabeds. *Appl. Sci.* **2023**, *13*, 7519. [CrossRef]
27. Celli, D.; Li, Y.; Ong, M.C.; Di Risio, M. The role of submerged berms on the momentary liquefaction around conventional rubble mound breakwaters. *Appl. Ocean Res.* **2019**, *85*, 1–11. [CrossRef]
28. Jeng, D.S.; Ye, J.H.; Zhang, J.S.; Liu, L.F. An integrated model for the wave-induced seabed response around marine structures: Model verifications and applications. *Coast. Eng.* **2013**, *72*, 1–19. [CrossRef]
29. Jisheng, Z.; Linlong, T.; Jinhai, Z.; Rui, H.; Yakun, G. Effects of Soil-Resistance Damping on Wave-Induced Pore Pressure Accumulation around a Composite Breakwater. *J. Coast. Res.* **2018**, *34*, 573–585. [CrossRef]
30. Young, Y.L.; White, J.A.; Xiao, H.; Borja, R.I. Liquefaction potential of coastal slopes induced by solitary waves. *Acta Geotech.* **2009**, *4*, 17–34. [CrossRef]
31. Chang, K.-T.; Jeng, D.-S. Numerical study for wave-induced seabed response around offshore wind turbine foundation in Donghai offshore wind farm, Shanghai, China. *Ocean Eng.* **2014**, *85*, 32–43. [CrossRef]
32. Li, Y.; Ong, M.C.; Tang, T. Numerical analysis of wave-induced poro-elastic seabed response around a hexagonal gravity-based offshore foundation. *Coast. Eng.* **2018**, *136*, 81–95. [CrossRef]
33. Lin, Z.; Pokrajac, D.; Guo, Y.; Jeng, D.-S.; Tang, T.; Rey, N.; Zheng, J.; Zhang, J. Investigation of nonlinear wave-induced seabed response around mono-pile foundation. *Coast. Eng.* **2017**, *121*, 197–211. [CrossRef]
34. Qi, W.-G.; Gao, F.-P. Physical modeling of local scour development around a large-diameter monopile in combined waves and current. *Coast. Eng.* **2014**, *83*, 72–81. [CrossRef]
35. Sui, T.; Zhang, C.; Guo, Y.; Zheng, J.; Jeng, D.; Zhang, J.; Zhang, W. Three-dimensional numerical model for wave-induced seabed response around mono-pile. *Ships Offshore Struct.* **2016**, *11*, 667–678. [CrossRef]
36. Zhao, H.; Alam, A.; San, C.-Y.; Eguchi, S.; Chen, Q.; Lian, Q.; Ma, D. Molecular mechanisms of brain-derived neurotrophic factor in neuro-protection: Recent developments. *Brain Res.* **2017**, *1665*, 1–21. [CrossRef]
37. Sui, T.; Zheng, J.; Zhang, C.; Jeng, D.-S.; Zhang, J.; Guo, Y.; He, R. Consolidation of unsaturated seabed around an inserted pile foundation and its effects on the wave-induced momentary liquefaction. *Ocean Eng.* **2017**, *131*, 308–321. [CrossRef]
38. Cui, L.; Jeng, D.-S.; Liu, J. Numerical analysis of the seabed liquefaction around a fixed gravity-based structure (GBS) of an offshore platform and protection. *Ocean Eng.* **2022**, *249*, 110844. [CrossRef]
39. Qin, C.; Duan, L.; Wang, D.; Duan, B.; Fan, M.; Wang, H. Three-dimensional numerical simulation of wave-induced seabed response around a dumbbell-shaped cofferdam. *Appl. Ocean Res.* **2023**, *134*, 103501. [CrossRef]
40. Chen, W.; Liu, C.; He, R.; Chen, G.; Jeng, D.; Duan, L. Stability of the foundation trench of the immersed tunnel subjected to combined wave and current loading. *Appl. Ocean Res.* **2021**, *110*, 102627. [CrossRef]
41. Han, S.; Jeng, D.-S.; Tsai, C.-C. Response of a Porous Seabed around an Immersed Tunnel under Wave Loading: Meshfree Model. *J. Mar. Sci. Eng.* **2019**, *7*, 369. [CrossRef]
42. Li, Y.; Ong, M.C.; Tang, T. A numerical toolbox for wave-induced seabed response analysis around marine structures in the OpenFOAM® framework. *Ocean Eng.* **2020**, *195*, 106678. [CrossRef]
43. Cuéllar, P.; Mira, P.; Pastor, M.; Fernández Merodo, J.A.; Baeßler, M.; Rücker, W. A numerical model for the transient analysis of offshore foundations under cyclic loading. *Comput. Geotech.* **2014**, *59*, 75–86. [CrossRef]
44. Kudella, M.; Oumeraci, H.; de Groot, M.B.; Meijers, P. Large-Scale Experiments on Pore Pressure Generation underneath a Caisson Breakwater. *J. Waterw. Port Coast. Ocean Eng.* **2006**, *132*, 310–324. [CrossRef]

45. Liao, C.; Chen, J.; Zhang, Y. Accumulation of pore water pressure in a homogeneous sandy seabed around a rocking mono-pile subjected to wave loads. *Ocean Eng.* **2019**, *173*, 810–822. [CrossRef]
46. Sumer, S.K.; Sumer, B.M.; Dixen, F.H.; Fredsøe, J. Pore pressure buildup in the subsoil under a caisson breakwater. In Proceedings of the Eighteenth International Offshore and Polar Engineering Conference, Ottawa, ON, Canada, 6–11 July 2008; pp. 664–671.
47. Wang, Z.; Yang, Y.; Yu, H.-S. Effects of principal stress rotation on the wave–seabed interactions. *Acta Geotech.* **2017**, *12*, 97–106. [CrossRef]
48. Zhu, J.F.; Zhao, H.Y.; Jeng, D.S. Effects of principal stress rotation on wave-induced soil response in a poro-elastoplastic sandy seabed. *Acta Geotech.* **2019**, *14*, 1717–1739. [CrossRef]
49. Qi, W.-G.; Shi, Y.-M.; Gao, F.-P. Uplift soil resistance to a shallowly-buried pipeline in the sandy seabed under waves: Poro-elastoplastic modeling. *Appl. Ocean Res.* **2020**, *95*, 102024. [CrossRef]
50. Qi, W.-G.; Gao, F.-P. A modified criterion for wave-induced momentary liquefaction of sandy seabed. *Theor. Appl. Mech. Lett.* **2015**, *5*, 20–23. [CrossRef]
51. Qi, W.-G.; Gao, F.-P. Wave induced instantaneously-liquefied soil depth in a non-cohesive seabed. *Ocean Eng.* **2018**, *153*, 412–423. [CrossRef]
52. Zhou, M.-Z.; Jeng, D.-S.; Qi, W.-G. A new model for wave-induced instantaneous liquefaction in a non-cohesive seabed with dynamic permeability. *Ocean Eng.* **2020**, *213*, 107597. [CrossRef]
53. Wu, S.; Jeng, D.S. Effects of dynamic soil permeability on the wave-induced seabed response around a buried pipeline. *Ocean Eng.* **2019**, *186*, 106132. [CrossRef]
54. Wu, S.; Jeng, D.S.; Seymour, B.R. Numerical Modelling of consolidation-induced solute transport in unsaturated soil with dynamic hydraulic conductivity and degree of saturation. *Adv. Water Resour.* **2020**, *135*, 103466. [CrossRef]
55. Arulandan, K.; Sybico, J.; Houlsby, G.T.; Schofield, A.N. Post-liquefaction settlement of sands. *Géotechnique* **2009**, *59*, 739–749.
56. Ha, I.S.; Park, Y.H.; Kim, M.M. Dissipation Pattern of Excess Pore Pressure After Liquefaction in Saturated Sand Deposits. *Transp. Res. Rec.* **2003**, *1821*, 59–67. [CrossRef]
57. Haigh, S.K.; Eadington, J.; Madabhushi, S.P.G. Permeability and stiffness of sands at very low effective stresses. *Géotechnique* **2012**, *62*, 69–75. [CrossRef]
58. Shahir, H.; Mohammadi-Haji, B.; Ghassemi, A. Employing a variable permeability model in numerical simulation of saturated sand behavior under earthquake loading. *Comput. Geotech.* **2014**, *55*, 211–223. [CrossRef]
59. Ueng, T.-S.; Wang, Z.-F.; Chu, M.-C.; Ge, L. Laboratory tests for permeability of sand during liquefaction. *Soil Dyn. Earthq. Eng.* **2017**, *100*, 249–256. [CrossRef]
60. Wang, B.; Zen, K.; Chen, G.Q.; Zhang, Y.B.; Kasama, K. Excess pore pressure dissipation and solidification after liquefaction of saturated sand deposits. *Soil Dyn. Earthq. Eng.* **2013**, *49*, 157–164. [CrossRef]
61. Zhou, M.; Liu, H.; Jeng, D.-S.; Qi, W.; Fang, Q. Modelling the wave-induced instantaneous liquefaction in a non-cohesive seabed as a nonlinear complementarity problem. *Comput. Geotech.* **2021**, *137*, 104275. [CrossRef]
62. Kunisch, K.; Rösch, A. Primal-Dual Active Set Strategy for a General Class of Constrained Optimal Control Problems. *SIAM J. Optim.* **2002**, *13*, 321–334. [CrossRef]
63. Zhou, M.-Z.; Qi, W.-G.; Jeng, D.-S.; Gao, F.-P. A non-Darcy flow model for a non-cohesive seabed involving wave-induced instantaneous liquefaction. *Ocean Eng.* **2021**, *239*, 109807. [CrossRef]
64. Fukumoto, Y.; Ohtsuka, S. 3-D direct numerical model for failure of non-cohesive granular soils with upward seepage flow. *Comput. Part. Mech.* **2018**, *5*, 443–454. [CrossRef]
65. Skempton, A.W.; Brogan, J.M. Experiments on piping in sandy gravels. *Géotechnique* **1994**, *44*, 449–460. [CrossRef]
66. Scholtès, L.; Chareyre, B.; Michallet, H.; Catalano, E.; Marzougui, D. Modeling wave-induced pore pressure and effective stress in a granular seabed. *Contin. Mech. Thermodyn.* **2015**, *27*, 305–323. [CrossRef]
67. Turcotte, B.R. *Laboratory Evaluation of Wave Tank Parameters for Wave-Sediment Interaction*; Cornell University: Ithaca, NY, USA, 1984.
68. Zhou, X.-L.; Zhang, J.; Wang, J.-H.; Xu, Y.-F.; Jeng, D.-S. Stability and liquefaction analysis of porous seabed subjected to cnoidal wave. *Appl. Ocean Res.* **2014**, *48*, 250–265. [CrossRef]
69. Dean, R.; Dalrymple, R.A. *Water Wave Mechanics for Engineers and Scientists*; World Scientific: Singapore, 1984.
70. Biot, M.A. General Theory of Three-Dimensional Consolidation. *J. Appl. Phys.* **1941**, *12*, 155–164. [CrossRef]
71. Verruijt, A. Elastic storage of aquifers. In *Flow Through Porous Media*; Academic Press: New York, NY, USA, 1969.
72. Cheng, A.H.D.; Liu, P.L.F. Seepage Force on a Pipeline Buried in a Poroelastic Seabed under Wave Loadings. *Appl. Ocean Res.* **1986**, *8*, 22–32. [CrossRef]

Journal of
Marine Science and Engineering

MDPI

Article

Dynamic Response Analysis and Liquefaction Potential Evaluation of Riverbed Induced by Tidal Bore

Dongzi Pan [1] and Ying Li [2,*]

[1] Zhejiang Key Laboratory of Estuary and Coast, Zhejiang Institute of Hydraulics and Estuary, Hangzhou 310020, China; pandz@zjwater.gov.cn
[2] College of Geomatics, Zhejiang University of Water Resources and Electric Power, Hangzhou 310018, China
* Correspondence: liying@zjweu.edu.cn; Tel.: +86-571-8643-8014

Abstract: Tidal bores, defined by sudden upstream surges of tidal water in estuaries, exert significant hydrodynamic forces on riverbeds, leading to complex sedimentary responses. This study examines the dynamic response and liquefaction potential of riverbeds subjected to tidal bores in macro-tidal estuaries. An analytical model, developed using the generalized Biot theory and integral transform methods, evaluates the dynamic behavior of riverbed sediments. Key factors such as permeability, saturation, and sediment properties are analyzed for their influence on momentary liquefaction. The results indicate that fine sand reduces liquefaction risk by facilitating pore water discharge, while silt soil increases sediment instability. Additionally, the study reveals that pressure gradients induced by tidal bores can trigger momentary liquefaction, with the maximum liquefaction depth predicted based on horizontal pressure gradients being five times that predicted based on vertical pressure gradients. This research highlights the critical role of sediment characteristics in riverbed stability, providing a comprehensive understanding of the interactions between tidal bores and riverbed dynamics. The findings contribute to the development of predictive models and guidelines for managing the risks of tidal bore-induced liquefaction in coastal and estuarine environments.

Keywords: tidal bore; riverbed; liquefaction; integral transform method; analytical solution

1. Introduction

Tidal bores are remarkable natural phenomena in estuaries and bays [1], characterized by a sudden upstream surge of tidal water against the ebb flow [2,3]. This occurs due to the deformation of tidal waves in shallow water, creating a steep front or a series of wave trains that move upstream along rivers or through convergent bays [4,5]. Recent studies have identified tidal bores in 117 rivers across 25 countries [6]. As depicted in Figure 1, notable examples include the Qiantang River in China, the Amazon River in Brazil, Cook Inlet in Alaska, the Hooghly River in India, the Batang Lupar River in Malaysia, the Kampar River in Indonesia, and the Sittang River in Myanmar [7]. During a tidal bore, water levels can rise by over 3 m within seconds, with flow velocities reaching up to 12 m/s as the river rapidly transitions from ebb to flood tide [8,9]. These powerful waves exert increased pressure on the riverbed, affecting pore water pressure within the sediment and potentially leading to liquefaction. The complex interaction between hydrodynamic forces and sedimentary responses highlights the power and complexity of tidal bores, making them a compelling subject for research.

Since the nineteenth century, tidal bore research has advanced from descriptive studies to dynamic mechanism analyses [10]. Early work focused on in situ observation [11], while contemporary research integrates field investigations [12–14], laboratory experiments [2,15–17], and numerical simulations [17–19] to examine the dynamics and turbulent structures of the tidal bore front [20–22]. Despite these advancements, limitations in observational and simulation techniques hinder our understanding of the interactions between tidal bores and riverbeds, necessitating further research to improve analytical accuracy and reliability in this field.

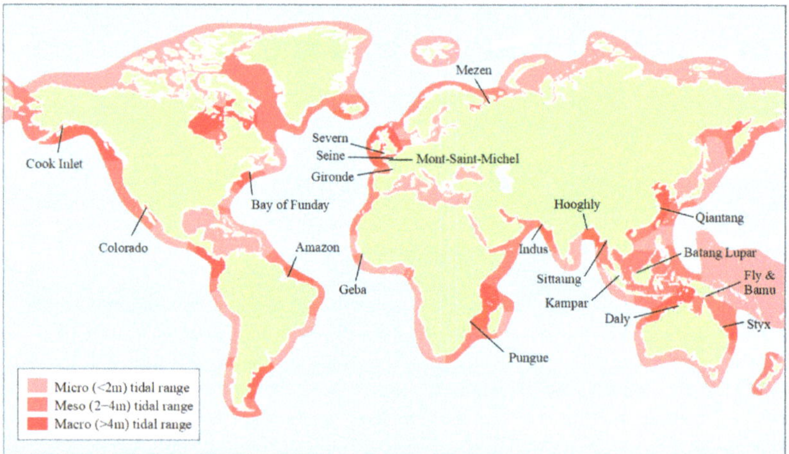

Figure 1. Distribution of typical tidal bores around the world.

Liquefaction, as defined by the Geotechnical Earthquake Engineering Committee of the American Society of Civil Engineers, is the act or process of transforming any substance into a liquid [23]. In cohesionless soils, the shift from a solid to a liquefied state results from increased pore pressure and a decrease in effective stress [24]. Seabed liquefaction under wave loading has been extensively studied, with comprehensive reviews by Jeng [25] and Lin et al. [26]. In natural environments, two mechanisms of wave-induced liquefaction exist [27–29]: momentary liquefaction and residual liquefaction. Momentary liquefaction primarily arises from the attenuation and phase lag of oscillatory pore pressure along the seabed depth [30], associated with the momentary volumetric strain of sediments. During this process, pore pressure does not accumulate; however, a significant upward seepage force is generated at wave troughs due to vertical pressure gradients, causing sediment particle movement and overturning [31–33]. Residual liquefaction refers to the instability and failure induced by the cumulative rise in pore water pressure and the reduction in effective stress within seabed sediments under wave action [34–36], related to the permanent compaction deformation of sediments.

Recent research has also explored tidal flat liquefaction caused by tidal bores from a sedimentological perspective [37]. Tessier and Terwindt [38] investigated significant soft sediment deformations in the tidal channel of Mont-Saint-Michel Bay, characterized by folds and tobacco-pouch structures, attributing these features to the liquefaction of unconsolidated sediments triggered by the passage of a tidal bore. In the Turnagain Arm area of Alaska, Greb and Archer [39] found that tidal bores are a primary cause of soft sediment deformation structures similar to those formed by earthquake-induced sand liquefaction. Fan et al. [40] analyzed sedimentary structures in the northern tidal flat of the Qiantang River estuary, indicating that the rapid increase in water pressure and wave impact induced by tidal bores can cause liquefaction, resulting in features such as convolute bedding and drainage structures. Tidal bore-induced soft sediment deformations, primarily involving liquefaction, are evident in contorted beds and dewatering structures, with geometries ranging from 2D folds to 3D chaotic envelopes. Due to the occurrence of liquefaction, tidal bore-induced sedimentary signatures are more conspicuous than regular tidal facies and are easily recognized within tidal channel infilling successions.

The dynamic response and liquefaction potential of soils in coastal and estuarine environments are strongly influenced by factors such as permeability, saturation, and sediment properties, including grain size and fines content [41]. High permeability facilitates the rapid dissipation of pore water pressure, reducing the risk of liquefaction, whereas low permeability increases susceptibility. Saturation further affects this process, with unsaturated soils more prone to increased pressure gradients [42]. Additionally, sediment characteristics play a critical role: coarse-grained soils generally have a lower liquefaction potential, but the presence of fines can increase susceptibility by restricting drainage [43]. Understanding the interaction of these factors is key to predicting soil behavior under dynamic conditions, such as those caused by tidal bores.

Analytical solutions for the response of seabeds to wave loading have garnered significant attention since the 1940s [25], leading to the development of various computational models. These models include uncoupled (drained) models, consolidation (quasi-static) models, dynamic models, and poro-elastoplastic models [41]. The core of seabed response analysis under wave loading involves examining the coupled effects of soil stress and pore fluid pressure induced by wave trains [44]. Building on Biot's pioneering work on consolidation theory [45–47], analytical solutions for seabed response under wave loading have been proposed for both two-dimensional [31,48] and three-dimensional spaces [42,49]. A set of analytical solutions for wave-induced seabed response under cnoidal waves was developed to evaluate the impact of high-order nonlinearity in shallow water [44,50], with a parametric study revealing that this nonlinearity significantly influences shallow water waves and amplifies the effects of soil characteristics and liquefaction potential. All the aforementioned analytical studies were formulated based on the assumption of simple harmonic waves. However, the transient response of the seabed under the abrupt application of wave loads has not been accounted for in these analyses.

Wave action can lead to localized liquefaction and erosion of the seabed, which is particularly evident in transient responses. Selecting appropriate integral transforms facilitates the conversion of differential and integral equations into solvable algebraic equations. Consequently, several analytical or semi-analytical solutions for seabed transient response under wave action have utilized integral transform methods [51,52]. For layered seabeds, the state-space method can be extended to improve computational efficiency [53,54]. In macro-tidal estuary regions, a tidal bore is a typical transient wave. It is characterized by a sudden rise in water level at the front of the tidal wave. When a breaking bore occurs, the front becomes nearly vertical and propagates upstream as a form of moving load. However, previous studies have often been limited by inadequate consideration of the complex interactions between tidal bores and the riverbed, particularly in terms of transient response and the liquefaction potential. These limitations highlight the need for more comprehensive models that can accurately capture the dynamic behavior of the riverbed under such extreme conditions. The purpose of this study is to address these gaps by developing a more detailed analysis of the transient response of the riverbed and to better understand the mechanisms leading to momentary liquefaction induced by pressure gradient.

The structure of this paper is as follows: Section 2 provides a detailed description of the theoretical model and methodologies used in this study, including the governing equations and the integral transform method. Section 3 presents the results of the study, with subsections focusing on model verification, riverbed response analysis, seepage characteristics, and liquefaction analysis. The discussion and limitations of these findings are addressed in Section 4. Finally, Section 5 concludes the study, summarizing the key contributions and outlining directions for future research. The findings of this study are expected to provide new insights into the stability of riverbeds under extreme hydrodynamic conditions, contributing to improved predictive models and risk assessment frameworks for coastal engineering and management practices.

2. Materials and Methods

2.1. Theoretical Model

The generalized Biot theory provides a comprehensive framework for describing the interaction between skeleton deformation and pore fluid flow in multiphase porous media under transient loading conditions [55]. This study employs a theoretical model to analyze the dynamic response of a riverbed subjected to tidal forces, based on the generalized Biot theory. The model operates under the following assumptions:

- The riverbed is considered horizontal, homogeneous, highly saturated, isotropic in permeability, and of finite thickness.
- The compressibility coefficients of the soil skeleton and pore water are constant, while the soil particles are assumed to be incompressible.
- The stress–strain relationship of the soil skeleton adheres to Hooke's law.
- The seepage of pore water complies with Darcy's law.
- The energy loss during tidal bore propagation is disregarded.
- The water pressure exerted on the riverbed surface is equivalent to the water pressure on the surface of an impermeable, rigid horizontal riverbed at the same depth.

2.1.1. Governing Equations

In a two-dimensional Cartesian coordinate system (x, z), we analyze the propagation of an incident tidal bore at a water depth d over a submerged porous riverbed with a thickness h, as illustrated in the side view sketch in Figure 2. Following Zienkiewicz et al. [55], the equilibrium equations for a unit total volume can be expressed as

$$\sigma_{ij,j} + \rho_0 g_i = \rho_0 \ddot{u}_i + \rho_w \ddot{w}_i, \tag{1}$$

where σ_{ij} is the total stress component, u_i is the displacement of the solid matrix, w_i is the average relative displacement of the fluid to the solid, ρ_0 represents the combined density, $\rho_0 = (1 - n)\,\rho_s + n\,\rho_w$, ρ_w, and ρ_s denotes the fluid and solid density, respectively, n is the porosity of the solid phase, and g_i refers to the body force acceleration, typically defined by the relevant components of gravity.

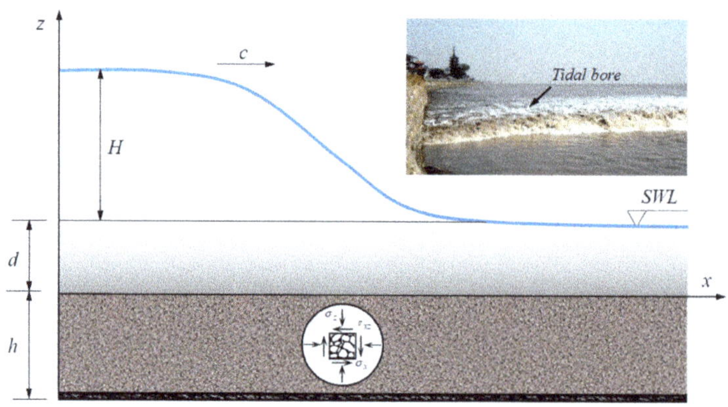

Figure 2. Schematic diagram of the interaction between a tidal bore and the riverbed. Inset: the Qiantang River tidal bore at Yanguan, China; c represents the celerity of the tidal bore, d is the water depth before the bore arrives, h is the thickness of the riverbed, H is the height of the tidal bore, σ_x and σ_z correspond to the normal stresses in the x and z directions, respectively, τ_{xz} is the shear stress, and SWL indicates the still water line.

The equilibrium equations for the fluid phase, incorporating viscous resistance as defined by Darcy's law and assuming isotropic permeability, are established as follows:

$$-p_{,i} + \rho_w g_i = \rho_w \ddot{u}_i + \frac{\rho_w}{n} \ddot{w}_i + \frac{\rho_w g}{k_z} \dot{w}_i, \tag{2}$$

where p is the pore pressure in the riverbed; k_z is the permeability coefficient in the z direction, considering a seabed with isotropic permeability $k_z = k_x$; and k_x is the permeability coefficient in the x directions.

The continuity equation for mass conservation in the fluid is as follows:

$$n\beta\,\dot{p} + \dot{u}_{i,i} + \dot{w}_{i,i} = 0, \tag{3}$$

where β is the volumetric compressibility coefficient of the fluid, which can be expressed as follows for a highly saturated riverbed [56]

$$\beta = \frac{1}{K_w} + \frac{1 - S_r}{P_{w0}}, \tag{4}$$

where K_w is the true modulus of elasticity of water, P_{w0} is the absolute water pressure, and S_r is the degree of saturation.

In plane strain, the constitutive relationship is expressed as follows:

$$\sigma_{ij} = \lambda\delta_{ij}\theta + G(u_{i,j} + u_{j,i}) - \delta_{ij}p, \tag{5}$$

where λ and G are the Lamé constants, δ_{ij} is the Kronecker delta denotation, and θ is the volume strain, $\theta = u_{i,i}$.

2.1.2. Integral Transform Method

Omitting the static gravity terms and focusing only on excess pressures and stresses above the static state, apply the Laplace transform to generalized Biot's Equations (1) through (3) and constitutive Equation (5), respectively [41]

$$\hat{f}(x,z,s) = \int_0^{+\infty} f(x,z,t)e^{-st}dt, \tag{6}$$

where $f(x, z, t)$ is the original function in the time domain; $\hat{f}(x,z,s)$ is the Laplace transformed function; and s is the complex frequency variable in the Laplace domain. After simplification, the generalized Biot's equations can be written in the following form:

$$G\nabla^2\hat{u}_x + (\lambda + G)\frac{\partial\hat{\theta}}{\partial x} - (1 - \vartheta)\frac{\partial\hat{p}}{\partial x} - s^2(\rho_0 - \rho_w\vartheta)\hat{u}_x = 0, \tag{7}$$

$$G\nabla^2\hat{u}_z + (\lambda + G)\frac{\partial\hat{\theta}}{\partial z} - (1 - \vartheta)\frac{\partial\hat{p}}{\partial z} - s^2(\rho_0 - \rho_w\vartheta)\hat{u}_z = 0, \tag{8}$$

$$\nabla^2\hat{p} - \frac{n\beta\rho_w s^2}{\vartheta}\hat{p} - \frac{1 - \vartheta}{\vartheta}\rho_w s^2\hat{\theta} = 0, \tag{9}$$

where

$$\vartheta = \frac{\rho_w s^2 n}{\rho_w s^2 + nbs}, \tag{10}$$

$$b = \frac{\rho_w g}{k_z}. \tag{11}$$

From Equations (7) through (9), the following equations can be derived

$$\nabla^4\hat{p} + \beta_1\nabla^2\hat{p} + \beta_2\hat{p} = 0, \tag{12}$$

where $M = \lambda + 2G$,

$$\beta_1 = -\frac{s^2(Mn\beta\rho_w + \rho_w - 2\rho_w\vartheta + \vartheta\rho_0)}{M\vartheta}, \tag{13}$$

$$\beta_2 = \frac{(\rho_0 - \rho_w\vartheta)n\beta s^4\rho_w}{M\vartheta}. \tag{14}$$

Apply the following Fourier transforms to Equations (7), (8), and (12):

$$\widetilde{f}(\xi,z,s)=\int_{-\infty}^{+\infty}\hat{f}(x,z,s)e^{i\xi x}dx, \tag{15}$$

where $\hat{f}(x,z,s)$ is the original function, $\widetilde{f}(\xi,z,s)$ is the Fourier transformed function, ξ is the frequency variable, and i is the imaginary unit.

After converting the partial differential equation into an ordinary differential equation, the operator method is used to obtain the following general solution:

$$\mathbf{Y}_1(\xi,z,s)=\mathbf{T}_1(\xi,z,s)\mathbf{X}_0(\xi,z,s), \tag{16}$$

where

$$\mathbf{Y}_1=\left[\widetilde{p}(\xi,z,s)\widetilde{\theta}(\xi,z,s)\ \widetilde{u}_z(\xi,z,s)\xi(i\widetilde{u}_x)(\xi,z,s)\right]^T, \tag{17}$$

$$\mathbf{X}_0=\left[A_1e^{\gamma_1 z}A_2e^{-\gamma_1 z}A_3e^{\gamma_2 z}A_4e^{-\gamma_2 z}A_5e^{\gamma_3 z}A_6e^{-\gamma_3 z}\right]^T, \tag{18}$$

$$\mathbf{T}_1=\begin{bmatrix}1&1&1&1&0&0\\-G\kappa_1&-G\kappa_1&-G\kappa_2&-G\kappa_2&0&0\\\gamma_1\alpha_1&-\gamma_1\alpha_1&\gamma_2\alpha_2&-\gamma_2\alpha_2&1&1\\\chi_1&-\chi_1&\chi_2&-\chi_2&\gamma_3&-\gamma_3\end{bmatrix}, \tag{19}$$

$$\gamma_j=\sqrt{\xi^2-L_j^2}, \tag{20}$$

$$L_1^2=\frac{1}{2}(\beta_1+\sqrt{\beta_1^2-4\beta_2}), \tag{21}$$

$$L_2^2=\frac{1}{2}(\beta_1-\sqrt{\beta_1^2-4\beta_2}), \tag{22}$$

$$L_3^2=\frac{(\rho_w-\rho_0)\vartheta s^2}{G}, \tag{23}$$

$$\kappa_j=\frac{\vartheta L_j^2+n\beta\rho_w s^2}{G(1-\vartheta)\rho_w s^2}, \tag{24}$$

$$\alpha_j=\frac{\kappa_j(\lambda+G)+(1-\vartheta)/G}{L_3^2-L_j^2}, \tag{25}$$

$$\chi_j=G\kappa_j+\gamma_j^2\alpha_j. \tag{26}$$

Using Equation (16) and the constitutive Equation (5), the total stress components can be expressed in the Laplace–Fourier transform domain as follows:

$$\mathbf{Y}_2(\xi,z,s)=\mathbf{T}_2(\xi,z,s)\mathbf{X}_0(\xi,z,s), \tag{27}$$

where

$$\mathbf{Y}_2=\left[\widetilde{\sigma}_x(\xi,z,s)\widetilde{\sigma}_z(\xi,z,s)i\xi\widetilde{\tau}_{xz}(\xi,z,s)\right]^T, \tag{28}$$

$$\mathbf{T}_2=\begin{bmatrix}c_1&c_1&c_2&c_2&-2G\gamma_3&2G\gamma_3\\d_1&d_1&d_2&d_2&2G\gamma_3&-2G\gamma_3\\g_1&-g_1&g_2&-g_2&\gamma_3^2+\xi^2&\gamma_3^2+\xi^2\end{bmatrix}, \tag{29}$$

where

$$c_j=-(\lambda\kappa_j+2\chi_j)G, \tag{30}$$

$$d_j=-(M\kappa_j-2\chi_j)G, \tag{31}$$

$$g_j=\gamma_j(\xi^2\alpha_j+\chi_j)G. \tag{32}$$

At this stage, the analytical solution of the dynamic response control equation for the riverbed in the Laplace–Fourier transform domain is obtained. In Equations (16) and (27),

A_j (j = 1, ..., 6) in $\mathbf{X}_0(\xi, z, s)$ are undetermined constants, which will be solved based on the specific problem in the next subsection.

2.2. Solving the Boundary Value Problem

According to [57], the tidal bore-induced dynamic water pressure at the water–riverbed interface can be expressed as [52]

$$p_b = \rho g H\{1 - \tanh b(x - ct)\}/2, \tag{33}$$

where

$$b = \left(3H/4d^3\right)^{1/2}. \tag{34}$$

The boundary conditions for the riverbed response problem under the action of tidal bore can be described as follows:

$$z = 0,\ \sigma_z + p = 0,\ \tau_{xz} = 0,\ p = p_b, \tag{35}$$

$$z = -h,\ u_x = u_z = 0,\ \frac{\partial p}{\partial z} = 0. \tag{36}$$

Applying the Laplace transform followed by the Fourier transform to Equations (35) and (36) yields

$$\widetilde{\sigma}_z(\xi, 0, s) = -\widetilde{p}_b, \widetilde{p}(\xi, 0, s) = \widetilde{p}_b, \widetilde{\tau}_{xz}(\xi, 0, s) = 0, \tag{37}$$

$$\widetilde{u}_x(\xi, -h, s) = 0, \widetilde{u}_z(\xi, -h, s) = 0, \frac{\partial \widetilde{p}}{\partial z}(\xi, -h, s) = 0, \tag{38}$$

where

$$\widetilde{p}_b = 2\pi\rho g H\left\{\frac{1}{s}\delta(\xi) + \sum_{k=1}^{\infty}\frac{(-1)^k}{s + 2kbc}\delta(\xi - 2kbi)\right\}. \tag{39}$$

The detailed derivation process of Equation (39) can be found in Appendix A.

Substitute Equations (37) and (38) into Equations (16) and (27), solve the system of equations, and obtain the expressions for the coefficients A_j (j = 1, ..., 6)

$$A_j = \widetilde{p}_b B_j \quad (j = 1, \cdots, 6), \tag{40}$$

where coefficients B_j are functions of ξ and s; the specific forms can be found in Appendix B.

The solution for the riverbed response under tidal bore in the Laplace–Fourier transform domain is given as follows:

$$\widetilde{f}(\xi, z, s) = 2\pi\rho g H\left\{\frac{1}{s}\delta(\xi) + \sum_{k=1}^{\infty}\frac{(-1)^k}{s + 2kbc}\delta(\xi - 2kbi)\right\}\boldsymbol{\Gamma}(\xi, z, s), \tag{41}$$

where $\boldsymbol{\Gamma}(\xi, z, s)$ is a rational function of ξ, s and z, and the specific expression varies depending on the response variable.

By applying the inverse Fourier transform to Equation (41) and utilizing the sifting property of the δ function, we obtain the following result:

$$\hat{f}(x, z, s) = \rho g H\left\{\frac{\boldsymbol{\Gamma}(0, z, s)}{s} + \sum_{k=1}^{\infty}\frac{(-1)^k\boldsymbol{\Gamma}(2kbi, z, s)}{s + 2kbc}e^{2kbx}\right\}. \tag{42}$$

By applying the inverse Laplace transform, using the numerical method for inverse Laplace transform proposed by Durbin [58], a corresponding program was developed to perform the calculations and obtain the results.

Meanwhile, let

$$\hat{\mathbf{n}}_1(x, z, s) = \frac{\boldsymbol{\Gamma}(0, z, s)}{s}, \tag{43}$$

$$\hat{\mathbf{n}}_2(x,z,s) = \frac{(-1)^k \mathbf{\Gamma}(2kbi,z,s)}{s+2kbc} e^{2kbx}. \tag{44}$$

Considering Equation (43), it can be written as a reduced rational function

$$\hat{\mathbf{n}}_1(x,z,s) = \frac{\mathbf{W}_{11}(x,z,s)}{\mathbf{W}_{12}(x,z,s)}. \tag{45}$$

where coefficients $\mathbf{W}_{11}(x, z, s)$ and $\mathbf{W}_{12}(x, z, s)$ are irreducible proper rational functions.

Assuming the denominator has N zeros, $s_1, s_2, \ldots s_N$, with $s = 0$ clearly being one of them, let us assume $s_1 = 0$. Then, according to Heaviside's first expansion theorem for the inverse Laplace transform [59], the following result can be obtained:

$$\mathbf{n}_1(x,z,t) = \mathbf{\Gamma}(0,z,0) + \sum_{n=2}^{N} \frac{\mathbf{W}_{11}(x,z,s_n)}{\mathbf{W}'_{12}(x,z,s_n)} \exp[s_n t]. \tag{46}$$

where

$$\mathbf{W}'_{12}(x,z,s) = \frac{d}{ds}[\mathbf{W}_{12}(x,z,s)]. \tag{47}$$

Similarly, from Equation (44), we obtain the following result:

$$\mathbf{n}_2(x,z,t) = \sum_{k=1}^{\infty} (-1)^k \mathbf{\Gamma}(2kbi,z,-2kbc) e^{2kb(x-ct)} + \sum_{k=1}^{\infty} \sum_{n=2}^{N} \frac{\mathbf{W}_{21}(x,z,s_n)}{\mathbf{W}'_{22}(x,z,s_n)} \exp[s_n t]. \tag{48}$$

where coefficient $\mathbf{W}_{21}(x, z, s)$ and $\mathbf{W}_{22}(x, z, s)$ are irreducible proper rational functions.

The second term on the right-hand side of Equations (46) and (48) decays rapidly in the time domain, i.e.,

$$\left. \sum_{n=2}^{N} \frac{\mathbf{W}_{11}(x,z,s_n)}{\mathbf{W}'_{12}(x,z,s_n)} \exp[s_n t] \right|_{t\to\infty} \to 0, \tag{49}$$

$$\left. \sum_{k=1}^{\infty} \sum_{n=2}^{N} \frac{\mathbf{W}_{21}(x,z,s_n)}{\mathbf{W}'_{22}(x,z,s_n)} \exp[s_n t] \right|_{t\to\infty} \to 0. \tag{50}$$

Therefore, the steady-state solution for the dynamic response of the riverbed under tidal bore action is given by

$$f^*(x,z,t) = \rho g H \left\{ \mathbf{\Gamma}(0,z,0) + \sum_{k=1}^{\infty} (-1)^k \mathbf{\Gamma}(2kbi,z,-2kbc) e^{2kb(x-ct)} \right\}. \tag{51}$$

2.3. Liquefaction Triggered Criterion

The propagation of tidal bores leads to sudden changes in water depth and rapid fluctuations in physical variables such as velocity and pressure fields. When a tidal bore arrives, the flow quickly shifts from ebb to flood in a very short time. For instance, in the Qiantang River in China, a tidal bore can cause a sudden rise of 2–4 m within just 5 s. The resulting pressure gradient is strong enough to cause the momentary failure of riverbed sediments, potentially leading to the liquefaction of the bed material. This study examines the momentary liquefaction of riverbed triggered by the vertical and horizontal pressure gradients.

When fluid flows through a porous medium, momentary liquefaction can occur if the vertical or horizontal pressure gradient exceeds a certain threshold. Bear [60] proposed that when the vertical resultant force acting on the particles composing a porous medium becomes zero, localized momentary liquefaction can occur

$$-\frac{1}{\rho_w g}\left(\frac{\partial p}{\partial z}\right) = \frac{s_z}{k_z} > \frac{\gamma'_s}{\gamma_w}, \tag{52}$$

where S_z represents the vertical seepage velocity, γ_s' stands for the effective unit weight of the soil particles, and γ_w is the unit weight of water.

Effective unit weight is the unit weight of the soil when buoyancy from water is considered. It is calculated by subtracting the buoyant force from the gravitational force acting on the soil particles within a unit volume of soil. For most silty soil,

$$\frac{\gamma'_s}{\gamma_w} \approx 0.7. \tag{53}$$

Meanwhile, Madsen [61] suggested that when the horizontal pressure gradient exceeds the intergranular stress, localized momentary liquefaction can also occur

$$-\frac{1}{\rho_w g}\left(\frac{\partial p}{\partial x}\right)_{crit} = \left(\frac{S_x}{k_x}\right)_{crit} = \frac{\gamma'_s}{\gamma_w}\tan\phi, \tag{54}$$

where S_x represents the horizontal seepage velocity and φ is the internal friction angle.

The internal friction angle of soil represents the soil's inherent frictional properties. It is generally considered to comprise two components: the surface friction between soil particles and the interlocking force generated by the embedding and interlocking of particles. Thus, it is not only related to the material properties of the soil particles but also to the stress conditions. It is typically, for clay, $0 \leq \varphi \leq 20°$; for saturated silty soil, it is $0 \leq \varphi \leq 35°$. For the silt in the Qiantang River, a particular value of $\varphi = 35°$ was chosen for the calculations.

$$\frac{\gamma'_s}{\gamma_w}\tan\phi \approx 0.5. \tag{55}$$

3. Results

3.1. Verification

Assuming that the soil skeleton is incompressible, meaning a rigid soil bed, the dynamic water pressure on the bed surface and the seepage velocity under the influence of a tidal bore can be approximately expressed as [57]

$$p_b/\rho g d = \varepsilon\{1 - \tanh bX\}/2, \tag{56}$$

$$\frac{S_x}{k_x} = \frac{\varepsilon\sqrt{3\varepsilon}}{4}\mathrm{sech}^2 bX, \tag{57}$$

$$\frac{S_z}{k_z} = \frac{3h\varepsilon^2}{4d}\tanh bX \,\mathrm{sech}^2 bX, \tag{58}$$

where $X = x - ct$, $\varepsilon = H/d$.

The results presented in Equation (51) can easily degenerate to the case of a rigid soil bed. A comparison between the degenerate solution and the analytical solution by Packwood and Peregrine [57] is shown in Figure 3. The calculation parameters are set as follows: tidal bore height H = 1.2 m, water depth d = 3.0 m, and soil bed thickness h = 3.0 m. For the rigid soil bed, $G \to \infty$, $\beta \to 0$. As illustrated in the Figure 3, the two solutions show good agreement.

3.2. Riverbed Responses

The degradation solutions of pore water pressure and seepage velocity caused by tidal bores have been validated using the analytical solution provided by Packwood and Peregrine [57]. Here, we provide more general and broader results regarding the riverbed response induced by tidal bores. In the example of riverbed response to be presented, the following common values are assumed: water depth d = 3 m, tidal bore height H = 3 m, riverbed thickness h = 9 m, permeability coefficient $k_x = k_z = 10^{-3}$ m/s, Poisson's ratio

$\mu = 1/3$, porosity $n = 0.3$, and shear modulus $G = 10^7$ N/m^2. Given Poisson's ratio μ and the shear modulus G, the Lame constant λ can be calculated using the following formula:

$$\lambda = \frac{2G\mu}{1-2\mu}. \tag{59}$$

The response of riverbed pore water pressure and effective stresses to tidal bore events is significantly influenced by the degree of saturation. Figure 4 presents the contours of excess pore water pressure within the riverbed for two saturation conditions: fully saturated ($S_r = 1$) and nearly fully saturated ($S_r = 0.99$). The results indicate that in the fully saturated scenario, the pore spaces are entirely filled with water, leading to higher excess pore water pressures near the riverbed surface. This uniform pressure distribution is attributed to the incompressibility of water. Conversely, in the nearly saturated case, the presence of a minimal amount of air in the pores reduces the overall pore pressure and results in more dispersed pressure contours. This suggests a more gradual dissipation of energy within the soil matrix. Unlike waves, the pore water pressure induced by tidal bores is always positive; it decreases with increasing depth. In an unsaturated riverbed, the gradients of pore pressure change both horizontally and vertically are greater than those in a saturated riverbed.

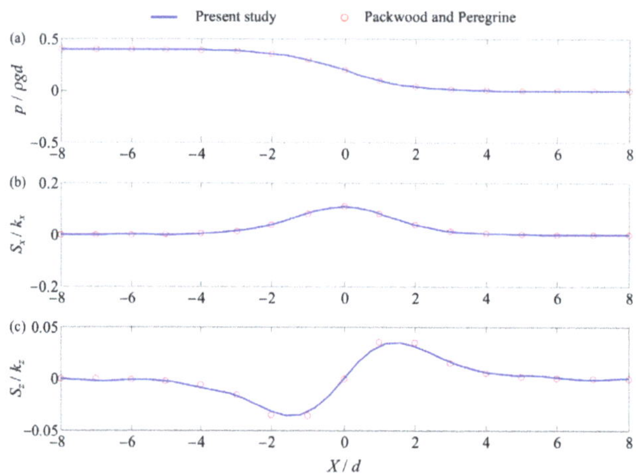

Figure 3. Comparison between the degenerate solution and the existing analytical solution [57]. (**a**) Dynamic water pressure, (**b**) horizontal seepage velocity, and (**c**) vertical seepage velocity.

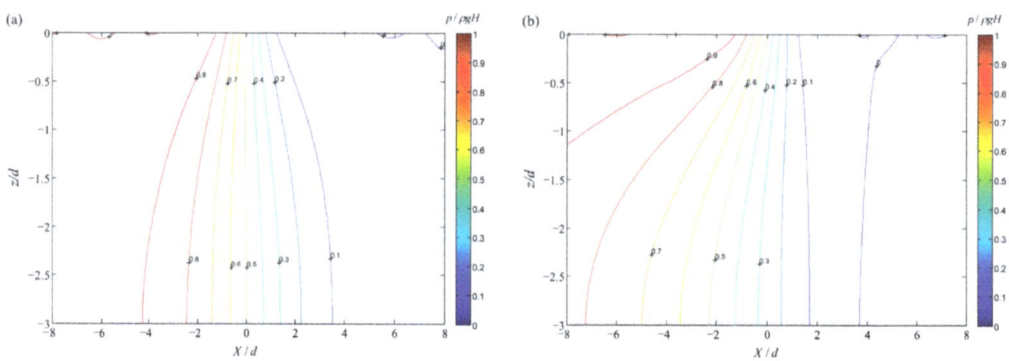

Figure 4. Contours of tidal bore induced pore pressure within the riverbed. (**a**) $S_r = 1$ and (**b**) $S_r = 0.99$, S_r is the degree of saturation.

Figure 5 illustrates the contours of effective stress for two saturation conditions: fully saturated and nearly fully saturated. In the fully saturated scenario, the stress distribution is more localized, with concentrated regions of effective stresses occurring directly beneath the front of the tidal bore. In contrast, the nearly saturated condition results in a more dispersed stress pattern, with higher magnitudes of both compressive and tensile stresses extending across a broader area of the riverbed. The presence of air in the pores reduces pore water pressure, leading to an increase in soil compressibility and a wider distribution of stresses. When the riverbed is fully saturated, the effective stress distribution induced by the tidal bore is generally symmetrical about the front at $X = 0$. The impact of saturation on effective normal stress is greater than that on shear stress.

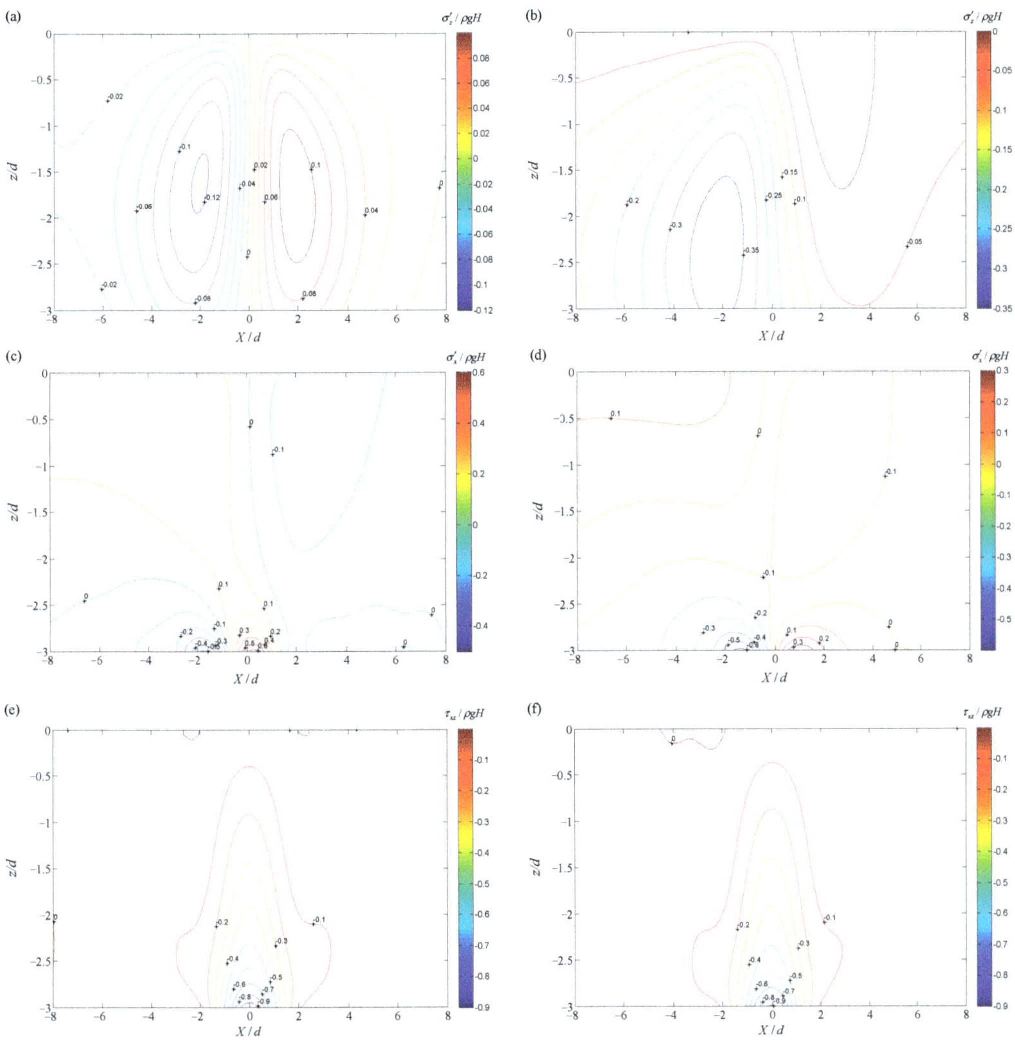

Figure 5. Contours of tidal bore-induced effective stresses within the riverbed. Vertical effective stress: (**a**) $S_r = 1$ and (**b**) $S_r = 0.99$; horizontal effective stress: (**c**) $S_r = 1$ and (**d**) $S_r = 0.99$; and shear stress: (**e**) $S_r = 1$ and (**f**) $S_r = 0.99$. S_r is the degree of saturation.

3.3. Seepage Characteristics

Figure 6 illustrates the time histories of tidal bore-induced horizontal and vertical seepage velocities within the fully saturated riverbed at three elevations: $z = 0$, $z = -0.25\,h$, and $z = -0.5\,h$. The dynamic water pressure (Figure 6a) is uniformly affected across all elevations. However, the horizontal and vertical seepage velocity (Figure 6b,c) shows a pronounced decrease with depth, being highest at the seabed surface and minimal at $z = -0.5\,h$.

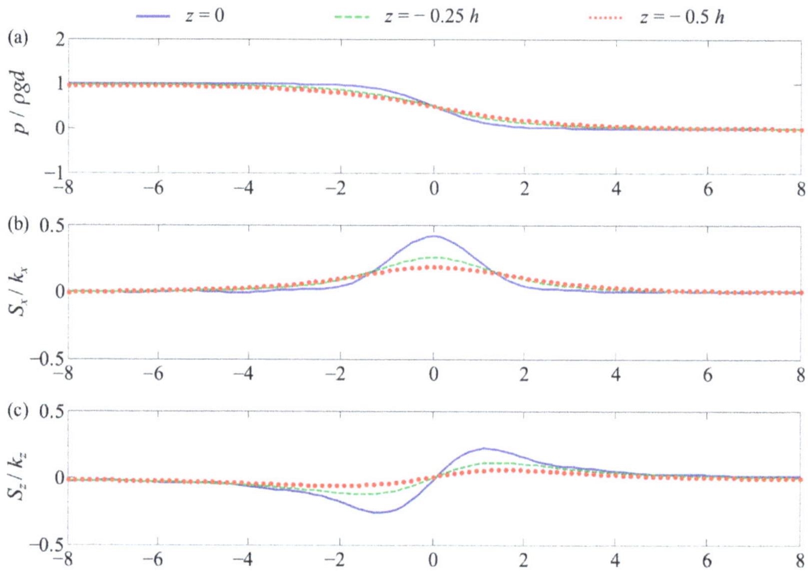

Figure 6. Time histories of the horizontal and vertical seepage velocities induced by the tidal bore at three different elevations (0, $-0.25\,h$, and $-0.5\,h$). (**a**) Dynamic water pressure at the interface of water and riverbed, (**b**) horizontal seepage velocity, and (**c**) vertical seepage velocity.

Figure 7 illustrates the contours of both vertical and horizontal seepage velocities induced by a tidal bore within the riverbed for two different degrees of saturation: fully saturated and nearly fully saturated. In the fully saturated case (Figure 7a,c), both vertical and horizontal seepage velocities are concentrated near the surface, showing higher peak velocities and sharper distributions, with the vertical seepage displaying alternating upward and downward flow zones and the horizontal seepage exhibiting intense and localized flow. In contrast, when the saturation is slightly reduced (Figure 7b,d), the seepage velocity distributions become broader and more diffuse, with lower peak velocities extending deeper into the riverbed. This difference highlights the impact of saturation on seepage dynamics, where a fully saturated riverbed exhibits more intense and localized seepage, while reduced saturation leads to a more widespread but weaker seepage response.

3.4. Liquefaction Analysis

To examine the influence of tidal bore and soil parameters on the momentary liquefaction of the riverbed, the following five sets of parameters are considered:

- Tidal bore height: 0.6 m to 4 m;
- Water depth: 2 m to 5 m;
- Bed thickness: 0.6 m to 10 m;
- Degree of saturation: $G\beta$ from 0 to 0.6;
- Permeability coefficient: $k_x = k_z$, from 10^{-6} m/s to 10^{-2} m/s.

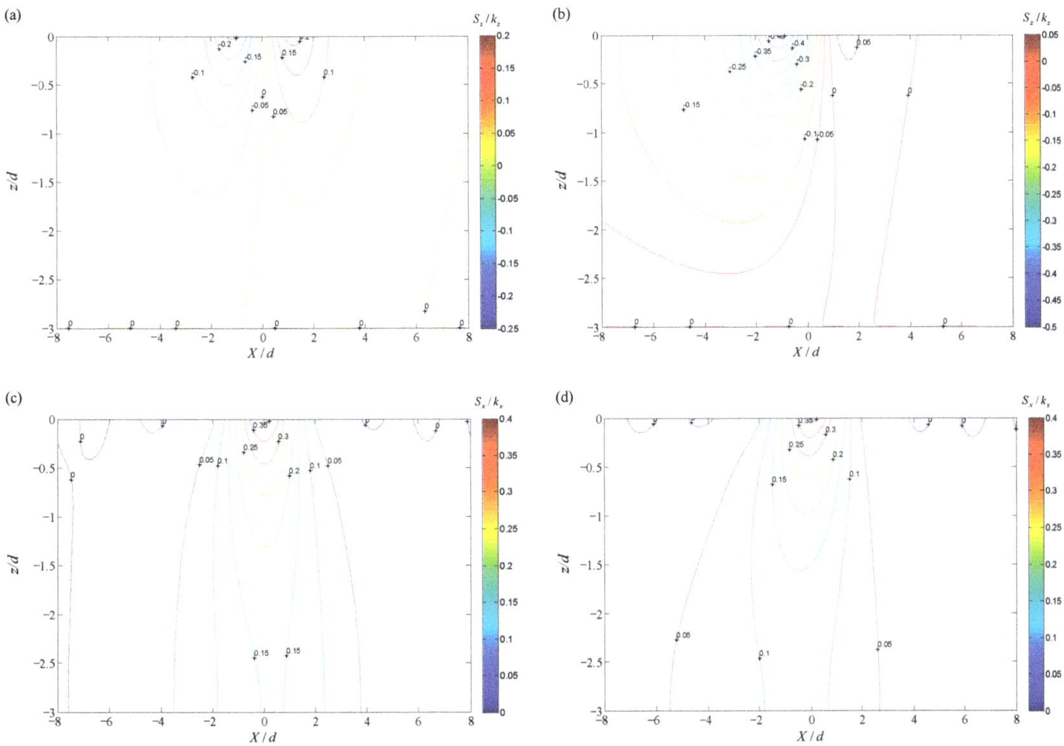

Figure 7. Contours of tidal bore-induced seepage velocity within the riverbed. Vertical seepage velocity: (**a**) $S_r = 1$ and (**b**) $S_r = 0.99$; horizontal seepage velocity: (**c**) $S_r = 1$ and (**d**) $S_r = 0.99$. S_r is the degree of saturation.

The other parameters remain constant, including Poisson's ratio $\mu = 1/3$, porosity $n = 0.3$, and shear modulus $G = 10^7$ N/m^2. Figure 8 illustrates the boundary between the momentary liquefaction zone and the stable zone within the riverbed, as predicted by two different models: Equation (53) by Bear [60] and Equation (55) by Madsen [61].

According to Bear's criterion, momentary liquefaction occurs when the vertical resultant force on the particles becomes zero, resulting in a shallow liquefaction zone with a depth of $z_l = 0.1\ d$. In contrast, Madsen's criterion suggests that liquefaction occurs when the horizontal pressure gradient exceeds the intergranular stress, leading to a deeper liquefaction zone extending to $z_l = 0.5\ d$. The maximum liquefaction depth predicted based on horizontal pressure gradients is five times that predicted based on vertical pressure gradients. This difference in liquefaction depth highlights the varying influences of vertical and horizontal seepage forces, with Bear's criterion predicting shallow liquefaction under rapidly changing vertical seepage forces, while Madsen's criterion indicates deeper liquefaction under significant horizontal pressure gradients. For conservative reasons, the liquefaction criterion proposed by Madsen [61] is adopted in the subsequent parameter analysis.

Figure 9 illustrates the relationship between maximum liquefaction depth (z_l/d) and tidal bore height (H/d) for a partially saturated riverbed composed of isotropic silt soil and fine sand. The results indicate a distinction in liquefaction behavior between the two soil types. Silt soil exhibits a greater liquefaction depth across all tidal bore heights compared to fine sand, highlighting its higher susceptibility to liquefaction. This behavior can be attributed to the finer particle size and lower permeability of silt, which result in higher pore pressure gradient under loading conditions such as those induced by tidal bores. When the water depth is 2 m and the tidal bore height is less than 1.4 m, no liquefaction occurs in the fine sand. However, when the tidal bore height exceeds 1.4 m, the liquefaction depth increases sharply. This indicates that when the relative tidal bore height is low, it is insufficient to impact the fine sandy seabed.

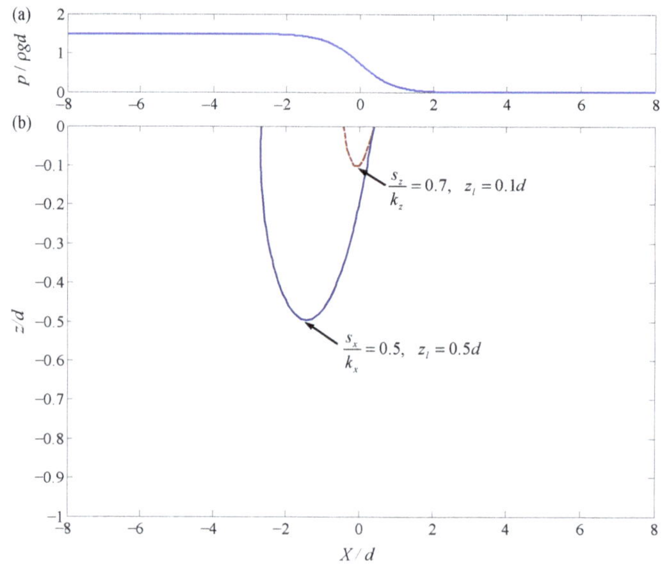

Figure 8. (**a**) Dynamic water pressure at the water–riverbed interface. (**b**) The boundary between the liquefaction zone and the stable zone based on Equations (53) and (55), respectively.

Figure 10 illustrates the relationship between the maximum liquefaction depth (z_l/d) and the water depth (d/h) before the arrival of the tidal bore for a partially saturated riverbed composed of isotropic silt soil and fine sand, with a constant tidal bore height. The data reveal a pronounced decrease in liquefaction depth as the initial water depth increases relative to the tidal bore height for both soil types. Silt soil, due to its finer particle size and lower permeability, initially leads to a higher liquefaction depth. However, as the water depth increases, the relative impact of the tidal bore on the silt soil's stability lessens, leading to reduced liquefaction. As the water depth increases, the liquefaction depth in fine sand decreases more rapidly than in silt, indicating that fine sand exhibits relatively better stability under deeper water conditions.

Figure 11 illustrates the relationship between the liquefaction depth z_l/d and the relative riverbed thickness h/d, comparing isotropic silt soil and fine sand. As h/d increases, the liquefaction depth for both soil types also increases, indicating that thicker riverbeds are more prone to deeper liquefaction zones. For silt riverbeds, when h/d is less than 1, the liquefaction depth increases significantly with the increase in riverbed thickness; when h/d exceeds 1, the impact of riverbed thickness on liquefaction depth diminishes, tending towards a stable depth. The pattern is similar for fine sand riverbeds, but the critical value is around $h/d = 2.5$. This suggests that there is a most unstable thickness in the riverbed, resulting from the interaction between incident tidal bore and reflected stress waves from the riverbed bottom. Therefore, in engineering site selection, these sensitive areas should be avoided, and structural foundations should be placed in relatively safer locations.

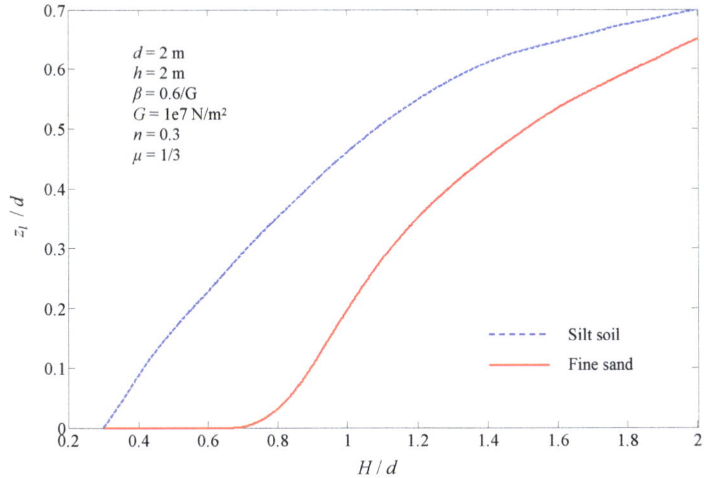

Figure 9. Maximum liquefaction depth z_l/d as a function of tidal bore height *H/d* for a partially saturated riverbed composed of isotropic silt soil and fine sand.

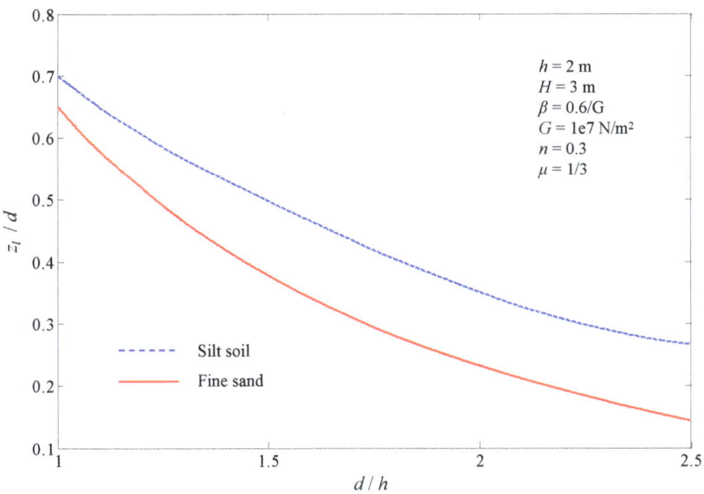

Figure 10. Maximum liquefaction depth z_l/d as a function of water depth *d/h* before the arrival of the tidal bore for a partially saturated riverbed composed of isotropic silt soil and fine sand.

Figure 12 illustrates the relationship between the maximum liquefaction depth z_l/d and the soil stiffness parameter $G\beta$, comparing the liquefaction behavior of a riverbed composed of isotropic silt soil and fine sand. The results indicate that as $G\beta$ increases, the liquefaction depth for both types of soil also increases, though the rate of increase differs. The liquefaction depth for silt soil rises significantly with increasing $G\beta$, suggesting that the liquefaction potential of silt soil is highly sensitive to changes in soil stiffness. In contrast, the liquefaction depth for fine sand increases more gradually, indicating that its liquefaction potential is less affected by variations in soil stiffness. As the saturation degree increases, the $G\beta$ value decreases and the soil stiffness increases, leading to a reduction in liquefaction depth, meaning that the more saturated the riverbed, the more stable it becomes. However, when the saturation degree $S_r \to 1$, the soil stiffness $G\beta \to 0$ and the maximum liquefaction depth of fine sand riverbeds does not vary significantly with changes in saturation.

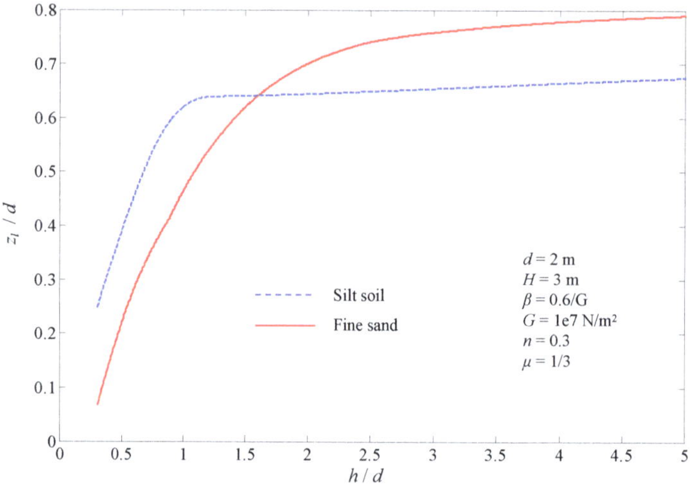

Figure 11. Maximum liquefaction depth z_l/d as a function of relative riverbed thickness h/d for a partially saturated riverbed composed of isotropic silt soil and fine sand.

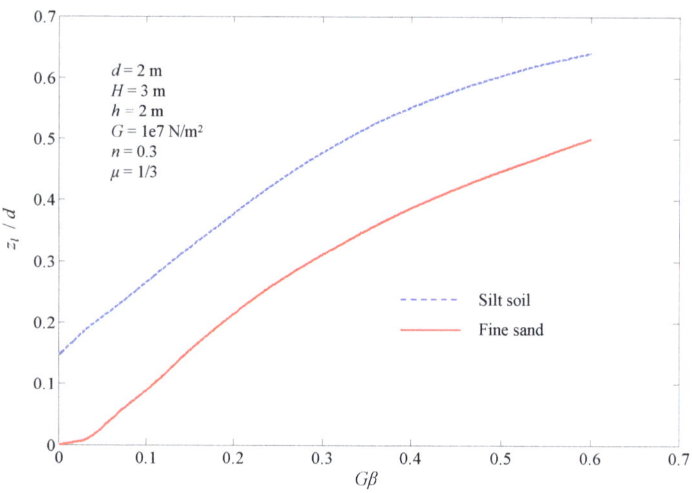

Figure 12. Maximum liquefaction depth z_l/d as a function of soil stiffness $G\beta$ for a riverbed composed of isotropic silt soil and fine sand.

Figure 13 illustrates the relationship between the maximum liquefaction depth z_l/d and the soil permeability k_x, examining the liquefaction behavior of partially saturated isotropic riverbeds ($k_z = k_x$). The results show that the liquefaction depth initially increases with increasing permeability, reaches a peak, and then decreases, indicating that there is a specific permeability range where the liquefaction depth is maximized. Specifically, at lower permeability values typical of silt soil ($k_z < 1.6 \times 10^{-4}$ m/s), the liquefaction depth increases as permeability increases due to the limited drainage capacity, which leads to a higher pore water pressure gradient under tidal bore, thereby enhancing liquefaction potential. However, as it continues to increase to values typical of fine sand, the liquefaction depth decreases because higher permeability facilitates drainage, reducing the pore water pressure gradient and thus lowering the risk of liquefaction. When the permeability coefficient is greater than 7×10^{-3} m/s, no liquefaction occurs in the riverbed because the pore water is easily discharged.

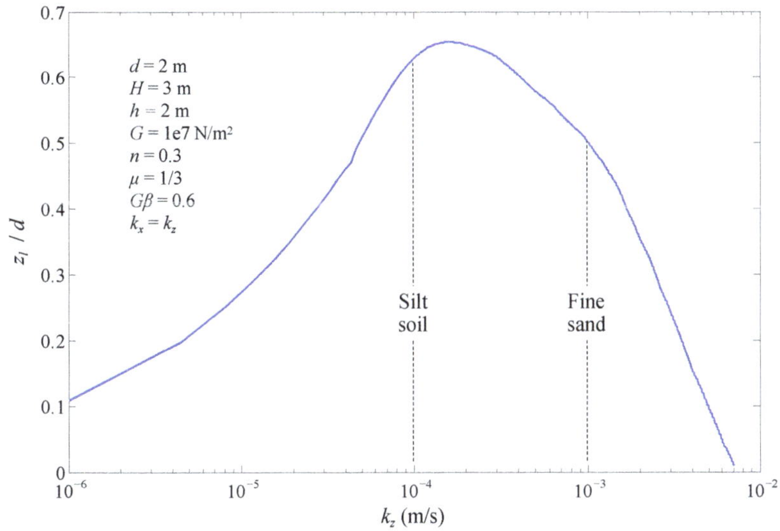

Figure 13. Maximum liquefaction depth z_l/d as a function of soil permeability k_z for a partially saturated isotropic riverbed.

4. Discussion

4.1. Liquefaction Potential Evaluation

The results of this study reveal the critical impact of tidal bores on the liquefaction potential of riverbeds, particularly in macro-tidal estuaries, where the sudden influx of water creates significant hydrodynamic forces that challenge the structural integrity of the sediment layers. By applying the generalized Biot theory and integral transform methods, this study not only captures the transient response of riverbed sediments under these dynamic pressures but also offers a quantitative framework for predicting the conditions under which liquefaction is likely to occur. The study demonstrates that momentary liquefaction is not merely a theoretical possibility but a realistic outcome influenced by key factors such as the permeability coefficient and the degree of saturation of the riverbed. This understanding bears similarity to the seabed liquefaction phenomenon induced by wave loading [29,52].

Fine sand facilitates pore water discharge, significantly reducing liquefaction risk, whereas silt soil increases sediment instability. This finding emphasizes the critical need to consider local soil characteristics when evaluating the vulnerability of riverbeds to tidal bore events. In riverbeds with low permeability, the retention of pore water can result in pore pressure gradients reaching critical levels, thereby increasing the risk of liquefaction. Conversely, in riverbeds with high permeability, the efficient dissipation of pore water pressure can significantly mitigate this risk, identifying a potential area for targeted engineering interventions aimed at enhancing riverbed stability.

Additionally, the study's insights extend beyond the specific context of tidal bores, offering broader implications for the management of coastal and riverine environments subjected to extreme hydrodynamic events. The consistency of these findings with previous research [25,27] reinforces the critical role of soil properties in determining sediment stability under rapid hydrodynamic changes. At the same time, the study advances the field by providing a more detailed analytical approach that considers the specific interactions between tidal bores and riverbed sediments. This approach not only enhances the understanding of sediment behavior in extreme conditions but also lays the groundwork for future studies that might explore more complex scenarios, including the effects of sediment layering, heterogeneity, and multi-bore interactions. Such advancements will be crucial in developing more effective strategies for mitigating the impacts of tidal bores and preserving the integrity of vulnerable estuarine and riverine ecosystems.

4.2. Addressing Limitations

Although this study offers a comprehensive analysis, several limitations indicate the need for further research to fully understand the complex dynamics of tidal bore-induced liquefaction. One key limitation is the assumption of homogeneity and isotropy in riverbed properties. This oversimplifies real-world conditions, where riverbeds typically display significant stratification and variability in material properties, such as grain size, permeability, and sedimentary structure. As highlighted by Hazirbaba [62], Zuo and Baudet [63], and Porcino et al. [64], such heterogeneities can lead to diverse responses to dynamic forces. While this study focused on clean sand and silty soils, the role of fines content in sand–silt mixtures was not explicitly considered. Another limitation lies in the use of a constant shear modulus G. Liquefaction involves large deformations, and G typically varies with shear strain and the number of applied cycles. While this study employed a constant G, we have acknowledged that strain-dependent variations in G are crucial for accurately modeling soil behavior under cyclic loading. Future work will incorporate these variations to enhance the precision of liquefaction predictions, especially in layered sediments where mechanical behavior may differ significantly from homogeneous assumptions.

Additionally, this study does not account for the effects of turbulent flow and sediment transport, which are critical in natural tidal bore environments [7]. Turbulent flow can generate complex stress patterns, leading to localized high-pressure zones that may trigger unexpected liquefaction. Sediment transport processes, such as erosion and deposition, also alter riverbed morphology over time, affecting stability in subsequent tidal events. The exclusion of these processes limits the model's applicability in actual scenarios. To address this, we have expanded our discussion on the assumptions made and emphasized the importance of validating the model through field studies and in situ experiments. These additions strengthen the conclusions and highlight the need for empirical validation to support theoretical findings.

Moreover, the study primarily focuses on the dynamic response to a single tidal bore event, overlooking the long-term effects of repeated events. In many estuarine environments, riverbeds are subjected to daily tidal bores, leading to cumulative effects that may weaken sediment structures over time. The current model does not fully capture the progressive weakening of the riverbed under sustained cyclic loading. Future research should explore these long-term effects to better predict the potential for failure in riverbed stability.

Finally, we have addressed the limitations of the analytical solution, particularly when coefficients change over time, and have introduced numerical methods, such as the Gauss elimination method and the Newmark time integration scheme. These approaches offer alternatives when analytical solutions are impractical, providing a pathway for future refinements of the model. A numerical analysis method that combines stochastic processes with the finite element method (FEM) can also be employed to assess the quantitative impact of the random distribution of soil material properties [65].

5. Conclusions

This study presents a comprehensive analysis of the dynamic response and liquefaction potential of riverbeds induced by tidal bores, with a focus on macro-tidal estuaries. The analysis highlights the importance of permeability and saturation in determining the stability of riverbed sediments under the dynamic pressures generated by tidal bores. It also provides valuable insights into the role of different soil types in influencing liquefaction risk in estuarine environments.

The key findings demonstrate that fine sand reduces liquefaction risk by facilitating pore water discharge, while silt soil increases the likelihood of sediment instability. Additionally, the degree of saturation was found to significantly affect the distribution of effective stresses and seepage characteristics within the riverbed. Partially saturated riverbeds exhibit higher pore pressure gradients, which raise the risk of liquefaction compared to fully saturated conditions.

The assumptions of homogeneity and isotropy in riverbed properties, along with the exclusion of turbulent flow, sediment transport, and repeated tidal bores, limit the applicability of the study in complex natural environments. Future research should develop models that account for the layered structure of riverbeds, turbulent flow, sediment transport, and long-term effects of tidal bores to improve liquefaction risk predictions. Field validation in estuarine environments with stratified riverbeds and frequent tidal bores is also crucial. Gathering high-resolution data on sediment properties, flow dynamics, and riverbed behavior will further refine these models and provide more accurate risk assessments.

Author Contributions: Conceptualization, writing—original draft preparation and methodology, D.P. and Y.L.; validation and data curation, Y.L.; funding acquisition and writing—review and editing, D.P. All authors have read and agreed to the published version of the manuscript.

Funding: This research was funded by the Zhejiang Provincial Natural Science Foundation of China, grant numbers LZJWZ22E090002 and LZJWY22E090006; the National Natural Science Foundation of China, grant numbers 42176214 and 41876095; and the Key Program of the President of the Zhejiang Institute of Hydraulics and Estuary, grant number ZIHE21Z001.

Institutional Review Board Statement: Not applicable.

Informed Consent Statement: Not applicable.

Data Availability Statement: The original contributions presented in the study are included in the article, further inquiries can be directed to the corresponding author.

Conflicts of Interest: The authors declare no conflict of interest.

Appendix A

According to Equation (33), the bed surface dynamic pressure under tidal bore conditions, without considering the turbulent effects on the free surface, can be expressed as follows:

$$p_b = \rho g H\{1 - \tanh b(x - ct)\}/2. \tag{A1}$$

Let $l_1 = 1, l_2 = \tanh b(x - ct)$; applying the Laplace transform followed by the Fourier transform to Equation (A1) yields

$$\widetilde{p_b} = \frac{1}{2}\rho g H\Big\{\widetilde{l_1}(\xi, 0, s) - \widetilde{l_2}(\xi, 0, s)\Big\}, \tag{A2}$$

where

$$\widetilde{l_1}(\xi, 0, s) = \int_{-\infty}^{+\infty}\int_0^{+\infty} l_1 e^{i\xi x}e^{-st}dtdx = \int_{-\infty}^{+\infty}\int_0^{+\infty} e^{i\xi x}e^{-st}dtdx = \frac{2\pi}{s}\delta(\xi), \tag{A3}$$

$$\widetilde{l_2}(\xi, 0, s) = \int_{-\infty}^{+\infty}\int_0^{+\infty} l_2{}^{i\xi x}e^{-st}dtdx = \int_{-\infty}^{+\infty}\int_0^{+\infty} \tanh b(x - ct)e^{i\xi x}e^{-st}dtdx. \tag{A4}$$

and $\delta(x)$ is the Dirac delta function, satisfying the following sifting property:

$$\int_{-\infty}^{+\infty} \delta(\xi - \xi_0)\phi(\xi)d\xi = \phi(\xi_0). \tag{A5}$$

Expanding $\tanh b(x-ct)$ into a q-series yields

$$\tanh b(x-ct) = -1 - 2\sum_{k=1}^{\infty} (-1)^k q^{2k},\ q = e^{b(x-ct)}. \tag{A6}$$

Substituting Equation (A6) into Equation (A4) yields

$$\begin{aligned}\widetilde{l}_2(\xi,0,s) &= \int_{-\infty}^{+\infty}\left\{\int_0^{+\infty}\left[-1-2\sum_{k=1}^{\infty}(-1)^k e^{2bk(x-ct)}\right]e^{-st}dt\right\}e^{i\xi x}dx \\ &= -\frac{2\pi}{s}\delta(\xi) - 2\int_{-\infty}^{+\infty}\left\{\int_0^{+\infty}\sum_{k=1}^{\infty}(-1)^k e^{2bk(x-ct)}\, e^{-st}dt\right\}e^{i\xi x}dx \\ &= -\frac{2\pi}{s}\delta(\xi) - 2\int_{-\infty}^{+\infty}\left\{\sum_{k=1}^{\infty}\frac{(-1)^k e^{2kbx}}{s+2kbc}\right\}e^{i\xi x}dx\end{aligned}. \tag{A7}$$

From Equation (A5) yields

$$\frac{1}{2\pi}\int_{-\infty}^{+\infty} 2\pi\delta(\xi-2kbi)\, e^{-i\xi x}dx = \int_{-\infty}^{+\infty}\delta(\xi-2kbi)\, e^{-i\xi x}dx = e^{-i\xi x}\Big|_{\xi=2kbi} = e^{2kbx}. \tag{A8}$$

Combining Equations (A7) and (A8) yields

$$\widetilde{l}_2(\xi,0,s) = -\frac{2\pi}{s}\delta(\xi) - 4\pi\sum_{k=1}^{\infty}\frac{(-1)^k}{s+2kbc}\delta(\xi-2kbi). \tag{A9}$$

Substituting Equations (A9) and (A3) into Equation (A2) yields

$$\widetilde{p}_b = 2\pi\rho g H\left\{\frac{1}{s}\delta(\xi) + \sum_{k=1}^{\infty}\frac{(-1)^k}{s+2kbc}\delta(\xi-2kbi)\right\}. \tag{A10}$$

Appendix B

$$B_j(\xi,s) = \frac{\Sigma_j(\xi,s)}{\Sigma_0(\xi,s)},(j=1,\cdots,6) \tag{A11}$$

$$\begin{aligned}\Sigma_j &= (E_{j,0}\cosh(\gamma_1 h) + E_{j,1}\sinh(\gamma_1 h))\cosh(\gamma_2 h)\cosh(\gamma_3 h) \\ &+(E_{j,2}\sinh(\gamma_1 h) + E_{j,3}\cosh(\gamma_1 h))\cosh(\gamma_2 h)\sinh(\gamma_3 h) \\ &+(E_{j,4}\sinh(\gamma_1 h) + E_{j,5}\cosh(\gamma_1 h))\sinh(\gamma_2 h)\cosh(\gamma_3 h) \\ &+(E_{j,6}\cosh(\gamma_1 h) + E_{j,7}\sinh(\gamma_1 h))\sinh(\gamma_2 h)\sinh(\gamma_3 h) \\ &+E_{j,8}\cosh(\gamma_1 h) + E_{j,9}\sinh(\gamma_1 h) + E_{j,10}\cosh(\gamma_2 h) \\ &+E_{j,11}\sinh(\gamma_2 h) + E_{j,12}\cosh(\gamma_3 h) + E_{j,13}\sinh(\gamma_3 h)\end{aligned}, \tag{A12}$$

where

$$\begin{aligned}&E_{0,0} = \gamma_1\gamma_2\gamma_3(\xi^2+\gamma_3^2)(a_1-a_2)(d_1-d_2) + 2\gamma_3(\gamma_1 g_2 b_2 + \gamma_2 g_1 b_1) \\ &E_{0,2} = -\gamma_2 b_1(\xi^2+\gamma_3^2)(d_1-d_2) - 2\gamma_1\gamma_2\gamma_3^2 g_1(a_1-a_2) \\ &E_{0,4} = -2\gamma_3(\gamma_1 g_1 b_2 + \gamma_2 g_2 b_1) \\ &E_{0,6} = \gamma_1 b_2(\xi^2+\gamma_3^2)(d_1-d_2) + 2\gamma_1\gamma_2\gamma_3^2 g_2(a_1-a_2) \\ &E_{0,8} = 2\gamma_1\gamma_2\gamma_3 b_2(\xi^2+\gamma_3^2)(a_1-a_2) + \gamma_1\gamma_3 g_2(d_1-d_2) \\ &E_{0,10} = -2\gamma_1\gamma_2\gamma_3 b_1(\xi^2+\gamma_3^2)(a_1-a_2) - \gamma_1\gamma_3 g_1(d_1-d_2) \\ &E_{0,12} = -2\gamma_3(\gamma_1 g_2 b_1 + \gamma_2 g_1 b_2)\end{aligned},$$

$$
\begin{array}{l}
E_{0,1} = E_{0,3} = E_{0,5} = E_{0,7} = E_{0,9} = E_{0,11} = E_{0,13} = 0 \\
E_{1,0} = \gamma_1\gamma_2\gamma_3 d_2(\xi^2 + \gamma_3^2)(a_2 - a_1) + 2\gamma_1\gamma_3 b_2 g_2 \\
E_{1,2} = \gamma_2 b_1 d_2(\xi^2 + \gamma_3^2) \quad E_{1,4} = -2\gamma_2\gamma_3 b_1 g_2 \\
E_{1,6} = 2\gamma_1\gamma_2\gamma_3^2 g_2(a_1 - a_2) - \gamma_1 b_2 d_2(\xi^2 + \gamma_3^2) \\
E_{1,8} = 2\gamma_1\gamma_2\gamma_3 b_2(\xi^2 + \gamma_3^2)(a_1 - a_2) - \gamma_1\gamma_3 d_2 g_2 \\
E_{1,10} = \gamma_2\gamma_3 d_2 g_1 \qquad E_{1,11} = \gamma_2\gamma_3 d_1 g_2 \\
E_{1,12} = -2\gamma_2\gamma_3 b_2 g_1 \qquad E_{1,13} = -\gamma_2 b_2 d_1(\xi^2 + \gamma_3^2) \\
E_{1,1} = E_{1,0}\ E_{1,3} = E_{1,2}\ E_{1,5} = E_{1,4}\ E_{1,7} = E_{1,6}\ E_{1,9} = E_{1,8}
\end{array},
$$

$$E_{2,j} = (-1)^j E_{1,j},$$

$$
\begin{array}{l}
E_{3,0} = \gamma_1\gamma_2\gamma_3 d_1(\xi^2 + \gamma_3^2)(a_1 - a_2) + 2\gamma_2\gamma_3 b_1 g_1 \\
E_{3,1} = -2\gamma_1\gamma_3 b_2 g_1 \qquad E_{3,3} = \gamma_1 b_2 d_1(\xi^2 + \gamma_3^2) \\
E_{3,2} = 2\gamma_1\gamma_2\gamma_3^2 g_1(a_2 - a_1) - \gamma_2 b_1 d_1(\xi^2 + \gamma_3^2) \\
E_{3,8} = \gamma_1\gamma_3 d_1 g_2 \qquad E_{3,9} = \gamma_1\gamma_3 d_2 g_1 \\
E_{3,10} = 2\gamma_1\gamma_2\gamma_3 b_1(\xi^2 + \gamma_3^2)(a_2 - a_1) - \gamma_2\gamma_3 d_1 g_1 \\
E_{3,12} = -2\gamma_1\gamma_3 b_1 g_2 \quad E_{3,13} = -\gamma_1 b_1 d_2(\xi^2 + \gamma_3^2) \\
E_{3,4} = E_{3,1}\ E_{3,5} = E_{3,0}\ E_{3,6} = E_{3,3}\ E_{3,7} = E_{3,2}\ E_{3,11} = E_{3,10}
\end{array},
$$

$$E_{4,j} = (-1)^j E_{3,j},$$

$$
\begin{array}{l}
E_{5,0} = -\gamma_1 b_2 d_1 g_2 - \gamma_2 b_1 d_2 g_1 \qquad E_{5,1} = \gamma_1\gamma_2\gamma_3 g_1 d_2(a_1 - a_2) \\
E_{5,4} = \gamma_1 b_2 d_2 g_1 + \gamma_2 b_1 d_1 g_2 \qquad E_{5,5} = \gamma_1\gamma_2\gamma_3 g_2 d_1(a_2 - a_1) \\
E_{5,8} = \gamma_1\gamma_2 b_2 d_1(\xi_2 + \gamma_3^2)(a_1 - a_2) \\
E_{5,9} = 2\gamma_1\gamma_2\gamma_3 b_2 g_1(a_2 - a_1) \\
E_{5,10} = \gamma_1\gamma_2 b_1 d_2(\xi_2 + \gamma_3^2)(a_1 - a_2) \\
E_{5,11} = 2\gamma_1\gamma_2\gamma_3 b_1 g_2(a_1 - a_2) \\
E_{5,12} = \gamma_2 b_2 d_1 g_1 + \gamma_1 b_1 d_2 g_2 \\
E_{5,2} = E_{5,1}\ E_{5,3} = E_{5,1}\ E_{5,6} = E_{5,5} \\
E_{5,7} = E_{5,4}\ E_{5,13} = E_{5,12}
\end{array},
$$

$$E_{6,j} = (-1)^{j+1} E_{5,j}$$

Notation of Physical Quantity

1. c, the tidal bore celerity
2. d, the water depth
3. $f(x,z,t)$, the original function in the time domain
4. , the Laplace transformed function
5. , the Fourier transformed function
6. G, the shear modulus
7. g_i, the body force acceleration
8. h, the riverbed thickness
9. H, the tidal bore height
10. K_w, the true modulus of elasticity of water
11. k_x, the permeability coefficient in the x direction
12. k_z, the permeability coefficient in the z direction
13. n, the solid phase porosity
14. p, the pore pressure in the riverbed
15. p_b, the water pressure at the water–riverbed interface
16. P_{w0}, the absolute water pressure
17. s, the complex frequency variable in the Laplace domain
18. S_r, the saturation degree
19. S_x, the horizontal seepage velocity
20. S_z, the vertical seepage velocity
21. u_i, the solid matrix displacement
22. w_i, the average relative displacement of the fluid to the solid

23. x, the horizontal coordinate
24. z, the vertical coordinate
25. β, the volumetric compressibility coefficient of the fluid
26. $\gamma_s{}'$, the effective unit weight of the soil particles
27. γ_w, the unit weight of water
28. δ_{ij}, the Kronecker delta denotation
29. θ, the volume strain
30. λ, the Lamé constant
31. ξ, the frequency variable in the Fourier domain
32. ρ_0, the combined density
33. ρ_s, the solid density
34. ρ_w, the fluid density
35. σ_{ij}, the total stress components
36. σ_x, the normal stresses in the x directions
37. σ_z, the normal stresses in the z directions
38. τ_{xz}, the shear stress
39. φ, the internal friction angle

References

1. Lynch, D.K. Tidal bores. *Sci. Am.* **1982**, *247*, 146–157. [CrossRef]
2. Li, Y.; Pan, D.-Z.; Chanson, H.; Pan, C.-H. Tidal bore progressing on a small slope. *Exp. Therm. Fluid Sci.* **2017**, *88*, 513–518. [CrossRef]
3. Li, Y.; Pan, D.-Z.; Chanson, H.; Pan, C.-H. Real-time characteristics of tidal bore propagation in the Qiantang River Estuary, China, recorded by marine radar. *Cont. Shelf Res.* **2019**, *180*, 48–58. [CrossRef]
4. Bonneton, P.; Bonneton, N.; Parisot, J.-P.; Castelle, B. Tidal bore dynamics in funnel-shaped estuaries. *J. Geophys. Res. Oceans* **2015**, *120*, 923–941. [CrossRef]
5. Filippini, A.G.; Arpaia, L.; Bonneton, P.; Ricchiuto, M. Modeling analysis of tidal bore formation in convergent estuaries. *Eur. J. Mech. B Fluids* **2019**, *73*, 55–68. [CrossRef]
6. Colas, A. *Mascaret, Prodige de la Marée*; Editions YEP: Anglet, France, 2017.
7. Pan, D.; Li, Y.; Pan, C. A review on field observations of tidal bore-induced flow and sediment dynamic processes. *Ocean Eng.* **2023**, *41*, 169–181.
8. Li, Y.; Pan, D. Tidal bore impact pressures on a trestle pier in the Qiantang River Estuary, China. *Adv. Mech. Eng.* **2022**, *14*, 16878132221131435. [CrossRef]
9. Pan, D.; Li, Y. Tidal bore scour around a spur dike. *J. Mar. Sci. Eng.* **2022**, *10*, 1086. [CrossRef]
10. Chanson, H. *Tidal Bores, Aegir, Eagre, Mascaret, Pororoca Theory and Observations*; World Scientific: Singapore, 2012.
11. Moore, R.N. *Report on the Bore of the Tsien-Tang Kiang*; Hydrographic Office: London, UK, 1888.
12. Reungoat, D.; Lubin, P.; Leng, X.; Chanson, H. Tidal bore hydrodynamics and sediment processes: 2010–2016 field observations in France. *Coast. Eng. J.* **2018**, *60*, 484–498. [CrossRef]
13. Tu, J.; Fan, D.; Voulgaris, G. Field observations of turbulence, sediment suspension, and transport under breaking tidal bores. *Mar. Geol.* **2021**, *437*, 106498. [CrossRef]
14. Wisha, U.J.; Wijaya, Y.J.; Hisaki, Y. Real-time properties of hydraulic jump off a tidal bore, its generation and transport mechanisms: A case study of the Kampar River Estuary, Indonesia. *Water* **2022**, *14*, 2561. [CrossRef]
15. Leng, X.; Chanson, H. Turbulent advances of a breaking bore: Preliminary physical experiments. *Exp. Therm. Fluid Sci.* **2015**, *62*, 70–77. [CrossRef]
16. Li, Y.; Chanson, H. Physical modeling of vanishing bores in open-channel flows. *Can. J. Civ. Eng.* **2022**, *49*, 289–294. [CrossRef]
17. Simon, B.; Lubin, P.; Chanson, H. Hydrodynamic shock in rivers: Physical and numerical modelling of flow structures in tsunami-like bores. *Phys. Fluids* **2023**, *35*, 106607. [CrossRef]
18. Gualtieri, C.; Chanson, H. Physical and numerical modelling of air-water flows: An introductory overview. *Environ. Model. Softw.* **2021**, *143*, 105109. [CrossRef]
19. Leng, X.; Simon, B.; Khezri, N.; Lubin, P.; Chanson, H. CFD Modelling of tidal bores: Development and validation challenges. *Coast. Eng. J.* **2018**, *60*, 423–436. [CrossRef]
20. Shi, R.; Wuthrich, D.; Chanson, H. Air-water properties of unsteady breaking bores Part 1: Novel Eulerian and Lagrangian velocity measurements using intrusive and non-intrusive Techniques. *Int. J. Multiph. Flow* **2023**, *159*, 104338. [CrossRef]
21. Shi, R.; Wuthrich, D.; Chanson, H. Air-water properties of unsteady breaking bore Part 2: Void fraction and bubble statistics. *Int. J. Multiph. Flow* **2023**, *159*, 104337. [CrossRef]
22. Wuthrich, D.; Shi, R.; Chanson, H. Strong free-surface turbulence in breaking bores: A physical study on the free-surface dynamics and air-water interfacial features. *J. Fluid Mech.* **2021**, *924*, A20. [CrossRef]

23. Committee on Soil Dynamics of the Geotechnical Engineering Division, ASCE. Definition of terms related to liquefaction. *J. Geotech. Eng. Div.* **1978**, *104*, 1197–1200.
24. Popescu, R.; Chakrabortty, P. Mechanism of seismic liquefaction for heterogeneous soil. *Soil Dyn. Earthq. Eng.* **2024**, *176*, 10833. [CrossRef]
25. Jeng, D.-S. Wave-induced sea floor dynamics. *Appl. Mech. Rev.* **2003**, *56*, 407–429. [CrossRef]
26. Lin, J.; Jeng, D.-S.; Zhao, H.; Gao, Y.; Liu, J.; Guo, Y. Recent advances of seabed liquefaction around the vicinity of marine structures. *Ocean Eng.* **2023**, *280*, 114660. [CrossRef]
27. Sumer, B.M. *Liquefaction Around Marine Structures*; World Scientific: Singapore, 2014.
28. Guo, Z.; Jeng, D.-S.; Guo, W. Simplified approximation of wave-induced liquefaction in a shallow porous seabed. *Int. J. Geomech.* **2014**, *14*, 06014008. [CrossRef]
29. Qi, W.G.; Gao, F.P. A modified criterion for wave-induced momentary liquefaction of sandy seabed. *Theor. Appl. Mech. Lett.* **2015**, *5*, 20–23. [CrossRef]
30. Sui, T.; Yang, M.; Peng, L.; Chen, J.; Zhang, C.; Zheng, J. Wave-induced residual response and liquefaction of a nonhomogeneous layered seabed. *Front. Mar. Sci.* **2024**, *11*, 1360641. [CrossRef]
31. Yamamoto, T.; Koning, H.L.; Sellmeijer, H.; Hijum, E.V. On the response of a poro-elastic bed to water waves. *J. Fluid Mech.* **1978**, *87*, 193–206. [CrossRef]
32. Tsai, C.P. Wave-induced liquefaction potential in a porous seabed in front of a breakwater. *Ocean Eng.* **1995**, *22*, 1–18. [CrossRef]
33. Ulker, M.B.C.; Rahman, M.S.; Guddati, M.N. Wave-induced dynamic response and instability of seabed around a breakwater. *Ocean Eng.* **2010**, *37*, 1522–1545. [CrossRef]
34. Seed, H.B.; Rahman, M.S. Wave-induced pore pressure in relation to ocean floor stability of cohesionless soils. *Mar. Georesour. Geotec.* **1978**, *3*, 123–150. [CrossRef]
35. Sumer, B.M.; Kirca, V.S.O.; Fredsøe, J. Experimental validation of a mathematical model for seabed liquefaction under waves. *Int. J. Offshore Polar Eng.* **2012**, *22*, 133–141.
36. Sui, T.; Zhang, C.; Jeng, D.-S.; Guo, Y.; Zheng, J.; Zhang, W.; Shi, J. Wave- induced seabed residual response and liquefaction around a mono-pile foundation with various embedded depth. *Ocean Eng.* **2019**, *173*, 157–173. [CrossRef]
37. Tessier, B.; Furgerot, L.; Mouazé, D. Sedimentary signatures of tidal bores: A brief synthesis. *Geo.-Mar. Lett.* **2016**, *37*, 325–331. [CrossRef]
38. Tessier, B.; Terwindt, J.H.J. An example of soft-sediment deformations in an intertidal environment-the effect of a tidal bore. *C. R. Acad. Sci. Sér. II* **1994**, *319*, 217–233.
39. Greb, S.F.; Archer, A.W. Soft-sediment deformation produced by tides in a meizoseismic area, Turnagain Arm, Alaska. *Geology* **2007**, *35*, 425–438. [CrossRef]
40. Fan, D.D.; Cai, G.F.; Shang, S.; Wu, Y.J.; Zhang, Y.W.; Gao, L. Sedimentation processes and sedimentary characteristics of tidal bores along the north bank of the Qiantang Estuary. *Chin. Sci. Bull.* **2012**, *57*, 1578–1589. [CrossRef]
41. Pan, D. Dynamic Response of Seabed and Pipeline Due to Wave Loading. Ph.D. Thesis, Zhejiang University, Hangzhou, China, 2007.
42. Hsu, J.R.C.; Jeng, D.S.; Tsai, C.P. Short-crested wave-induced soil response in a porous seabed of infinite thickness. *Int. J. Numer. Anal. Methods Geomech.* **1993**, *17*, 553–576. [CrossRef]
43. Arab, A.; Belkhatir, M. Fines content and cyclic preloading effect on liquefaction potential of silty sand: A laboratory study. *Acta Polytech. Hung.* **2012**, *9*, 47–64.
44. Hsu, C.-J.; Tsai, C.-C.; Chen, Y.-Y. Wave-induced seabed momentary liquefaction in shallow water. *Appl. Ocean Res.* **2021**, *115*, 102819. [CrossRef]
45. Biot, M.A. General theory of three-dimensional consolidation. *J. Appl. Phys.* **1941**, *12*, 155–164. [CrossRef]
46. Biot, M.A. Theory of propagation of elastic waves in a fluidsaturated porous solid. I. Low-frequency range. *J. Acoust. Soc. Am.* **1956**, *282*, 168–178. [CrossRef]
47. Biot, M.A. Mechanics of deformation and acoustic propagation in porous media. *J. Appl. Phys.* **1962**, *33*, 1482–1498. [CrossRef]
48. Madsen, O.S. Wave-induced pore pressures and effective stresses in a porous bed. *Geotechnique* **1978**, *28*, 377–393. [CrossRef]
49. Hsu, C.-J.; Hung, C. Three-dimensional wave-induced dynamic response in anisotropic poroelastic seabed. *Water* **2019**, *12*, 1465. [CrossRef]
50. Hsu, C.-J.; Chen, Y.-Y.; Tsai, C.-C. Wave-induced seabed response in shallow water. *Appl. Ocean Res.* **2019**, *89*, 211–223. [CrossRef]
51. Wang, L.; Pan, D.; Ling, D. Analysis on integral transform of the wave-induced response in seabed and its application. *Chin. J. Geotech. Eng.* **2006**, *28*, 847–852.
52. Tong, L.; Liu, P.L.F. Transient wave-induced pore-water-pressure and soil responses in a shallow unsaturated poroelastic seabed. *J. Fluid Mech.* **2022**, *938*, A36. [CrossRef]
53. Pan, D.; Wang, L.; Ling, D. State-space solution of the wave-induced responses in multilayered seabed and its application. *Ocean Eng.* **2006**, *24*, 23–30.
54. Pan, D.; Wang, L.; Pan, C. State-space solution of seabed–geotextile systems subjected to cyclic wave loadings. *Geotext. Geomembr.* **2007**, *25*, 242–250. [CrossRef]
55. Zienkiewicz, O.C.; Chang, C.T.; Bettess, P. Drained, undrained, consolidating and dynamic behaviour assumptions in soils. *Geotechnique* **1980**, *30*, 385–395. [CrossRef]

56. Verruijt, A. Elastic Storage of Aquifers. In *Flow Through Porous Media*; de Wiest, R.J.M., Ed.; Academic Press: Cambridge, MA, USA, 1969.
57. Packwood, A.R.; Peregrine, D.H. Loss of water wave energy due to percolation in a permeable sea bottom. *Coast. Eng.* **1980**, *3*, 221–242. [CrossRef]
58. Durbin, F. Numerical inversion of Laplace transforms: An efficient improvement to Dubner and Abate's method. *Comput. J.* **1974**, *17*, 371–376. [CrossRef]
59. Cohen, A.M. *Numerical Methods for Laplace Transform Inversion*; Springer Science: New York, NY, USA, 2007.
60. Bear, J. *Dynamics of Fluids in Porous Media*; American Elsevier Publishing Company: New York, NY, USA, 1972.
61. Madsen, O.S. Stability of a sand bed under breaking waves. In Proceedings of the 14th International Conference on Coastal Engineering, Copenhagen, Denmark, 24–28 June 1974; pp. 776–794.
62. Hazirbaba, K. Pore Pressure Generation Characteristics of Sands and Silty Sands: A Strain Approach. Ph.D. Thesis, University of Texas, Austin, TX, USA, 2005.
63. Zuo, L.; Baudet, B.A. Determination of the transitional fines content of sand-non plastic fines mixtures. *Soils Found.* **2015**, *55*, 213–219. [CrossRef]
64. Porcino, D.D.; Diano, V.; Triantafyllidis, T.; Wichtmann, T. Predicting undrained static response of sand with non-plastic fines in terms of equivalent granular state parameter. *Acta Geotech.* **2020**, *15*, 867–882. [CrossRef]
65. Savvides, A.A.; Papadrakakis, M. A probabilistic assessment for porous consolidation of clays. *SN Appl. Sci.* **2020**, *2*, 2115. [CrossRef]

Journal of
*Marine Science
and Engineering*

MDPI

Article

Micromechanical Analysis of Lateral Pipe–Soil Interaction Instability on Sloping Sandy Seabeds

Yu Peng [1] and Liming Qu [2,*]

[1] Department of Civil and Environmental Engineering, The Hong Kong Polytechnic University, Hung Hom, Kowloon, Hong Kong 999077, China; yupeng@polyu.edu.hk
[2] Faculty of Geosciences and Environmental Engineering, Southwest Jiaotong University, Chengdu 610031, China
* Correspondence: qlming@swjtu.edu.cn; Tel.: +86-131-4591-3179

Abstract: The micromechanical mechanism of pipe instability under lateral force actions on sloping sandy seabeds is unclear. This study investigated the effects of slope angle and instability direction (upslope or downslope) on pipe–soil interaction instability for freely laid and anti-rolling pipes using coupled discrete element method and finite element method (DEM–FEM) simulations. The numerical results were analyzed at both macro- and microscales and compared with the experimental results. The findings revealed that the ultimate drag force on anti-rolling pipes increased with slope angle and was significantly larger than that on freely laid pipes for both downslope and upslope instabilities. Additionally, the rotation-induced upward traction force was proved to be the essential reason for the smaller soil deformation around freely laid pipes. Moreover, the shape differences in the motion trajectories of pipes were successfully explained by variations in the soil supporting force distributions under different slope conditions. Additionally, synchronous movement between the pipe and adjacent particles was identified as the underlying mechanism for the reduced particle collision and shear wear on pipe surfaces under a high interface coefficient. Furthermore, an investigation of particle-scale behaviors revealed conclusive mechanistic patterns of pipe–soil interaction instability under different slope conditions. This study could be useful for the design of pipelines in marine pipeline engineering.

Keywords: micromechanics; sloping seabed; pipe–soil interaction; lateral instability; discrete element method; finite element method

1. Introduction

Subsea pipelines are crucial components in offshore engineering, facilitating the transportation of oil, gas, and other resources from ocean to land [1–4]. These pipelines are usually installed on sloping seabeds and are often exposed to lateral hydrodynamic forces caused by ocean water flows or tides [5–7]. The interaction between the untrenched pipelines and granular soils under lateral hydrodynamic force actions on sloping seabeds is complex [6,8,9]. However, conventional pipeline designs are usually based on studies on horizontal seabeds, overlooking the lateral pipe–soil interaction instability under lateral oblique force actions on sloping seabeds [2,3,10]. It is evident that the slope condition introduces variations in forces and moments for pipe–soil interactions, leading to inaccurate estimation of pipe lateral instability and the intensity of pipe–soil interactions on the seabed [2,6]. Therefore, investigating pipe–soil interaction behavior and pipe stability under lateral oblique force actions on sloping seabeds is important for marine pipeline design and ensures engineering safety.

Previous studies have shown significant interest in understanding the behavior of marine pipe–soil interactions under lateral oblique force actions [1,10–14]. Researchers have conducted scaled model tests, including centrifugal tests, and field trials using mechanical-actuator methods to evaluate the instability of untrenched pipelines and explore the relationship between soil resistance and pipe displacements on soil seabeds [6,15–18]. These experimental tests have revealed that several factors, such as soil type, magnitude and direction of lateral force actions, pipe surface roughness, end constraint condition of pipes, and initial embedment, influence the untrenched pipe–soil interaction behaviors [6,8,19–21]. Among these factors, the slope angle and the direction of lateral force action play crucial roles in determining pipeline–soil interaction instability on seabeds [2,6]. However, the specific influence of slope on pipe–soil interactions under lateral oblique force actions has received limited attention thus far [3,6]. Additionally, over the past few decades, numerous numerical simulation studies have been conducted to enhance our understanding of progressive pipe–soil interactions under lateral oblique force actions [6,22,23]. These simulation studies have typically employed the Coulomb friction theory or other improved theories to develop empirical pipe–soil interaction models for predicting pipeline instability on seabeds [5,6,15,24]. Undoubtedly, these experimental tests and numerical simulations have contributed to the scientific understanding of pipe–soil interaction instability under lateral oblique force actions on sloping seabeds.

However, previous studies on untrenched pipe–soil interactions on sloping seabeds have primarily focused on soil resistance magnitude, macroscale pipe movement behaviors, and pipe instability [2,6,25]. Consequently, the mechanisms of microscale pipe–soil interactions on sloping seabeds remain unclear. Furthermore, previous continuous numerical methods, such as the finite element method (FEM), have failed to accurately simulate particle sliding and collision behaviors, as well as large deformation behaviors of the soil [6,22,23,26,27]. This limitation has reduced the precision of analyzing pipe–soil interaction in granular soil environments. Additionally, many pipe–soil interaction behaviors on underwater slopes observed in experimental studies, such as different soil deformation under downslope and upslope conditions, variable motion trajectories of pipes, and particle-slide induced shear wear on pipe surfaces, have not been systematically explained [3,5,6]. Currently, the discrete element method (DEM) is favored by many scientists for its ability to analyze microscopic granular soil behaviors and their interaction with structures [28–31]. The DEM also excels in discontinuity and large deformation analysis [28,32,33], making it a preferred choice for simulating various granular soil-related issues during pipe–granular soil interactions [34–37]. However, the DEM is weak in modeling the behaviors of continuous materials such as structures; this is due to the fact that in the DEM, structures are typically represented by rigid walls, which are unable to model continuous stress and deformation [35,38,39]. Therefore, investigating the stress, deformation, and shear wear of pipe structures (continuous materials) solely using the DEM has proven to be difficult. Therefore, it is desirable to combine the advantages of both the DEM and the FEM to analyze pipe–soil interactions under lateral oblique force actions, considering the strengths and weaknesses of each method. To the best of the authors' knowledge, no studies in the current literature have utilized coupled DEM–FEM simulations to investigate pipe–soil interaction instability under lateral oblique force actions on sloping seabeds.

Accordingly, this study aims to explore lateral pipe–soil interaction instability on sloping sandy seabeds from micro- to macroscales through a series of three-dimensional coupled DEM–FEM simulations. The research begins by comparing lateral soil resistance–displacement curves obtained from model tests and numerical simulations, followed by examining macroscale soil resistance distributions and soil deformation phenomena. Subsequently, the detailed behaviors of the pipes, including their motion trajectory, stress and deformation, and surface shear intensity, are investigated, taking into account different slope angles and instability directions (upslope or downslope). Additionally, several conclusive patterns of lateral pipe–soil interaction on sloping seabeds are identified through further analysis of microparticle behaviors. The findings of this study provide new insights into understanding the lateral movement behaviors of submarine pipes and shed light on the mechanisms of pipe–soil interaction instability on sloping sandy seabeds.

2. Simulations of Pipe–Soil Interaction Behaviors with Coupled DEM–FEM

2.1. Brief Description of Pipeline Lateral Instability on Sloping Seabeds

With the expansion of oil and gas exploration into deeper waters, the lateral hydrodynamic action exerted by ocean water flow has become a dominant factor affecting submarine pipelines [22]. The on-bottom stability of long-distance untrenched pipelines varies significantly and is highly dependent on the end constraint conditions [3,6]. Two commonly encountered end constraint conditions (rotation conditions) for model pipes are as follows: (1) Anti-rolling condition, where the pipe's rolling motion is restricted, but it can move freely parallel and perpendicular to the seabed surface. (2) Freely laid condition, where the pipe can rotate around its axis without any end constraint and can move freely in parallel and perpendicular directions to the seabed surface.

Unlike previous studies that primarily focused on the on-bottom stability of pipelines on horizontal seabeds, the on-bottom stability of pipelines on sloping seabeds is more complex. Figure 1 illustrates two different lateral on-bottom instability modes induced by the relative water flow actions on sloping seabeds: (a) Upslope instability, where the pipe moves upward along the sloping seabed (θ is positive). (b) Downslope instability, where the pipe moves downward along the sloping seabed (θ is negative). Additionally, the magnitude of the seabed slope angle should also be considered when analyzing the lateral on-bottom stability of pipelines. As mentioned earlier, the lateral on-bottom stability of a submarine pipeline under ocean wave/current involves complex flow–pipe–soil interaction, which refers to the interaction between hydrodynamic loading, the untrenched pipeline, and the surrounding soil. The ultimate lateral soil resistance (F_{Ru}) for pipe instability is mainly influenced by pipeline parameters, sand properties, and hydrodynamic load characteristics. Previous studies on pipe lateral instability [3,6] have provided expressions for the ultimate lateral soil resistance (F_{Ru}) parallel to the slope surface and the corresponding pipe–soil contact force (F_{Cu}) perpendicular to the slope surface, which can be expressed as follows:

$$F_{\text{Ru}} = F_{Du} - G\sin\theta \tag{1}$$

$$F_{\text{Cu}} = G\cos\theta - F_{Du}\tan\theta \tag{2}$$

where F_{Du} is the ultimate drag force on the pipe; G is the submerged weight of the pipe per meter; and θ is the slope angle.

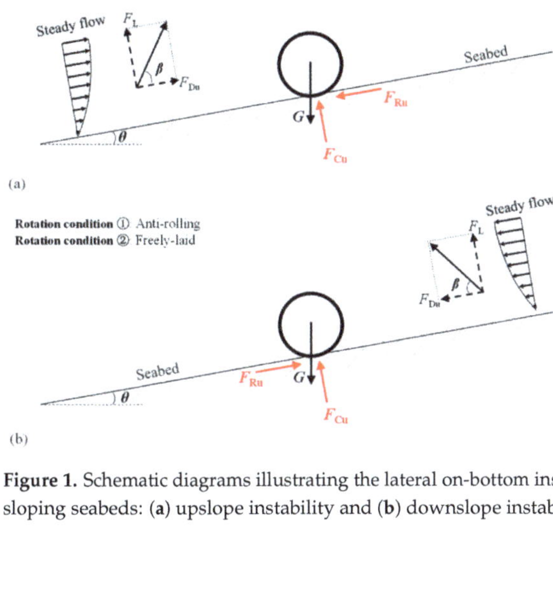

Figure 1. Schematic diagrams illustrating the lateral on-bottom instability of a submarine pipeline on sloping seabeds: (**a**) upslope instability and (**b**) downslope instability.

2.2. The Adopted Model Tests

This study followed the large-scale model tests conducted by Gao et al. [3]. Silty fine sand with a particle median diameter (d_{50}) of 0.11 mm was used in these model tests. The buoyant unit weight (γ') of the sandy soil was 9.02 kN/m^3 while the internal friction angle of the soil was 32°. The relative density of the sand was 16%. The pipe used in the tests had a diameter (D) of 0.35 m and a submerged weight of 0.801 kN/m. To simulate the lateral hydrodynamic action on the pipe during the experiments, a mechanical actuator simulation method was employed. The mechanical-loaded pipe–soil interaction facility (Figure 2) served as the main test facility for modeling pipe lateral instability on sloping sand beds. A displacement-controlled testing program was implemented in the mechanical actuator system. An inclined force was generated on the model pipe by a stepper motor through two cables: one connecting the pipe and the live pulley, and the other connecting the live pulley and the stepper motor via two fixed pulleys (Figure 2). Based on Jones's study (1976), the resultant hydrodynamic load on the pipe was found to be obliquely upwards with an inclination angle ranging approximately between 53° and 57° (around 55°). Thus, the length of the front cable was adjusted to achieve a specific inclined load angle (approximately 55°) in the model tests. Detailed analysis of the loads on the submarine pipeline was provided by Gao et al. [3,15].

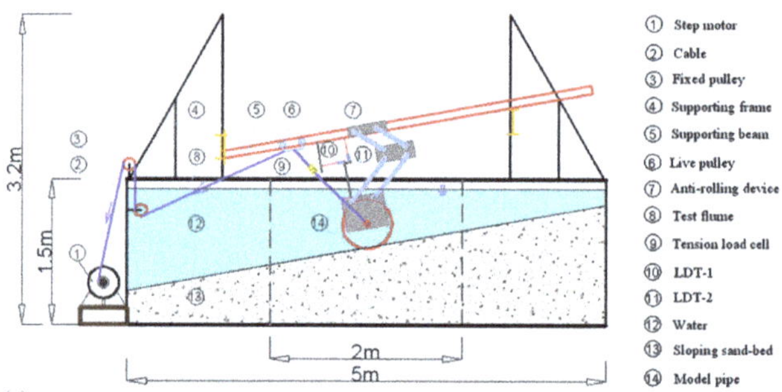

(a)

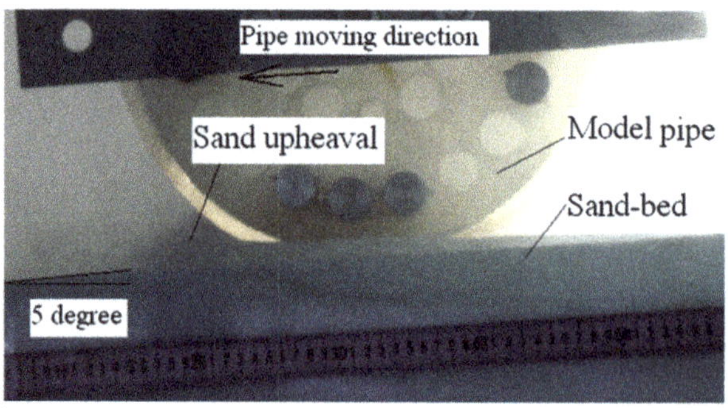

(b)

Figure 2. Experimental model testing setup for simulating pipe lateral instability on sloping seabeds (modified from Gao et al. [3]): (**a**) schematic diagram illustrating the experimental tests and (**b**) photo of the experimental tests.

Non-contact measurements of pipe displacements were performed using two laser displacement transducers (LDT-1 and LDT-2): LDT-1 measured the lateral displacement of the pipe (parallel to the seabed surface), while LDT-2 measured the settlement perpendicular to the seabed. A tension load cell installed on the front cable was used to measure the inclined load exerted on the model pipe. Experimental observations were recorded using a digital video camera placed behind a transparent glass wall. The additional settlement, lateral displacement of the test pipe, and corresponding load were measured simultaneously during the loss of lateral stability of the pipe. Further details of the adopted model tests can be found in Gao et al. [3].

2.3. Methodology of Coupled DEM–FEM

In this study, a coupled approach combining the discrete element method (DEM) and the finite element method (FEM) was employed to investigate the lateral instability of submarine pipes on sloping sandy seabeds. The numerical models in the three-dimensional DEM–FEM simulations had the same pipe size, slope angle, and pulling direction as the model tests conducted by Gao et al. [3]. Granular soil is suited for simulation using DEM, while solid structures are suited for continuous numerical simulations [40–45]. As in previous DEM–FEM coupling research [33], commercial DEM-FEM softwares were used in this study. In this study, the granular soil was represented by spherical balls in the DEM part, while the pipe was simulated using finite element meshes in the FEM part. This approach has been previously validated as an effective method for simulating soil–structure interaction [43,46,47]. In the model tests, the pipes exhibited only slight deformation compared to their displacement, meeting the requirement for implementation of the one-way transient DEM–FEM coupling method [48,49]. As in previous studies [33,46,50,51], this one-way transient coupling algorithm involved computing boundary data (i.e., particle forces on the surface of the structure) using the DEM and subsequently integrating it into the FEM simulation as a condition at each step, without providing feedback from the FEM to the DEM. During the DEM–FEM coupling, the particle forces on the boundary of the pipe (triangular geometry) were collected first in the DEM part, and then the vertex force per boundary triangle was computed as the average of the values corresponding to the surrounding triangles. Next, these forces on the vertex were used for mechanical transient structure analysis in the FEM part.

As shown in Figure 3a, a three-dimensional (3D) particle domain with several fixed boundaries was used to model a thin slice of soil perpendicular to the pipe axis, following the approach of Macaro [34]. The boundaries of the container consisted of several frictionless walls. As shown in Figure 3b, the domain thickness adopted in the simulations presented here was 100 mm. The pipe segment was implemented via finite element mesh (triangular geometry) as a circular cylinder (the number of vertices was 6888), and the mechanical parameters for the pipe were based on the elastic properties of steel. In DEM-based simulations, particle size in the numerical simulation is usually much larger than that in model tests while the span of particle size distribution is narrower to obtain an acceptable computational cost, especially for three-dimensional engineering structure–soil interaction simulations, as presented in previous studies [34,38,52–54]. The diameters of particles in this simulation were between 12 and 24 mm while the median particle diameter (d_{50}) was 16 mm, generated by the gravity deposition method [55]. In the gravity deposition method, as in previous studies [56,57], particles were released from the top of the container, and the slope angle for the sand deposit was controlled by adjusting the inclination angle of the deposition apparatus (wall boundary). Thus, the modeled mean grain size (d_{50}) in the DEM simulations met the requirement that the d_{50} should be smaller than 1/5 (approximately 1/6 in this study) of the domain thickness to neglect the boundary effect in the out-of-plane direction [34,38]. Additionally, the particle size in the DEM met the criterion that the ratio of the pipe diameter to d_{50} should be larger than 10 (approximately 22 in this study), according to previous research [34,38,58–60]. The total number of particles generated in the simulations ranged from 100,032 to 173,284 in different slope angle cases.

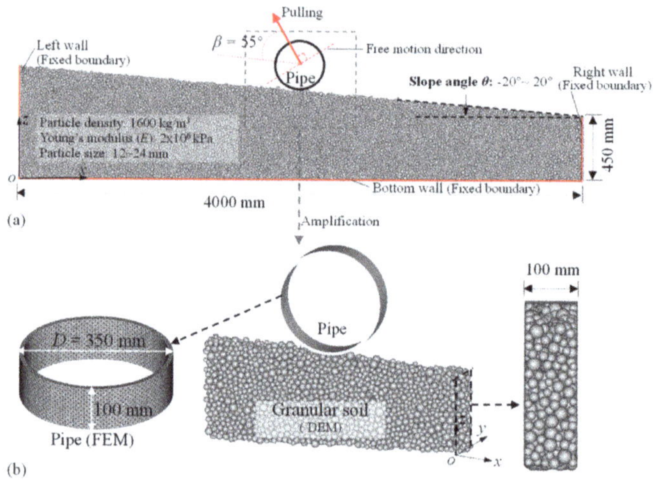

Figure 3. DEM–FEM simulation model of the pipe–sand interaction on sloping seabeds: (**a**) front view, (**b**) local details.

In accordance with previous simulations [33,61], the Hertzian spring–dashpot model was used for calculating the normal contact behavior, while the linear spring Coulomb limit model was used for calculating the tangential contact force behavior. The tangential to normal stiffness ratio was set to the commonly recommended value of 1.0. To model the pulling behaviors observed in the model test conducted by Gao et al. [3], where the pulling direction was at an angle of 55° to the slope surface, the pulling direction was controlled with the same angle of 55° in the numerical simulations, as illustrated in Figure 3. The pulling actions were velocity-controlled, with a pulling velocity of 1.74 mm/s, consistent with the velocity used in previous model tests [3,6]. The relative density of the sandy soil in the model tests was matched in the DEM-based simulations, with a relative density of 16% (void ratio of 0.83); this value was obtained using the commonly used method proposed by Deluzarche et al. [32]. In the method proposed by Deluzarche et al. [32], the relative density of granular soil (void ratio) is controlled by the friction coefficient of particles during the gravity deposition process of soil: a small void ratio can be obtained with a particle friction coefficient of 0.0 with vibration, while the largest void ratio can be obtained with a high friction coefficient. Based on former DEM-based studies on pipe–soil interactions [34–36,38,39,59], the elasticity modulus (E) was set at 2.0×10^6 kN/m^2 and 2×10^{11} kN/m^2 for the particles and the pipes, respectively, while the particle–particle friction coefficient was 0.5. Based on previous experiences in considering particle shape effects [62–67], the widely used linear spring rolling limit model proposed by Wensrich and Katterfeld [65] was used, and the rolling resistance (RR) of the particles was set at the typical value of 0.10 to model particle shape effect. In the numerical simulations, the critical timestep (6.7×10^{-5} s) was determined through extensive preliminary simulation analyses to ensure the stability of the results. The timestep in the simulations was 4.5×10^{-5} s, smaller than the critical step.

The calibration of the DEM models was performed on the basis of force–displacement curves obtained from the experimental tests (Figure 4), following previous studies on large-scale soil–structure interaction [68–71]. According to previous studies based on the DEM [47,72,73], direct calibration using force–displacement curves in large-scale physical experiments offers a comprehensive reflection of the intensity of structure–soil interaction on a physical level; this method reflects both the action and the properties of the structure, as well as the properties of the soil, compared with other calibration methods, such as calibration solely by parameters in small-scale soil property tests [33,68,69,73]. The friction among the particles was 0.5 while the interface friction between the pipe and the granular soil was 0.45 after the calibration: all within a rational range according to previous DEM simulations [74–76]. Table 1 gives the main parameters adopted in the coupled DEM–FEM simulations.

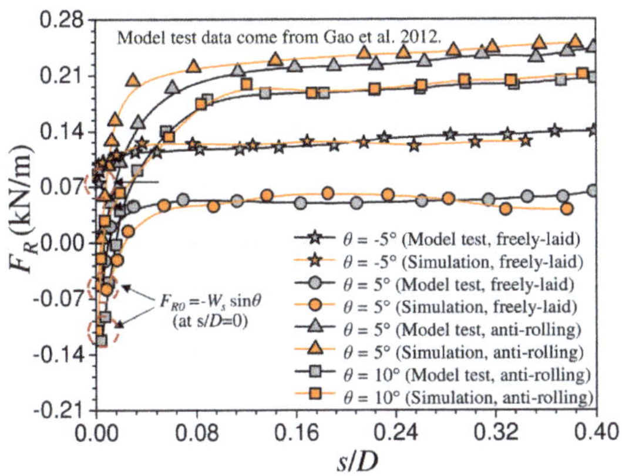

Figure 4. Relationships between lateral soil resistance F_R and lateral displacement s/D in the model tests and numerical simulations [3].

Table 1. Main parameters used in the DEM–FEM simulations.

Parameters	Value	Unit
Particles		
Particle density (ρ)	1600	kg/m^3
Young's modulus (E)	2×10^6	Pa
Friction coefficient (f)	0.50	\
Restitution coefficient	0.30	\
Rolling resistance (RR)	0.10	\
Timestep	4.5×10^{-5}	s
Poisson's ratio (υ)	0.30	\
Gravitational acceleration (g)	9.81	m/s^2
Pipes		
Young's modulus (E)	2×10^{11}	Pa
Friction coefficient (f)	0.45	\
Restitution coefficient	0.8	\
Poisson's ratio (υ)	0.3	\

In the coupled DEM–FEM numerical study, the effect of fluid (water) in the experimental tests was simplified (not directly modeled). The simulations were based on well-drained conditions (with negligible excess pore water pressure) for the following three reasons [3,15]: a. the adopted model tests were based on sandy soil, a kind of soil that facilitates the dissipation of excess pore pressure; b. the pipe was moving on the surface of the seabed (not buried in the soil), a fine drainage boundary; c. the steady moving velocity of the pipes was moderate (1.74 mm/s) on the soil surface. However, it should be noted that the drained condition in the DEM–FEM simulations (which accounts for the effect of water by using the buoyant unit weight of particles) is a simplification of real experimental tests.

Note that unacceptable computational efficiency will result if particle details in a DEM simulation are the same as those in model tests [34,38,52–54]. Therefore, it is widely accepted that, in DEM-based simulations, many factors such as particle size, shape, and size distribution are simplified compared to those observed in model tests, in order to achieve acceptable computational efficiency [77–82]. Consequently, rather than aiming for precise numerical values, simulations based on the DEM often prioritize providing valuable insights into the micromechanical mechanisms and trends in soil behavior.

3. Results and Analyses

3.1. Soil Resistance and Macro Pipe–Soil Interactions

The evaluation of the ultimate drag force on pipes provides a comprehensive reflection of the intensity of pipe–soil interaction for pipe instability analyses on sloping seabeds [2,6,19]. Relationships between ultimate drag force F_{Du} and slope angle θ in the numerical simulations are shown in Figure 5. Obviously, the ultimate drag force F_{Du} in a downslope condition was much smaller than that in an upslope condition, which is consistent with the common sense that moving downward is easier than moving upward on slopes. When comparing the ultimate drag force F_{Du} in anti-rolling and freely laid conditions, the F_{Du} was significantly larger (more than double) in anti-rolling conditions. In freely laid conditions, as the slope angle increased to $|\theta| \geq 10°$, "static instability" occurred, indicating that the freely laid pipe is relatively unstable on slopes. In anti-rolling conditions, as the slope angle (θ) increased, the F_{Du} increased significantly at first and then gradually tended to be stable, without the occurrence of "static instability" for $|\theta| \leq 20°$. This demonstrates that the end constraint (anti-rolling) noticeably enhances the stability of pipes on slopes. Therefore, both the slope effect and the end constraint effect on pipe–soil interactions warrant attention for evaluating pipe instability.

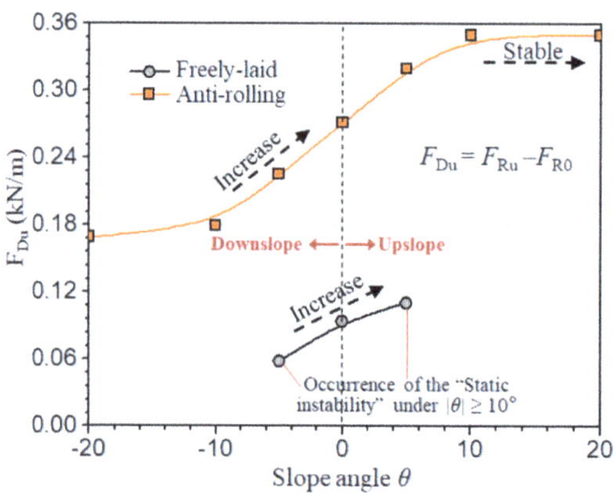

Figure 5. Relationships between the ultimate drag force (F_{Du}) and the slope angle θ as they affect pipe instability in the numerical simulations.

The soil deformation and the distribution of soil resistance (or supporting force) around pipes provide valuable insights into the manner of pipe–soil interaction on granular soil slopes. Figure 6 illustrates the progressive soil supporting force distributions and soil deformation around freely laid pipes under different slope angle conditions. In all cases, as the pipes rotated, a concave deformation beneath the pipe was formed due to particle slides under the weight effect of the pipes. In the case of a pipe experiencing downslope instability on a sloping seabed (see Figure 6a), the main supporting forces were consistently located on the downslope side of the pipes. In the case of a pipe experiencing instability on a flat seabed (see Figure 6b), the supporting forces were located on both sides of the pipes in the stationary conditions ($S/D = 0$), but the main supporting forces shifted to the pipe moving direction after noticeable pipe movement. In the case of a pipe experiencing upslope instability on a sloping seabed (see Figure 6c), the main supporting force zones underwent a location transition from being on the downslope side, to being on both sides, and, finally, to being on the upslope side, as the pipes rotated. Furthermore, in all cases, no significant soil upheaval in front of the pipe's movement direction was observed, and the concave deformation beneath the pipe noticeably decreased after pipe rotation. These observed soil deformations will be further analyzed in subsequent sections of this paper.

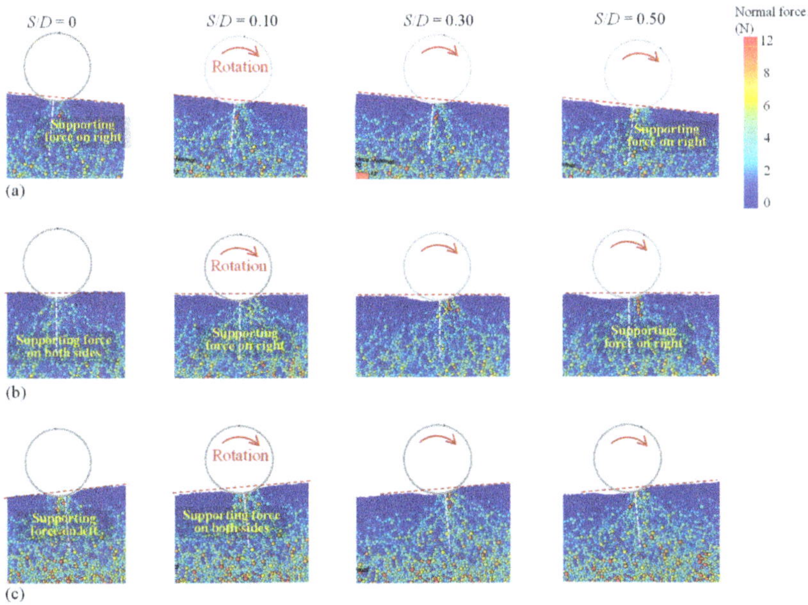

Figure 6. Progressive soil supporting force distributions and soil deformation around freely laid pipes at different slope angle conditions: (**a**) $\theta = -5°$ (downslope), (**b**) $\theta = 0°$ (flat), and (**c**) $\theta = 5°$ (upslope).

Figure 7 illustrates the distribution of soil resistance and progressive soil deformation of anti-rolling pipes on slopes under different slope angle conditions. To better illustrate how the soil upheaval in the pipe moving direction can be especially small in downslope conditions, the progressive soil supporting force and soil deformation distributions under $\theta = -10°$ are presented in Figure 7a. On the whole, the zone of supporting force distributions was similar to that in Figure 6, but the magnitude of the supporting forces was larger, which aligns with the larger ultimate drag force observed under anti-rolling conditions in Figure 5. Additionally, soil upheaval emerged in front of the pipe's movement direction, and the concave deformation of the slope surface and the soil upheaval increased with the slope angle θ in Figure 7, which is consistent with the increased ultimate drag force as slope angles increased in Figure 5. When comparing the settlement and soil deformation around freely laid pipes (see Figure 6) with that occurring around anti-rolling pipes (see Figure 7), it is notable that the settlement and soil deformation around anti-rolling pipes was significantly larger. This difference can be attributed to the influence of rotation-induced traction force on granular soil slopes, which will be further analyzed in the following sections.

It is important to note that the magnitude of soil deformation around pipes on a sloping seabed provides valuable insights into the intensity of pipe–soil interactions and is closely related to the magnitude of the ultimate drag force. Comparing Figures 6 and 7 reveals that the soil deformation around pipes in freely laid conditions is significantly smaller than that in anti-rolling conditions, consistent with findings from previous experimental studies [3,6]. However, previous studies have provided only vague interpretations for the smaller settlement of pipes in freely laid conditions, attributing it to a smaller ultimate drag force affecting pipe instability. In fact, the smaller settlement of pipes in freely laid conditions can be clearly explained. To clarify the underlying reasons, it is crucial to prioritize the recognition that the soil deformation around pipes is primarily induced by the gravitational force of the pipes (G), based on the force analyses in Figure 1. As shown in Figure 8, in the freely laid condition, the rotation of a pipe on the concave soil surface would induce an obliquely upward force (F_T), partially counteracting the pipe's gravity (G). Consequently, the soil deformation around the pipe decreases after significant pipe rotation on the concave soil surface. In the anti-rolling condition, where there is no pipe rotation and, therefore, no obliquely upward traction force to counteract the pipe's gravity (G) due to the pipe's end constraint, the soil deformation caused by the pipe's weight is pronounced. Therefore, the upward traction force induced by rotation is the main reason for the smaller soil deformation around pipes in freely laid conditions.

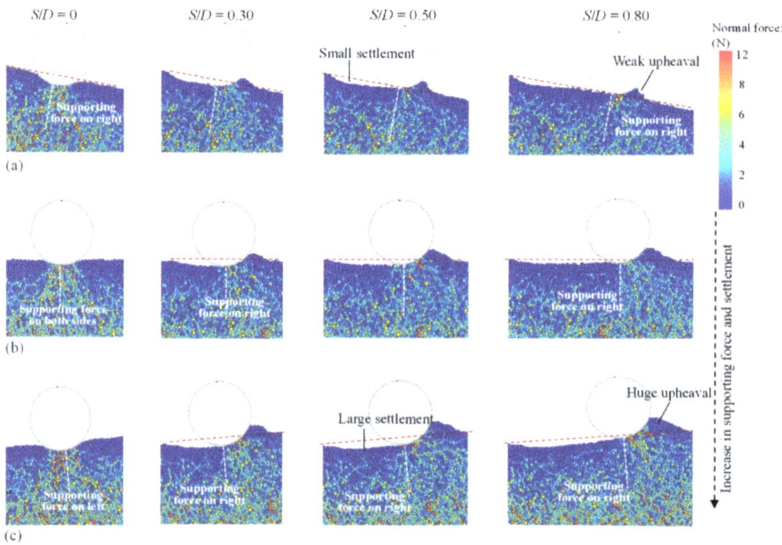

Figure 7. Progressive soil supporting force and soil deformation distributions around anti-rolling pipes at different slope angle conditions: (**a**) $\theta = -10°$ (downslope), (**b**) $\theta = 0°$ (flat), and (**c**) $\theta = 5°$ (upslope).

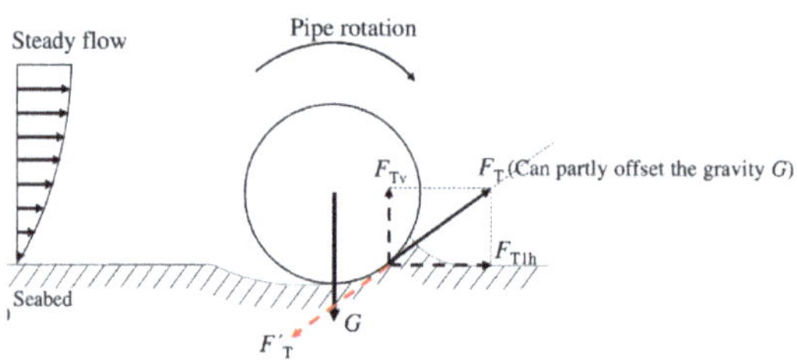

Figure 8. Schematic diagram illustrating the effect of free rotation on pipe–soil interaction on a concave soil surface.

3.2. Progressive Pipe Behaviors on Sloping Seabeds

The evaluation of pipe–soil interactions on sloping seabeds can benefit from analysis of the motion trajectory, stress, deformation, and shear wear characteristics of pipes [83–85]. However, the existing literature on DEM-based analysis lacks specific studies examining the influence of granular soil slopes on the stress, deformation, and shear wear characteristics of pipes from a cross-sectional perspective [1,86–88]. As previously discussed, the intensity of anti-rolling pipe–soil interactions is significantly stronger than that of freely laid pipe–

soil interactions on sloping seabeds. Consequently, this section primarily focuses on the behaviors of anti-rolling pipes.

The motion trajectory of pipes provides a comprehensive reflection of the intensity and manner of pipe–soil interactions on sloping seabeds. In scientific research, the motion trajectory of pipes is often compared using lateral displacement–settlement curves [6,15,18]. Previous studies [1,3] have pointed out the complexity of lateral displacement– settlement curves for pipes on sloping seabeds: sometimes a peak point exists in the lateral displacement–settlement curves, while at other times there is no peak point. To understand the pattern of pipe settlement development during lateral movement, lateral displacement–settlement curves for pipes under different slope angles (negative angle indicates downslope instability conditions) were produced and are presented in Figure 9. The lateral displacement and the settlements of pipes were normalized by the pipe diameter (D). By comparing the lateral displacement–settlement curves from the simulation results (see Figure 9), certain patterns can be observed. The pipe settlement continuously increased with lateral displacement when the slope angle was small ($\theta = -10^\circ$ in typical downslope instability conditions). In contrast, the pipe settlement initially decreased and then increased with the increase in lateral displacement, exhibiting an obvious peak at lower values when the slope angle was large ($\theta = 10^\circ$ and 20° in typical downslope instability conditions).

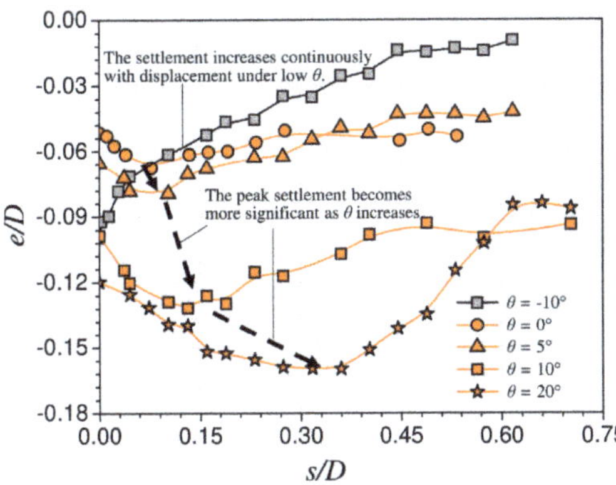

Figure 9. Normalized pipe settlement (e/D) vs. normalized lateral displacement (s/D) for anti-rolling pipes on sloping seabeds.

In fact, the observed shape rules in pipe settlement–displacement curves can be further elucidated by the differences in soil supporting force distributions under different slope conditions. Under the condition of no pipe movement on sloping seabeds (Figure 10a), the primary supporting force is concentrated on the downslope side of the pipe tip on the sloping seabed (as verified by the images in Figure 7c). If we only consider the effect of an obliquely upward driving force, the settlement of pipes will continuously decrease after the driving force is applied, due to the counteracting effect on the pipe's gravity. However, the change of soil support force distributions during lateral pipe movement also significantly affects the settlement of pipes. In the case of upslope instability (Figure 10b), the downslope supporting zone (contact zone) will be lost after the obliquely upward movement of pipes, resulting in an insufficient soil supporting force to counteract the pipe's gravity G. Consequently, the settlement of the pipe will increase due to the unbalanced gravitational force acting on the pipe, and the main soil supporting force will be transferred to the upslope side. In the case of upslope instability, the combination of the obliquely upward driving force and the change in soil support force distributions would result in a turning point on the pipe settlement–displacement curves. In the case of downslope instability (see Figure 10c), although the contact area on the upslope is reduced (much smaller contact area) after the pipes move downward, the main soil supporting force on the downslope side, which counteracts the gravity (G) of the pipes, experiences less change. As a result, the settlement of the pipe is mainly influenced by the obliquely upward driving force, leading to a continuous reduction in pipe settlement as the pipe moves laterally.

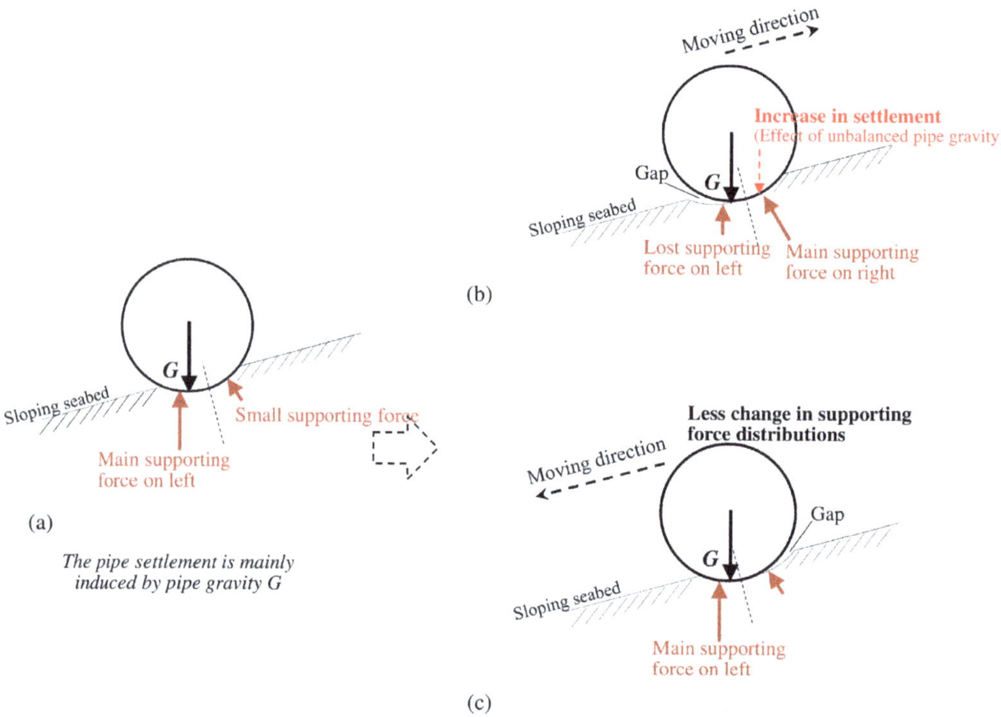

Figure 10. Schematic diagram illustrating the mechanism of the shape rules in pipe settlement–displacement curves for anti-rolling pipes on granular soil slopes: (**a**) stationary (Initial position), (**b**) upslope instability (after applying obliquely upward action), and (**c**) downslope instability (after applying obliquely downward action).

In the study conducted by Gao et al. [3], no experimental data on pipe stress and deformation were provided. This section aims to analyze the stress and deformation of pipes in cross-sections, which serves as a supplementary analysis to the experimental tests. Figure 11 illustrates the typical pipe stress and deformation in the central cross-sections for downslope instability and upslope instability. In all cases, the largest pipe deformations were observed in the lower part of the pipes in the pipe moving direction, which corresponds well to the main contact zone of pipe–soil interactions (see Figure 7). Due to the significant soil resistance at the pipe–soil interface, concave pipe deformation appeared in the lower part of the pipes in the pipe moving direction. In the downslope instability condition ($\theta = -5°$ in Figure 11a), the zone of concave deformation rotated slightly counterclockwise as the lateral pipe displacement increased, owing to the slight rise in soil upheaval in front of the pipe's movement direction. In the upslope instability condition ($\theta = 5°$ in Figure 11b), the stress and concave deformation of the pipes were larger than those in the downslope instability condition. Moreover, as a result of significant soil upheaval in front of the pipe's movement direction, the zone of concave deformation on the pipes exhibited noticeable rotation as lateral pipe displacement increased. The numerical simulation results indicate that the lower position ahead of the pipe's movement is relatively susceptible to deformation. In the upslope instability condition, the pipe deformation deserves more attention for failure analysis.

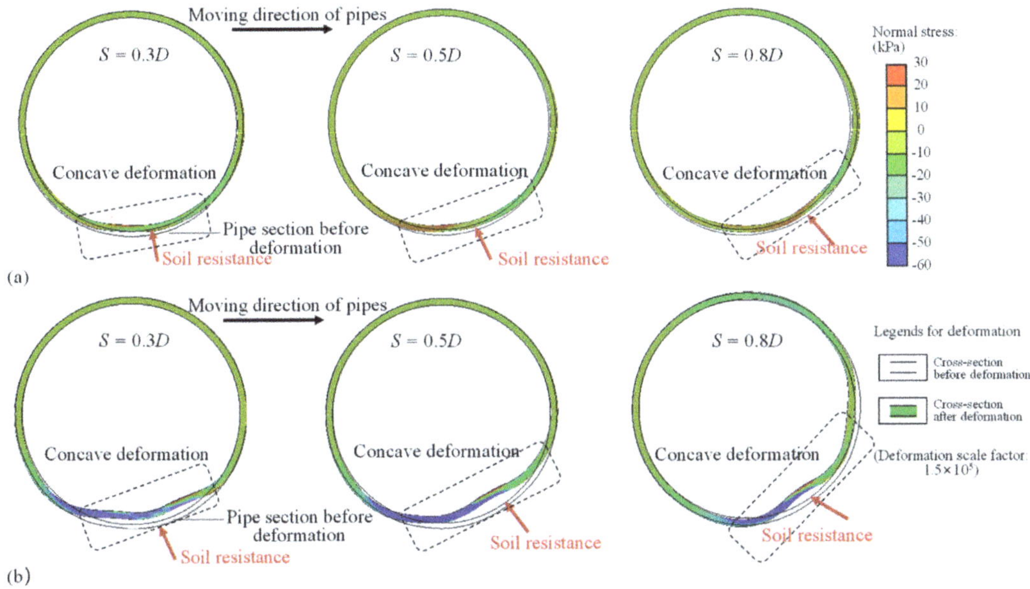

Figure 11. Comparison of the deformation of anti-rolling pipes in cross-sections: (**a**) $\theta = -5°$ (downslope instability) and (**b**) $\theta = 5°$ (upslope instability).

The tendency for shear wear to occur is a crucial factor in determining the placement of monitoring equipment and coatings to protect against seawater corrosion on pipelines. The pipe–soil interaction behaviors depicted in Figure 7 indicate a complex tendency for shear wear to occur on the outer surface of pipes. The intensity of shear wear mainly depends on the shear intensity [89]. As previously discussed, the intensity of pipe–soil interactions for freely laid pipelines is significantly weaker than that of anti-rolling pipe–soil interactions on sloping seabeds; thus, the shear intensities of freely laid pipe–soil interactions in all cases were found to be weak. Figure 12 presents the typical shear intensities on the pipe bottoms under different slope angles (negative angles represent downslope instability conditions) and different interface friction coefficients. In all cases, the main areas of high shear intensity were observed on the side of the pipe moving direction due to the predominant soil resistance on the side (the side in front of the pipe's movement direction). Moreover, the shear intensity increased with the slope angles (from downslope instability to upslope instability), which corresponds to the increased soil resistance in front of the pipe's movement direction with the increase in slope angles, as reflected in Figure 7. According to common sense, a higher interface friction coefficient leads to a higher interface friction force and, consequently, higher interface shear intensity. However, an interesting phenomenon was observed: the shear intensity on the bottom of the pipes increased with a decrease in the interface friction coefficient, as shown in Figure 12d–f. The reasons for the decrease in interface shear intensity with an increasing friction coefficient will be explained in detail in the following sections.

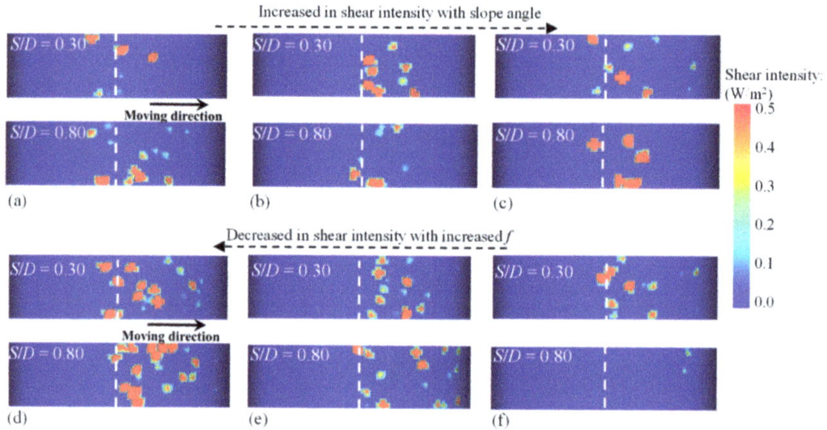

Figure 12. Comparison of the position and shear intensity on the bottom of anti-rolling pipes: (**a**) $\theta = -10°$ ($f = 0.45$), (**b**) $\theta = -5°$ ($f = 0.45$), (**c**) $\theta = 0°$ ($f = 0.45$), (**d**) $\theta = 5°$ ($f = 0.30$), (**e**) $\theta = 5°$ ($f = 0.45$), and (**f**) $\theta = 5°$ ($f = 0.60$).

3.3. Particle-Scale Soil Behaviors and Patterns of Pipe–Soil Interaction on Sloping Seabeds

Understanding particle-scale behaviors is crucial for comprehending the complex mechanisms of soil deformation and the interactions between structures and soil [90,91]. This section provides a more in-depth exploration of the interface shear behaviors, soil shear bands, soil disturbance and deformation zones, and patterns of pipe–soil interaction instability on sloping seabeds, with a focus on particle-scale behaviors.

The shear intensity on the pipe surfaces is directly related to the sliding and collision behaviors of particles at the pipe–soil interfaces. Figure 13 provides the particle collision behaviors on pipe–particle interfaces under different interface friction coefficients ($\theta = 5°$, for example). Numerous particle collisions were observed on the side of the pipe moving direction, which aligns with the zone of high shear intensity depicted in Figure 12. Additionally, the frequency of particle collisions and the sliding distances decreased with an increase in interface friction coefficients, consistent with the shear intensity shown in Figure 12d,e. This indicates that as the interface friction coefficients increase, the pipe and adjacent particles tend to move together to interact with the surrounding soil.

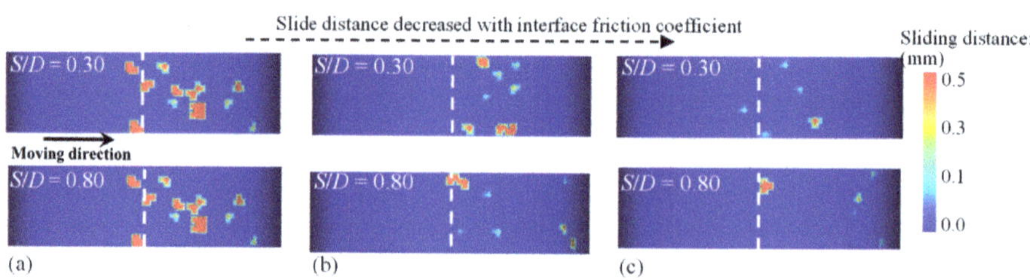

Figure 13. Comparison of the position and slide distance of particle collisions on the surface of anti-rolling pipes ($\theta = 5°$) under different interface friction coefficients: (**a**) $f = 0.30$, (**b**) $f = 0.45$, and (**c**) $f = 0.60$.

Furthermore, the particle rotation velocity distributions provide further evidence of the shear behaviors on the pipe surfaces. Previous studies [33,92] have shown that the zone of high particle rotation velocities in granular soil is indicative of the position of strong soil shear. Figure 14 explores the relationships between microparticle rotation velocity and shear bands or shear behaviors around pipes. In downslope instability conditions ($\theta = -5°$, for example), significant particle rotation occurred around the downslope zone of the pipes, as shown in Figure 14a, reflecting a high shear intensity in the zone. Additionally, the zone of high particle rotation velocities was relatively small in Figure 14a, corresponding to a limited soil upheaval in front of the pipe's movement direction. In upslope instability conditions (θ = 5°, for example), a soil shear band emerged below the interface shear zone in Figure 14b. This is associated with a pronounced soil upheaval in front of the pipe's movement direction. Interestingly, no notable particle rotation velocity was observed at the pipe–soil interface when the interface friction coefficient increased to f = 0.6 in Figure 14c; in this scenario, only a significant soil shear band at some distance from the interface (the soil shear band below the interface shear zone) was observed, suggesting that the particles on the pipe–soil interface move together with the pipes rather than undergoing relative sliding. The weak surface shear wear (see Figure 12) and fewer particle collisions (see Figure 13) under high interface friction coefficient conditions also support the synchronous movement of the pipe and adjacent particles. Clearly, the particle rotation behaviors based on discrete element method (DEM) simulations effectively reflect the differences in soil shear induced by slope and interface friction effects.

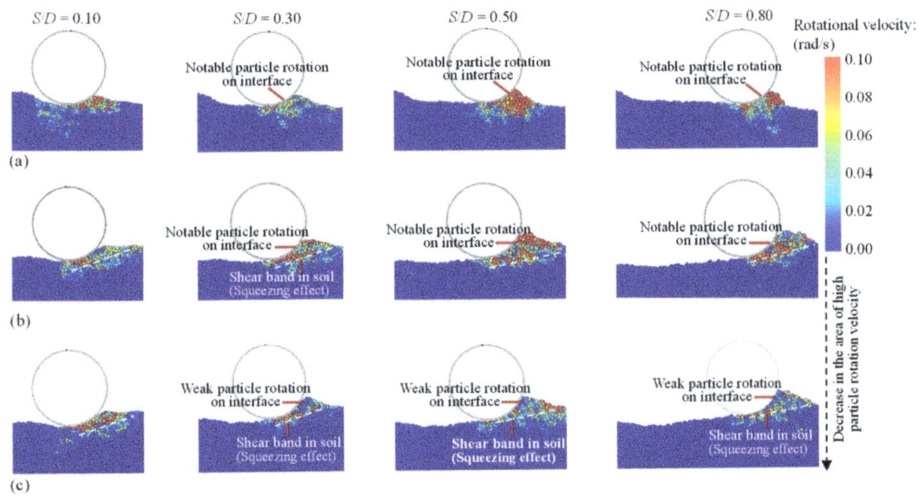

Figure 14. Comparison of the particle rotation velocity and soil shear behaviors in the progressive soil deformation process around anti-rolling pipes on sloping seabeds: (**a**) $\theta = -5°$ (f = 0.45), (**b**) θ = 5° (f = 0.45), and (**c**) θ = 5° (f = 0.60).

The trajectories and the translation velocity of particles at each tracing point in the pipe–soil interaction process are fundamental for in-depth interpretation of soil failure behaviors [33,93]. Figure 15 illustrates the trajectories of the particles and their translation velocities at each tracing point. In cases with freely laid pipes (Figure 15a), the short particle trajectories and low particle translation velocities indicate limited soil disturbance and weak pipe–soil interaction on sloping seabeds. In cases with anti-rolling pipes (Figure 15b–f), the long particle trajectories and high particle translation velocities indicate significant settlement of pipes on sloping seabeds; in these cases, the long and oblique trajectories of particles were a result of the strong squeezing movement of the pipes. Within the particle trajectory zones, the velocity of particle translation decreased with soil depth as a result of the increasing confining effect in deeper soil. Furthermore, the area of strong pipe–soil interaction zones increased with slope angles (from downslope instability to upslope instability), as increased by the increased area of particle trajectory zones. In summary, the end constraint condition of pipes, instability directions (upslope or downslope), and slope angles significantly influence the extent of soil disturbance on sloping seabeds.

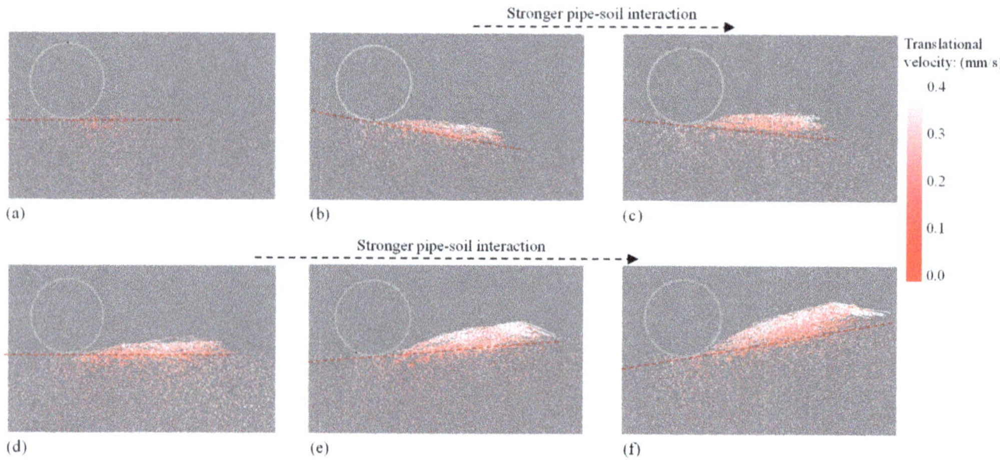

Figure 15. Comparison of particle trajectories and particle translation velocities during the progressive pipe–soil interaction process on sloping seabeds: (**a**) $\theta = 0°$ (freely laid), (**b**) $\theta = -10°$ (anti-rolling), (**c**) $\theta = -5°$ (anti-rolling), (**d**) $\theta = 0°$ (anti-rolling), (**e**) $\theta = 5°$ (anti-rolling), and (**f**) $\theta = 10°$ (anti-rolling).

3.4. Patterns of Pipe–Soil Interaction on Sloping Seabeds

In the field of marine pipeline engineering, pipelines are often found on sloping seabeds [1,11,14]. Therefore, it is crucial to analyze the patterns of lateral pipe–soil interactions under lateral oblique force actions on sloping seabeds. This study's macro- and microscale analyses of pipe–soil interaction yielded the mechanistic patterns of pipe–soil interactions on sloping seabeds, as shown in Figure 16. For freely laid pipes, without restraint, on sloping seabeds (Figure 16a), the pipes are able to rotate freely. As a result, the pipe–soil interaction is characterized by small pipe settlement and relatively mild soil disturbance, regardless of the instability directions (upslope or downslope). For anti-rolling pipes in downslope instability conditions (Figure 16b), the pipe–soil interaction is characterized by reduced pipe settlement and relatively weak soil upheaval as the pipe moves along the slope surface. For anti-rolling pipes in upslope instability conditions (Figure 16c), the pipe–soil interaction is characterized by an arc settlement trajectory and huge soil upheaval in front of the pipe's movement direction. In summary, the end constraint condition of pipes, instability directions (upslope or downslope), and slope angles significantly affect the patterns of interaction between pipes and granular soil on sloping seabeds. Therefore, applying conventional design theory while ignoring the effect of deformable slopes can lead to misjudging of the soil failure behaviors and the intensity of pipe–soil interactions under lateral oblique force actions, posing a hidden danger to engineering safety.

This study focused on the effect of upslope and downslope, as well as the effects of pipe rotation. However, it should be noted that the lateral on-bottom stability of a submarine pipeline subject to ocean waves/currents involves complex flow–pipe–soil interaction. Many factors could be further considered in future research. For example, loading speed plays a significant role in the hydrodynamic forces acting on the pipeline. Higher loading speeds, such as those caused by rapidly changing wave conditions or strong currents, can result in increased dynamic forces acting on the pipeline. This, in turn, may lead to larger soil deformations and potentially affect the lateral stability of the pipeline. In addition, soil liquefaction is another important factor that is worthy of consideration. In certain conditions, such as during seismic events or cyclic loading, some parts of the soil may undergo liquefaction, resulting in a loss of shear strength. This could significantly impact the lateral stability of the pipeline, potentially leading to instability and movement. Such further investigations would contribute to the development of more accurate models and design guidelines for submarine pipelines in various marine environments.

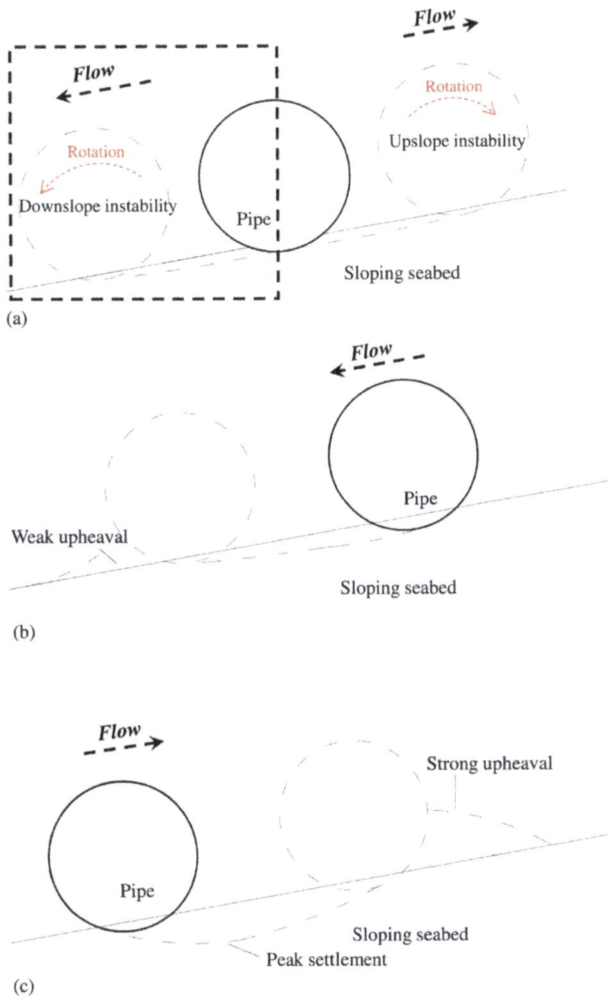

Figure 16. Schematics of the patterns of pipe–granular soil interaction under lateral oblique force actions on sloping seabeds: (**a**) freely laid pipe, (**b**) anti-rolling pipe (downslope instability), and (**c**) anti-rolling pipe (upslope instability).

4. Conclusions

This study investigated the underlying mechanisms behind the lateral pipe–soil interaction instability on sloping sandy seabeds, examining the effects of slope angle and instability direction on the behavior of pipe–soil interaction for freely laid and anti-rolling pipes subject to lateral oblique force actions, through a series of 3D coupled DEM–FEM simulations. The key findings can be summarized as follows:

(1) The pipe–soil interaction instability on downslopes and upslopes was compared using coupled DEM–FEM simulations. The ultimate drag force of anti-rolling pipes was found to be much larger than that of freely laid pipes on both downslope and upslope instabilities. In addition, progressive soil resistance and deformation distributions around pipes under different slope angles and instability directions were discovered and found to be increased with slope angles. Furthermore, the rotation-induced

obliquely upward traction force was proved to be the essential reason for the smaller soil resistance and smaller soil deformation around freely laid pipes.

(2) The effects of slope angle and instability direction on motion trajectory, stress, deformation, and shear wear of pipes were discovered and analyzed in detail. The difference in shapes of lateral displacement–settlement curves in upslope instability and downslope instability was compared and successfully explained by the difference in position distributions of soil supporting forces after obvious lateral pipe movement. Moreover, pipe stress and deformation under upslope instability conditions were found to be larger than those under downslope instability conditions, primarily concentrated around the contact position of pipe–soil interactions. In addition, the shear intensity on the pipe–soil interface was found to increase with slope angles and decrease with interface friction coefficients.

(3) The pipe–soil interaction in the upslope and downslope instabilities was further analyzed at particle scales. Particle trajectories, translation velocities, and rotation velocities were found to be advantageous in identifying granular soil behaviors, such as soil disturbance zones and shear bands. The synchronous movement of the pipe and adjacent particles, as indicated by soil shear bands, was identified as the mechanism behind the low particle collision and weak shear intensity on the pipe surfaces with high interface friction coefficients. As instability direction changed from downslope to upslope, the area of shear bands and soil disturbance exhibited a notable increase. Finally, several conclusive mechanistic patterns of pipe–soil interaction instability on sloping seabeds were identified by macro- to micro-analyses.

In conclusion, this study has highlighted the significant effects of slope angle and instability direction on the behavior of pipe–soil interaction for freely laid and anti-rolling pipes subject to lateral oblique force actions. The finding that slope angle and instability direction not only influence the magnitude of various pipe–soil interaction behaviors on slopes, but also shape the patterns of soil deformation and pipe movement, has significant implications for pipeline construction in marine engineering.

Author Contributions: Conceptualization, L.Q.; numerical simulations and data analyses L.Q. and Y.P.; writing—original draft preparation, Y.P.; All authors have read and agreed to the published version of the manuscript.

Funding: No special Funding was provided for this work.

Institutional Review Board Statement: Not applicable.

Informed Consent Statement: Not applicable.

Data Availability Statement: The data presented in this study are available on request from the corresponding author.

Conflicts of Interest: The authors declare no conflict of interest.

References

1. Wang, N.; Qi, W.; Gao, F. Predicting the Instability Trajectory of an Obliquely Loaded Pipeline on a Clayey Seabed. *J. Mar. Sci. Eng.* **2022**, *10*, 299. [CrossRef]
2. Yang, H.; Guo, Z.; Wang, L.; Dou, Y.; Liu, Z. Experimental study on wave-induced seabed response and force on the pipeline shallowly buried in a submerged sandy slope. *Ocean Eng.* **2022**, *251*, 111153. [CrossRef]
3. Gao, F.-P.; Han, X.-T.; Cao, J.; Sha, Y.; Cui, J.-S. Submarine pipeline lateral instability on a sloping sandy seabed. *Ocean Eng.* **2012**, *50*, 44–52. [CrossRef]
4. Wang, H.; Wang, K.; Guo, Z.; Rui, S.; Wang, Z. Resistance mechanism of partially embedded pipelines considering the influence of temperature fields. *Ocean Eng.* **2023**, *281*, 114720. [CrossRef]
5. Qi, W.-G.; Shi, Y.-M.; Gao, F.-P. Uplift soil resistance to a shallowly-buried pipeline in the sandy seabed under waves: Po-ro-elastoplastic modeling. *Appl. Ocean Res.* **2020**, *95*, 102024. [CrossRef]
6. Gao, F.-P.; Wang, N.; Li, J.; Han, X.-T. Pipe–soil interaction model for current-induced pipeline instability on a sloping sandy seabed. *Can. Geotech. J.* **2016**, *53*, 1822–1830. [CrossRef]
7. Jones, W.T. On-Bottom Pipe line Stability in Steady Water Currents. In Proceedings of the Offshore Technology Conference, Houston, TX, USA, 2–5 May 1976.

8. White, D.; Cathie, D. Geotechnics for Subsea Pipelines. In *Frontiers in Offshore Geotechnics II*; CRC Press: Boca Raton, FL, USA, 2010.
9. Rui, S.; Zhou, Z.; Jostad, H.P.; Wang, L.; Guo, Z. Numerical prediction of potential 3-dimensional seabed trench profiles considering complex motions of mooring line. *Appl. Ocean Res.* **2023**, *139*, 103704. [CrossRef]
10. Ameh, S.E.; Ameh, N.I. On-bottom stability design of submarine pipelines: The fundamentals. *J. Appl. Sci. Environ. Manag.* **2019**, *23*, 1985–1990. [CrossRef]
11. Zhang, W.; Askarinejad, A. Behaviour of buried pipes in unstable sandy slopes. *Landslides* **2019**, *16*, 283–293. [CrossRef]
12. Lyons, C. Soil Resistance to Lateral Sliding of Marine Pipelines. In Proceedings of the Fifth Annual Offshore Technology Conference, Houston, TX, USA, 30 April–2 May 1973.
13. Verley, R.; Lund, K.M. *A Soil Resistance Model for Pipelines Placed on Clay Soils*; American Society of Mechanical Engineers: New York, NY, USA, 1995.
14. Zhang, J.; Stewart, D.P.; Randolph, M.F. Modeling of Shallowly Embedded Offshore Pipelines in Calcareous Sand. *J. Geotech. Geoenviron. Eng.* **2002**, *128*, 363–371. [CrossRef]
15. Gao, F.-P.; Yan, S.-M.; Yang, B.; Luo, C.-C. Steady flow-induced instability of a partially embedded pipeline: Pipe–soil interaction mechanism. *Ocean Eng.* **2011**, *38*, 934–942. [CrossRef]
16. Verley, R.L.P.; Sotberg, T. A Soil Resistance Model for Pipelines Placed on Sandy Soils. *J. Offshore Mech. Arct. Eng.* **1994**, *116*, 145–153. [CrossRef]
17. Foray, P.; Bonjean, D.; Michallet, H.; Mory, M. Fluid-Soil-Structure Interaction in Liquefaction around a Cyclically Moving Cylinder. *J. Waterw. Port Coastal Ocean Eng.* **2006**, *132*, 289–299. [CrossRef]
18. White, D.; Cheuk, C. Modelling the soil resistance on seabed pipelines during large cycles of lateral movement. *Mar. Struct.* **2008**, *21*, 59–79. [CrossRef]
19. Xu, P.; Du, Z.; Zhang, T.; Chen, B. Vector form intrinsic finite element analysis of deepwater J-laying pipelines on sloping seabed. *Ocean Eng.* **2022**, *247*, 110709. [CrossRef]
20. Shi, Y.-M.; Wang, N.; Gao, F.-P.; Qi, W.-G.; Wang, J.-Q. Physical modeling of the axial pipe-soil interaction for pipeline walking on a sloping sandy seabed. *Ocean Eng.* **2019**, *178*, 20–30. [CrossRef]
21. Wu, S.; Jeng, D.-S. Effects of dynamic soil permeability on the wave-induced seabed response around a buried pipeline. *Ocean Eng.* **2019**, *186*, 106132. [CrossRef]
22. Chen, R.; Wu, L.; Zhu, B.; Kong, D. Numerical modelling of pipe-soil interaction for marine pipelines in sandy seabed subjected to wave loadings. *Appl. Ocean Res.* **2019**, *88*, 233–245. [CrossRef]
23. Lin, Z.; Guo, Y.; Jeng, D.-S.; Liao, C.; Rey, N. An integrated numerical model for wave–soil–pipeline interactions. *Coast. Eng.* **2016**, *108*, 25–35. [CrossRef]
24. Wagner, D.A.; Murff, J.D.; Brennodden, H.; Sveggen, O. Pipe-soil interaction model. *J. Waterw. Port Coast. Ocean Eng.* **1989**, *115*, 205–220. [CrossRef]
25. Liu, R.; Guo, S.; Yan, S. Study on the Lateral Soil Resistance Acting on the Buried Pipeline. *J. Coast. Res.* **2015**, *73*, 391–398. [CrossRef]
26. Mallikarachchi, H.E.; Pelecanos, L.; Soga, K. Nonlinear Finite-Element Analysis of Soil-Pipe Interaction for Laterally-Loaded Buried Offshore Pipelines. In *Numerical Methods in Geotechnical Engineering IX*; CRC Press: Boca Raton, FL, USA, 2018; Volume 2.
27. Liang, W.; Zhao, J. Multiscale modeling of large deformation in geomechanics. *Int. J. Numer. Anal. Methods Geomech.* **2019**, *43*, 1080–1114. [CrossRef]
28. Horabik, J.; Molenda, M. Parameters and contact models for DEM simulations of agricultural granular materials: A review. *Biosyst. Eng.* **2016**, *147*, 206–225. [CrossRef]
29. Chen, Q.; Peng, W.; Yu, R.; Tao, G.; Nimbalkar, S. Laboratory Investigation on Particle Breakage Characteristics of Calcareous Sands. *Adv. Civ. Eng.* **2021**, *2021*, 8867741. [CrossRef]
30. Yin, Z.-Y.; Wang, P.; Zhang, F. Effect of particle shape on the progressive failure of shield tunnel face in granular soils by coupled FDM-DEM method. *Tunn. Undergr. Space Technol.* **2020**, *100*, 103394. [CrossRef]
31. Peng, Y.; Yin, Z.-Y.; Ding, X. Micromechanical analysis of the particle corner breakage effect on pile load performance in coral sand. *Acta Geotech.* **2023**, *18*, 6353–6370. [CrossRef]
32. Deluzarche, R.; Cambou, B.; Fry, J. Modeling of rockfill behavior with crushable particles. In *Numerical Modeling in Micromechanics Via Particle Methods*; Routledge: Abingdon, UK, 2017; pp. 219–224.
33. Peng, Y.; Yin, Z.-Y. Micromechanical analysis of suction pile-granular soil interaction under inclined pulling load of mooring line: Mooring depth effect. *Mar. Struct.* **2023**, *92*, 103499. [CrossRef]
34. Macaro, G.; Utili, S.; Martin, C.M. DEM simulations of transverse pipe–soil interaction on sand. *Géotechnique* **2021**, *71*, 189–204. [CrossRef]
35. Yimsiri, S.; Soga, K. DEM analysis of soil-pipeline interaction in sand under lateral and upward movements at deep embedment. *Geotech. Eng.* **2006**, *37*, 83–94.
36. Jiang, M.J.; Zhang, W.C.; Liu, F.; Shen, Z.F. Investigating Particle-Size Effect on Uplift Mechanism of Pipes Buried in Sand Using Distinct Element Method. *Adv. Mater. Res.* **2012**, *361–363*, 505–509. [CrossRef]
37. Peng, Y.; Liu, H.; Li, C.; Ding, X.; Deng, X.; Wang, C. The detailed particle breakage around the pile in coral sand. *Acta Geotech.* **2021**, *16*, 1971–1981. [CrossRef]

38. Macaro, G.; Utili, S.; Martin, C. DEM analyses of pipe–soil interaction for offshore pipelines on sand. In *Geomechanics from Micro to Macro*; Soga, K., Kumar, K.K., Biscontin, G., Eds.; Taylor and Francis: London, UK, 2015.
39. Durmus, S.E. DEM Modelling of Soil Pipeline Interaction under Combined Vertical and Horizontal Loading. Master's Thesis, Politecnico di Milano, Milan, Italy, 2021.
40. Wang, Y.; Liu, J.; Guo, P.; Zhang, W.; Lin, H.; Zhao, Y.; Ou, Q. Simplified Analytical Solutions for Tunnel Settlement Induced by Axially Loading Single Pile and Pile Group. *J. Eng. Mech.* **2021**, *147*, 04021116. [CrossRef]
41. Liu, F.; Yi, J.; Cheng, P.; Yao, K. Numerical simulation of set-up around shaft of XCC pile in clay. *Geomech. Eng.* **2020**, *21*, 489–501.
42. Zheng, Z.; Zang, M.; Chen, S.; Zhao, C. An improved 3D DEM-FEM contact detection algorithm for the interaction simulations between particles and structures. *Powder Technol.* **2017**, *305*, 308–322. [CrossRef]
43. Cui, X.; Zhou, R.; Guo, G.; Du, B.; Liu, H. Effects of Differential Subgrade Settlement on Slab Track Deformation Based on a DEM-FDM Coupled Approach. *Appl. Sci.* **2021**, *11*, 1384. [CrossRef]
44. Qu, L.; Yang, C.; Ding, X.; Kouroussis, G.; Zheng, C. A continuum-based model on axial pile-head dynamic impedance in inhomo-geneous soil. *Acta Geotech.* **2021**, *16*, 3339–3353. [CrossRef]
45. Wu, Z.; Barosh, P.J.; Wang, L.; Hu, D.; Wang, W. Numerical modeling of stress and strain associated with the bending of an oil pipeline by a migrating pingo in the permafrost region of the northern Tibetan Plateau. *Eng. Geol.* **2008**, *96*, 62–77. [CrossRef]
46. Dratt, M.; Katterfeld, A. Coupling of FEM and DEM simulations to consider dynamic deformations under particle load. *Granul. Matter* **2017**, *19*, 49. [CrossRef]
47. Peng, Y.; Yin, Z.-Y.; Ding, X. Analysis of particle corner-breakage effect on pile penetration in coral sand: Model tests and DEM simulations. *Can. Geotech. J.* **2023**, *60*, 749–765. [CrossRef]
48. Kešner, A.; Chotěborský, R.; Linda, M.; Hromasová, M.; Katinas, E.; Sutanto, H. Stress distribution on a soil tillage machine frame segment with a chisel shank simulated using discrete element and finite element methods and validate by experiment. *Biosyst. Eng.* **2021**, *209*, 125–138. [CrossRef]
49. Orosz, Á.; Tamás, K.; Rádics, J.P.; Zwierczyk, P.T. Coupling Finite And Discrete Element Methods Using an Open Source and A Commercial Software. In Proceedings of the 32nd European Conference on Modelling and Simulation, Wilhelmshaven, Germany, 22–25 May 2018.
50. Castaneda, E.A.; Pineda Leon, R.; Cornejo, J. FEM and DEM Simulations of Tire-Soil and Drill-Soil Interactions in Off-Road Condi-tions for Mechanical Design Validation of a Space Exploration Rover. In Proceedings of the 2021 12th International Conference on Mechanical and Aerospace Engineering (ICMAE), Athens, Greece, 16–19 July 2021.
51. Munir, H.A.; Zakaria, A.; Ponniran, A.; Rahman, T.A.; Marimuthu, T. Investigation of The Dynamic Deflection of Conveyor Belts Via Simulation Modelling Methods on Idler Factor. *J. Phys. Conf. Ser.* **2022**, *2312*, 012027. [CrossRef]
52. Wang, P.; Yin, Z.-Y. Micro-mechanical analysis of caisson foundation in sand using DEM: Particle breakage effect. *Ocean Eng.* **2020**, *215*, 107921. [CrossRef]
53. Falagush, O.; McDowell, G.R.; Yu, H.-S. Discrete Element Modeling of Cone Penetration Tests Incorporating Particle Shape and Crushing. *Int. J. Géoméch.* **2015**, *15*, 04015003. [CrossRef]
54. Zhao, H.; Zhang, J.; Qiu, P.; Ji, S. Hierarchical Modeling Method for DEM Simulation and Its Application in Soil–Pile–Cap Interaction and Impact Case. *Int. J. Géoméch.* **2019**, *19*, 04019076. [CrossRef]
55. Dai, B.; Yang, J.; Zhou, C.; Luo, X. DEM investigation on the effect of sample preparation on the shear behavior of granular soil. *Particuology* **2016**, *25*, 111–121. [CrossRef]
56. Nakashima, H.; Shioji, Y.; Kobayashi, T.; Aoki, S.; Shimizu, H.; Miyasaka, J.; Ohdoi, K. Determining the angle of repose of sand under low-gravity conditions using discrete element method. *J. Terramech.* **2011**, *48*, 17–26. [CrossRef]
57. Peng, Y.; Yin, Z.-Y. Micromechanical analysis of suction pile-granular soil interaction under inclined pulling load: Effect of pulling angle. *Ocean Eng.* **2023**, *270*, 113615. [CrossRef]
58. Arroyo, M.; Butlanska, J.; Gens, A.; Calvetti, F.; Jamiolkowski, M. Cone penetration tests in a virtual calibration chamber. *Géotechnique* **2011**, *61*, 525–531. [CrossRef]
59. Calvetti, F.; di Prisco, C.; Nova, R. Experimental and Numerical Analysis of Soil–Pipe Interaction. *J. Geotech. Geoenviron. Eng.* **2004**, *130*, 1292–1299. [CrossRef]
60. Yimsiri, S. Distinct Element Analysis of Soil-Pipeline Interaction in Sand under Upward Movement at Deep Embedment Condition. In Proceedings of the 16th International Conference on Soil Mechanics and Geotechnical Engineering, Osaka, Japan, 12–16 September 2005.
61. Zhou, C.; Qian, J.; Yin, Z. Microscopic investigation of the influence of complex stress states on internal erosion and its impacts on critical hydraulic gradients. *Int. J. Numer. Anal. Methods Géoméch.* **2022**, *46*, 3377–3401. [CrossRef]
62. Jensen, R.P.; Edil, T.B.; Bosscher, P.J.; Plesha, M.E.; Ben Kahla, N. Effect of Particle Shape on Interface Behavior of DEM-Simulated Granular Materials. *Int. J. Géoméch.* **2001**, *1*, 1–19. [CrossRef]
63. Khazeni, A.; Mansourpour, Z. Influence of non-spherical shape approximation on DEM simulation accuracy by multi-sphere method. *Powder Technol.* **2018**, *332*, 265–278. [CrossRef]
64. Suhr, B.; Six, K. Simple particle shapes for DEM simulations of railway ballast: Influence of shape descriptors on packing behaviour. *Granul. Matter* **2020**, *22*, 43. [CrossRef]

65. Liu, G.; Jing, G.; Ding, D.; Shi, X. Micro-Analysis of Ballast Angularity Breakage and Evolution by Monotonic Triaxial Tests. In Proceedings of the 7th International Symposium on Environmental Vibration and Transportation Geodynamics, Hangzhou, China, 28–30 October 2016.
66. Hossain, Z.; Indraratna, B.; Darve, F.; Thakur, P.K. DEM analysis of angular ballast breakage under cyclic loading. *Géoméch. Geoengin.* **2007**, *2*, 175–181. [CrossRef]
67. Wensrich, C.; Katterfeld, A. Rolling friction as a technique for modelling particle shape in DEM. *Powder Technol.* **2012**, *217*, 409–417. [CrossRef]
68. Tohidifar, H.; Jafari, M.K.; Moosavi, M. Downwards force–displacement response of buried pipelines during dip–slip faulting in sandy soil. *Can. Geotech. J.* **2021**, *58*, 377–397. [CrossRef]
69. Khanal, M.; Elmouttie, M.; Poulsen, B.; Olsson, A.; Adhikary, D. Effect of Loading Rate on Sand Pile Failure: 2D DEM Simulation. *Geotech. Geol. Eng.* **2017**, *35*, 889–896. [CrossRef]
70. Coetzee, C.; Els, D. Calibration of granular material parameters for DEM modelling and numerical verification by blade–granular material interaction. *J. Terramech.* **2009**, *46*, 15–26. [CrossRef]
71. Peng, Y.; Yin, Z.-Y.; Zhou, C.; Ding, X. Micromechanical analysis of capillary suction effect on bearing capacity of unsaturated fine granular foundation soil using coupled CFD-DEM method. *Comput. Geotech.* **2023**, *153*, 105092. [CrossRef]
72. Mondal, S.; Disfani, M.; Narsilio, G.A. Battered minipiles in fine-grained soils: Soil-structure interaction. *Comput. Geotech.* **2022**, *147*, 104762. [CrossRef]
73. Brunelli, A.; de Silva, F.; Piro, A.; Parisi, F.; Sica, S.; Silvestri, F.; Cattari, S. Numerical simulation of the seismic response and soil–structure interaction for a monitored masonry school building damaged by the 2016 Central Italy earthquake. *Bull. Earthq. Eng.* **2021**, *19*, 1181–1211. [CrossRef]
74. Qian, J.-G.; Li, W.-Y.; Yin, Z.-Y.; Yang, Y. Influences of buried depth and grain size distribution on seepage erosion in granular soils around tunnel by coupled CFD-DEM approach. *Transp. Geotech.* **2021**, *29*, 100574. [CrossRef]
75. Nitka, M.; Combe, G.; Dascalu, C.; Desrues, J. Two-scale modeling of granular materials: A DEM-FEM approach. *Granul. Matter* **2011**, *13*, 277–281. [CrossRef]
76. Nguyen, T.K.; Combe, G.; Caillerie, D.; Desrues, J. FEM × DEM modelling of cohesive granular materials: Numerical homogenisation and multi-scale simulations. *Acta Geophys.* **2014**, *62*, 1109–1126. [CrossRef]
77. Wang, X.; Yin, Z.-Y.; Su, D.; Wu, X.; Zhao, J. A novel approach of random packing generation of complex-shaped 3D particles with controllable sizes and shapes. *Acta Geotech.* **2021**, *17*, 355–376. [CrossRef]
78. Zhang, J.; Shao, X.; Wang, X.; Hu, F.; Zuo, H. Discrete element simulation of crushing behavior of calcareous sands during pile jacking. *Rock Soil Mech.* **2015**, *36*, 272–278.
79. Qu, T.; Wang, M.; Feng, Y. Applicability of discrete element method with spherical and clumped particles for constitutive study of granular materials. *J. Rock Mech. Geotech. Eng.* **2022**, *14*, 240–251. [CrossRef]
80. Wang, X.; Yin, Z.Y.; Xiong, H.; Su, D.; Feng, Y.T. A spherical-harmonic-based approach to discrete element modeling of 3D irregular particles. *Int. J. Numer. Methods Eng.* **2021**, *122*, 5626–5655. [CrossRef]
81. Du, L.; Liu, X.; Han, Y.; Deng, Z. Generation of irregular particle packing with prescribed statistical distribution, spatial arrangement, and volume fraction. *J. Rock Mech. Geotech. Eng.* **2022**, *15*, 375–394. [CrossRef]
82. Yin, Z.-Y.; Wang, P. Micro-mechanical analysis of caisson foundation in sand using DEM: Particle shape effect. *Appl. Ocean Res.* **2021**, *111*, 102630. [CrossRef]
83. Boot, J.; Naqvi, M.; Gumbel, J. A new method for the structural design of flexible liners for gravity pipes of egg-shaped cross section: Theoretical considerations and formulation of the problem. *Thin-Walled Struct.* **2014**, *85*, 411–418. [CrossRef]
84. Zhang, J.; Liang, Z.; Zhang, H.; Feng, D.; Xia, C. Failure analysis of directional crossing pipeline and design of a protective device. *Eng. Fail. Anal.* **2016**, *66*, 187–201. [CrossRef]
85. Li, Z.; Yu, Y.; Zhao, M.; Li, C.; Ma, J.; Su, O.; Xu, L. Buckling analysis of offshore buried pipeline under the combination of strike-slip faulting and pressure loading. *Ocean Eng.* **2022**, *265*, 112531. [CrossRef]
86. Cheuk, C.Y.; White, D.J.; Bolton, M.D. Uplift Mechanisms of Pipes Buried in Sand. *J. Geotech. Geoenviron. Eng.* **2008**, *134*, 154–163. [CrossRef]
87. Thusyanthan, N.; Mesmar, S.; Wang, J.; Haigh, S. Uplift Resistance of Buried Pipelines and DNV-RP-F110 Guidelines. In Proceedings of the Offshore Petroleum Technology (OPT), Amsterdam, The Netherlands, 24–25 February 2010.
88. Wu, J.; Kouretzis, G.; Suwal, L.; Ansari, Y.; Sloan, S.W. Shallow and deep failure mechanisms during uplift and lateral dragging of buried pipes in sand. *Can. Geotech. J.* **2020**, *57*, 1472–1483. [CrossRef]
89. Katinas, E.; Chotěborský, R.; Linda, M.; Kuře, J. Sensitivity analysis of the influence of particle dynamic friction, rolling resistance and volume/shear work ratio on wear loss and friction force using DEM model of dry sand rubber wheel test. *Tribol. Int.* **2021**, *156*, 106853. [CrossRef]
90. Wang, P.; Yin, Z.-Y.; Zhou, W.-H.; Chen, W.-B. Micro-mechanical analysis of soil–structure interface behavior under constant normal stiffness condition with DEM. *Acta Geotech.* **2021**, *17*, 2711–2733. [CrossRef]
91. Fu, Z.; Chen, S.; Liu, S. Discrete Element Simulations of Shallow Plate-Load Tests. *Int. J. Géoméch.* **2016**, *16*, 04015077. [CrossRef]

92. Hazeghian, M.; Soroush, A. DEM-aided study of shear band formation in dip-slip faulting through granular soils. *Comput. Geotech.* **2016**, *71*, 221–236. [CrossRef]
93. Zhou, C.; Qian, J.-G.; Yin, Z.-Y.; Xiong, H. Suffusion in gap-graded granular soils subjected to strain-controlled cyclic loading with coupled CFD-DEM method. *Transp. Geotech.* **2023**, *42*, 101098. [CrossRef]

Journal of
Marine Science
and Engineering

MDPI

Article

Very High Cycle Fatigue Life of Free-Spanning Subsea Pipeline Subjected to Vortex-Induced Vibrations

Qingyuan Song [1,2], Jun Liu [1,2] and Fuping Gao [1,2,*]

[1] Institute of Mechanics, Chinese Academy of Sciences, Beijing 100190, China; songqingyuan@imech.ac.cn (Q.S.); liujun@imech.ac.cn (J.L.)
[2] School of Engineering Science, University of Chinese Academy of Sciences, Beijing 100049, China
* Correspondence: fpgao@imech.ac.cn

Abstract: Free-spanning subsea pipelines subjected to vortex-induced vibrations (VIVs) are particularly prone to fatigue failure. Existing flume observations indicated that the VIVs of a near-bed cylinder may be triggered effectively in moderate shear flows. This may imply that the vibration cycles of a spanned pipeline could be up to tens of millions. As such, very high cycle fatigue (VHCF) can occur during engineering service. The free span length is a key parameter for determining the structural natural frequency and the corresponding reduced velocity (*Vr*). On the basis of the dimensionless vibration amplitude *A/D–Vr* curve and the recommended S-N curves for high-strength steel pipelines with cathodic protection under seawater environments, a prediction method is proposed for the fatigue life of a free span undergoing VIVs. A parametric study is then performed to evaluate the fatigue life of the spanned pipelines with a focus on VHCF. It is indicated that the minimum fatigue life emerges at certain flow with a moderate velocity for a given span length. With a further decrease or increase in the flow velocity, the fatigue life would be enhanced correspondingly, which could be within the VHCF regime. Such nonlinear variation of the fatigue life with the span length and the flow velocity is attributed to being involved in various VIV branches of the *A/D–Vr* curve.

Keywords: free-spanning pipeline; vortex-induced vibration; very high cycle fatigue; pipeline–seabed interaction; high-strength steel; seawater environment

1. Introduction

Subsea pipelines have been widely employed for transporting oil and gas from underwater wells to offshore platforms and onshore processing facilities. The pipeline free-span scenario could be permanent when generated by seabed unevenness, artificial supports for pipeline crossing, or with short-to-long-term evolution by seabed mobility and local scour [1]. Under the action of ocean currents, a partially embedded pipeline may also be suspended because of tunnel erosion [2,3]. As such, free spans are frequently encountered or even inevitable, especially for a long laid pipeline for oil and gas exploitation in deep waters.

In subsea environments, a free-spanning pipeline may suffer from vortex-induced vibrations (VIVs), as illustrated in Figure 1. For an elastically mounted cylinder, the transverse VIVs can be effectively triggered when the reduced velocity (*Vr*) approaches a critical value $Vr_{\rm cr} \approx 4.0$ under wall-free conditions [4]. Once the VIVs are initiated, the frequency of vortex shedding and that of the structural oscillation would collapse into an identical value, which is known as the "lock-in" or "synchronization" phenomenon. The maximum amplitude of the vibration principally depends on the combined mass damping parameter $K_{\rm s}$ [5]. In engineering practice, subsea pipelines are usually in the proximity of the seabed. For such a near-bed cylinder, the flow fields around the cylinder would be altered due to bed proximity effects, which would subsequently affect the VIV responses. Recently, Liu and Gao [6] investigated the bed proximity effects in triggering transverse vibration of a cylinder under low K_s conditions. It was found that as the cylinder became closer to the bed boundary (i.e., with a decrease in the gap-to-diameter ratio e/D from 2.0 to 0.1), the values of Vr_{cr} significantly decreased (from 4.0 to 2.85), which was accompanied by a decrease in the vibration amplitudes. Previous flume observations by Gao et al. [7] indicated that once VIVs are triggered, the local scour around the vibrating pipe will be greatly enhanced due to the cyclic vibrations (e.g., the scour depth could be up to approximately 1.2D, which can be regarded as the wall-free condition for VIV responses).

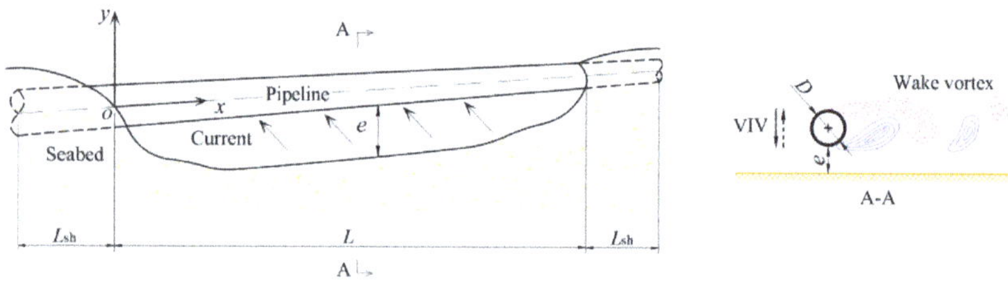

Figure 1. Schematic diagram of a spanning pipeline undergoing VIVs (not in scale).

In engineering practice, fatigue may take place under cyclic loading after a substantial period of service. The fatigue limit stress is generally defined as the highest stress at which specimens do not fail after 10^7 loading cycles. However, the existing studies have extended this knowledge by warning that fatigue failure may occur at lifetimes greater than $N = 10^7$ at a stress which is lower than the conventional fatigue limit [8]. Generally speaking, the fatigue of metallic materials can be divided into three regions: low cycle fatigue (LCF), high cycle fatigue (HCF) and very high cycle fatigue (VHCF), in which the number of cycles to failure is beyond 10^7. The fatigue behavior in the VHCF region is quite different compared with that in the conventional HCF region [9–13]. For instance, the fatigue behaviors of high-strength steels in the VHCF region show crack initiation at the inclusion site, while in the HCF region, the cracks are located preferentially at the surface [8]. The influence of hydrogen was found to be crucial for eliminating the fatigue limit in the extremely high cycle fatigue region [8,9].

For offshore structures subjected to typical wave and wind loading, significant fatigue failure may occur at $N \geq 10^7$ cycles (i.e., in the VHCF region) [14]. It is also extremely common that turbine blades experience more than $N = 10^7$ stress cycles by vibration. Nevertheless, fatigue test data are normally derived for a number of cycles less than 10^7, including those for subsea pipeline steels. For the scenario of free-spanning pipelines, the fatigue by VIVs is of great concern in engineering design [15]. The allowable span length of a suspended pipeline was investigated by considering VIV hysteresis effects [16]. The natural frequency of free spans was assessed by using a finite element model [17], and the random ocean current-induced fatigue life of free spans was further predicted on the basis of the Palmgren–Miner cumulative damage rule. Nevertheless, the correlation between the free span length and the corresponding service life has not been understood well, especially in the VHCF region.

In this study, the fatigue life of subsea-spanning pipeline undergoing VIVs is investigated, with a focus on VHCF. Based on the existing flume results for transverse VIV responses, a dimensionless vibration amplitude A/D–Vr curve with four branches is constructed in Section 2.1, and the maximum stress along the free span is then derived in Section 2.2. The bilinear stress range and number of cycles to failure (S-N) curves are recommended by considering the VHCF of high-strength steel pipelines with cathodic protection under seawater environments (see Section 3). Based on the constructed A/D–Vr curve and the recommended S-N curves, a flow chart is proposed in Section 4.1 to predict the fatigue life of a spanning pipeline undergoing VIVs. A parametric study is further performed in Section 4.2 to evaluate the VHCF life of the spanning pipeline. The correlations between the fatigue life and the flow velocity or the free span length are established. Finally, our concluding remarks are given in Section 5.

2. Vibration Amplitude and Stress Distribution along the Spanning Pipeline

2.1. Amplitude of VIVs

The vibration amplitude is correlated with the reduced velocity Vr, which is defined as

$$Vr = \frac{U}{f_n D} \tag{1}$$

where U is the characteristic velocity of the flow perpendicular to the cylinder's axis; f_n is the structural natural frequency; and D is the outer diameter of the cylinder [4]. The natural frequency (f_n) of a free-spanning pipeline is a key factor in identifying the reduced velocity (see Equation (1)). The natural frequency or the first eigen frequency of a free-spanning pipeline is influenced by many factors (such as the span length, elastic modulus of the pipe steel, boundary conditions, effective mass, and moment of inertia of the pipe), which can be evaluated with the following equation [15]:

$$f_n \approx \frac{C_1}{L_{\mathrm{eff}}^2}\left(\frac{(1+F_{\mathrm{CS}})EI}{m_e}\left[1+\frac{S_{\mathrm{eff}}}{P_{\mathrm{cr}}}+C_3\left(\frac{\delta}{D}\right)^2\right]\right)^{\frac{1}{2}} \tag{2}$$

where L_{eff} is the effective span length; m_e is the effective mass of the pipeline; E is the modulus of elasticity of the steel pipeline; I is the inertia moment of the steel pipeline; S_{eff} is the effective axial force (negative in compression); P_{cr} is the critical buckling load (positive sign); δ is the static deflection; C_1 and C_3 are the end boundary condition coefficients, which depend on the support conditions of the free-span boundary; F_{CS} denotes the stiffness of the concrete coating relative to the steel pipe stiffness; (1 + F_{CS}) is the stress concentration factor due to the concrete coating and localized bending; $F_{\mathrm{CS}} = k_c\left(\frac{(EI)_{\mathrm{conc}}}{EI}\right)^{0.75}$; $(EI)_{\mathrm{conc}}$ denotes the stiffness of concrete coating; and the empirical constant k_c accounts for the deformation or slippage in the corrosion coating and the cracking of the concrete coating. With reference to DNV GL [15], the effective span length L_{eff} can be calculated as follows:

$$\frac{L_{\mathrm{eff}}}{L} = \begin{cases} \frac{4.73}{-0.066\beta^2+1.02\beta+0.63} & \beta \geq 2.7 \\ \frac{4.73}{0.036\beta^2+0.61\beta+1.0} & \beta < 2.7 \end{cases} \tag{3}$$

In Equation (3), β is the non-dimensional soil stiffness (i.e., $\beta = \log_{10}\left[KL^4/((1+F_{\mathrm{CS}})EI)\right]$), where K is the soil stiffness (vertical or horizontal and static or dynamic). The values of β are generally in the range of $2.0 < \beta < 8.0$. For the case of transverse VIV responses, the dynamic vertical stiffness (K_{vd}) can be used (i.e., $K = K_{\mathrm{vd}}$), and the following expression can be applied for determination of the dynamic vertical stiffness; that is, $K_{\mathrm{vd}} = 0.88G/(1-\nu)$, where G is the secant shear modulus of soil and ν is the Poisson ratio of the soil [18]. The basic rules for the ideal end conditions are as follows: (1) pinned-pinned, which is used for free spans where each end is allowed to rotate about the pipeline axis, and (2) fixed-fixed, which should be used for free spans which are fixed in place by some sort of anchor at both ends. The following values of C_1 in Equation (2) have been widely used for these typical end conditions: (1) C_1 = 1.57 for the pinned-pinned end condition and (2) C_1 = 3.56 for the fixed-fixed end condition (see DNV GL [15] and Guo et al. [19]). The effective mass per meter (m_e) is the sum of the mass of the steel pipe per meter (m_p), the mass of the content inside the pipe per meter (m_c), and the added mass per meter (m_a):

$$m_e = m_p + m_c + m_a \tag{4a}$$

$$m_a = C_A \frac{\pi \rho_{\mathrm{water}} D^2}{4} \tag{4b}$$

where C_A is the added mass coefficient (for a circular pipeline, C_A = 1.0) and ρ_{water} is the density of the seawater. If the effective axial force and the static deflection are not taken into account, then Equation (2) can be simplified as follows:

$$f_n \approx \frac{C_1}{L_{eff}^2}\left(\frac{(1+F_{CS})EI}{m_e}\right)^{\frac{1}{2}} \quad (5)$$

In the past few decades, the VIVs of an elastically mounted cylinder have been investigated by quite a few researchers (see the comprehensive reviews of, for example, Williamson and Govardhan [5], Sarpkaya [20], Bearman [21] and Wu et al. [22]). In this study, the VIV amplitudes of wall-free cylinders in the region of the low mass damping parameter K_s (broadly $K_s < 0.05$; see Williamson and Govardhan [5]) were collected from some benchmark flume observations. The whole curve for the nonlinear variation of *A/D* with *Vr* was constructed as shown in Figure 2. Note that Wang et al. [23] experimentally studied the VIV of a neutrally buoyant circular cylinder in a recirculating open channel with the mass damping parameter $K_s \approx 3.46 \times 10^{-2}$. In the VIV experiment by de Oliveira Barbosa et al. [24], a freely vibrating cylinder was exposed to currents and placed near a plane boundary parallel to the cylinder axis, and $K_s \approx 1.45 \times 10^{-2}$. Daneshvar and Morton [25] studied the VIV of an elastically mounted circular cylinder in a water tunnel facility, where $K_s \approx 8.32 \times 10^{-3}$. Liu and Gao [6] recently investigated the wall proximity effects on the triggering of transverse VIVs of a circular cylinder with an accessorial low structural damping VIV simulation device, where $K_s \approx 3.14 \times 10^{-2}$. As indicated in Figure 2, in such a low mass damping system, four distinct branches of VIV responses can be identified and marked with various colors, namely the initial excitation branch, the upper branch, the lower branch, and the desynchronization branch (see also Khalak and Williamson [26]). In the initial excitation branch ($3.5 \leq Vr < 5.0$; see Figure 2), the cylinder starts to vibrate with a relatively small amplitude (note: transverse VIVs do not occur at $Vr < 3.5$); in the upper branch ($5.0 \leq Vr < 6.7$), the cylinder vibrates with a large amplitude; in the lower branch ($6.7 \leq Vr < 12$), the cylinder vibrates with a moderate amplitude; and in the desynchronization branch ($Vr \geq 12$), the vibration amplitude of a wall-free circular cylinder is further reduced. There exists a certain dispersion among the experimental data (e.g., at the upper branch; see Figure 2) which could be attributed to involving various values of K_s.

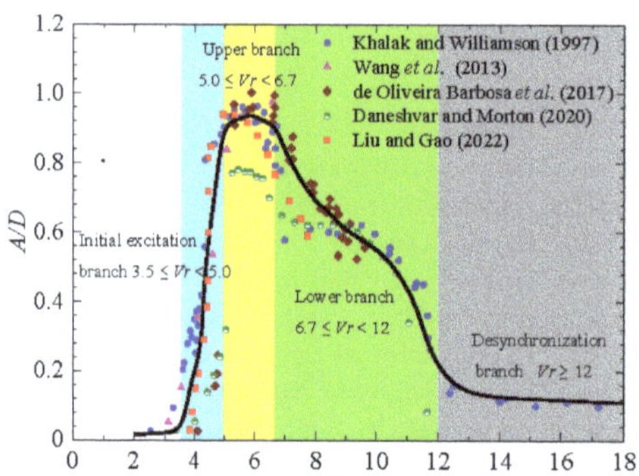

Figure 2. *A/D–Vr* curve for a wall-free circular cylinder with low mass damping [6,23–26].

The above analyses indicate that once the pipeline parameters and the flow velocity (*U*) are given, the corresponding reduced velocity (*Vr*) can be calculated with Equation (1) and Equation (5), and the vibration amplitude can be obtained with reference to Figure 2.

2.2. Stress Distribution along the Free Span

It is supposed that the ocean current is perpendicular to the axis of the free spans, as illustrated in Figure 1. As stated above, the shoulders for a free span can be simplified as the pinned-pinned or fixed-fixed end condition. The maximum deflection in the middle of the span ($w_{\max}$) under the action of a uniform load (q) can be calculated with Equation (6a) for the pinned-pinned end condition and with Equation (6b) for the fixed-fixed end condition [27]:

$$w_{\max} = \frac{5qL_{\text{eff}}^4}{384EI} \tag{6a}$$

$$w_{\max} = \frac{qL_{\text{eff}}^4}{384EI} \tag{6b}$$

where L_{eff} is the effective span length. But for the VIV responses, if the maximum deflection $w_{\max}$ is given ($w_{\max} = A$), then the corresponding maximum bending moment ($M_{\max}$) can be calculated with Equation (7a) for the pinned-pinned end condition and with Equation (7b) for the fixed-fixed end condition [27]:

$$M_{\max} = \frac{48EI}{5L_{\text{eff}}^2} w_{\max} \tag{7a}$$

$$M_{\max} = -\frac{32EI}{L_{\text{eff}}^2} w_{\max} \tag{7b}$$

Figure 3b,c shows the distributions of the bending moment along the free span under these two end conditions. It is indicated that the bending moment reaches its peak value in the middle of the span under the pinned-pinned end condition. But under the fixed-fixed end condition, the two ends connected with the span shoulders undertake the maximum bending moment. Consequently, the maximum normal stresses ($\sigma_{\max}$) correlated with the maximum deflections ($w_{\max}$) can be expressed as

$$\sigma_{\max} = \frac{24DE}{5L_{\text{eff}}^2} w_{\max} \tag{8a}$$

$$\sigma_{\max} = \frac{16DE}{L_{\text{eff}}^2} w_{\max} \tag{8b}$$

for the pinned-pinned and fixed-fixed end conditions, respectively.

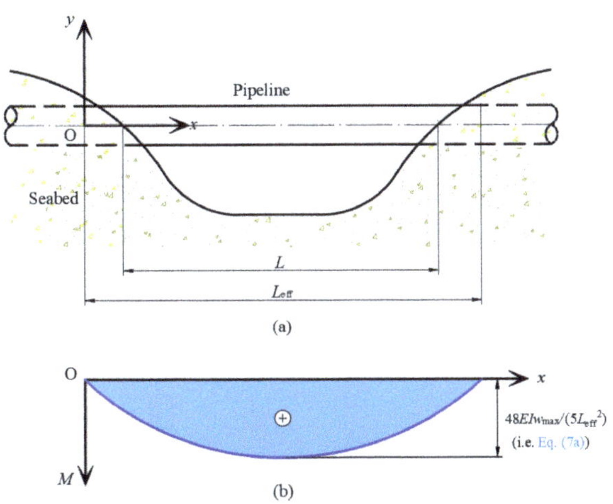

Figure 3. *Cont.*

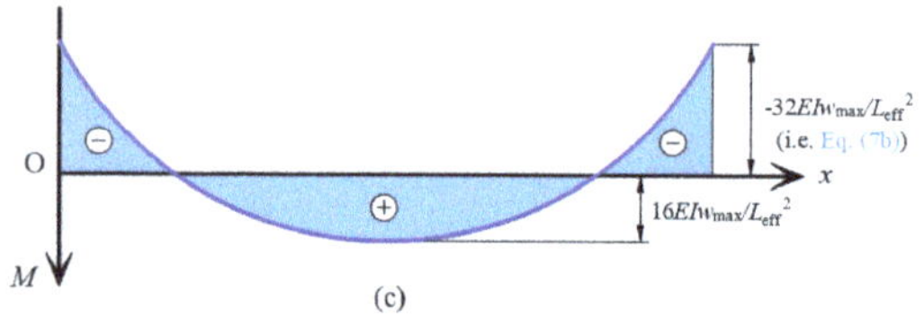

(c)

Figure 3. Distributions of bending moment along (**a**) the spanned pipeline under (**b**) the pinned-pinned end condition and (**c**) the fixed-fixed end condition.

3. Recommended S-N Curves

The fatigue design is generally based on the use of S-N curves, which were obtained from fatigue tests. The number of cycles to failure (N) at a stress range ($\Delta\sigma$) is defined by an S-N curve. As mentioned earlier, significant fatigue damage to offshore structures usually occurs at $N \geq 10^7$ cycles (VHCF region). For the VHCF mechanism of metallic materials, a few studies have already been carried out (e.g., Murakami et al. [9], Shiozawa et al. [10], Sakai et al. [28], and Hong et al. [29]). It has been generally recognized that cracks usually initiate from the surface of materials in the LCF and HCF regions, while the cracks preferably emerge in the interior of the materials with a fish eye morphology in the VHCF region. Cracks may initiate from several types of internal defects, which mainly include nonmetallic inclusions, process-related defects, second-phase particles, and microstructural inhomogeneities [30–33]. As illustrated in Figure 4a,b, an inclusion was found at the center of the fish eye zone. The rough area surrounding the inclusion within the fish eye zone, named the fine granular area (FGA), is the crack initiation region, which is critical to VHCF occurrence. The initiation and early growth of cracks could be attributed to the grain refinement caused by the dislocation interaction over a number of cyclic loads followed by the formation of micro-cracks, but the micro-cracks could also form irrespective of the grain refinement during cyclic loading [13].

Submarine pipelines may suffer from corrosion in engineering practices. A distinctive difference between corrosion fatigue and inert environment fatigue is that there is no safe stress range at which metallic materials have infinite fatigue life during engineering service in corrosive environments. The fatigue strength of metallic materials in corrosive environments decreases dramatically for the synergistic interactions between mechanical loading and the environmental effect. Cracks usually occur on the surfaces of metallic materials in very high cycle corrosion fatigue (VHC–CF) [34–37]. The surfaces of metallic materials may be degraded because of corrosion or erosion-corrosion, leading to the formation of pits, grooves, cracks, or rougher surfaces [38]. As illustrated in Figure 4c,d, surface corrosion pits are the origins of cracks for specimens tested in a seawater environment [36,37]. From the microscopic point of view, the corrosive media's attack on the material's surfaces can intensify the formation of microstructurally small cracks (MSCs), contributing to the acceleration of crack initiation and growth. The decrease in metal's ductility or fracture toughness caused by atomic hydrogen is called hydrogen embrittlement (HE), which can make cracks spread quickly with minimal macroscopic signs. In brief, corrosion fatigue is a combination of mechanical and environmental phenomena.

For the sake of service safety, an unfavorable scenario was considered in this case study, such as the weld joint happening to be installed in the middle of the span for the pinned-pinned boundary condition and at one end of the span for the fixed-fixed boundary condition. The service environment was considered to be in the seawater and with cathodic protection (CP).

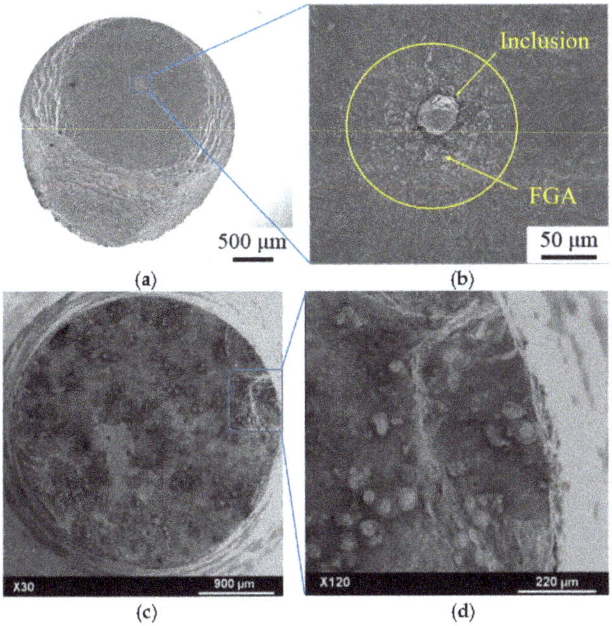

Figure 4. (**a**) The fracture surface of high-strength steel in an air environment. (**b**) A close-up of the crack initiation region in (**a**) [39]. (**c**) Specimen tested under seawater flow; (**d**) A close-up of the crack initiation regions in (**c**) [36].

The fatigue test data were normally derived for a number of cycles less than 10^7. As such, how to extrapolate the fatigue test data into the VHCF region was important for achieving a reliable assessment procedure [14]. Bilinear S-N curves (i.e., F3 and C1 (see Figure 5)) were recommended to define the number of cycles to failure at a stress range $\Delta\sigma$ (=$2\sigma_{max}$). Note that the F3 curve is for the circumferential butt weld made from one side without a backing bar, and the C1 curve is for the circumferential butt welds made from one side which were machined or ground flush to remove defects and weld overfill [14]. Meanwhile, the existing fatigue test data for the high-strength steels under various environments are also provided in Figure 5 for reference. For the pipe welds with different hot spot geometries, the structural stress concentration factor (SCF) should be accounted for. The SCF is defined as the ratio of local stress (hot spot stress) to nominal stress. It is recommended that SCF = 1.61 for the S-N curve F3 and SCF = 1.0 for the S-N curve C1 [14]. As shown in Figure 5, the S-N curves in the VHCF region became more horizontal as the number of cycles to failure N became larger than 10^6. The recommended S-N curves can be expressed in the following form [14]:

$$\mathrm{Log}_{10}(N) = \mathrm{Log}_{10}(\overline{a}) - m\mathrm{Log}_{10}\left(\Delta\sigma\left(\frac{t}{t_{\mathrm{ref}}}\right)^{k}\right) \tag{9}$$

where m is the negative inverse slope of the S-N curve; $\mathrm{Log}_{10}(\overline{a})$ is the intercept of the axis; $\mathrm{Log}_{10}(N)$ is the reference wall thickness of the pipe for welded connections; k is the thickness exponent for the fatigue strength; t is the wall thickness of the pipe; and t_{ref} is the reference wall thickness of the pipe (0.025 m).

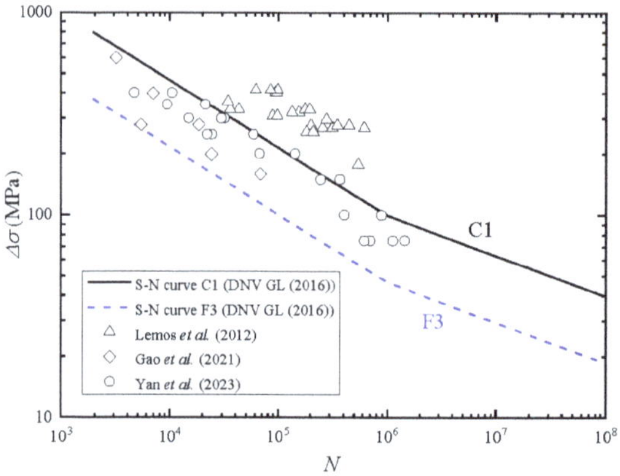

Figure 5. Recommended S-N curves for high-strength steel pipelines in seawater and with cathodic protection (adapted from DNV GL [14]) and the existing test data by Lemos et al. [40] for the saline environments saturated with carbon dioxide, Gao et al. [41] for NACE solution with saturated H_2S, and Yan et al. [42] for the saturated H_2S solution.

4. Fatigue Life Prediction and Discussions

4.1. A Flow Chart for Fatigue Life Prediction

A flow chart is proposed for the fatigue life prediction of a free-spanning pipeline undergoing VIVs, as illustrated in Figure 6. A brief description of the analysis procedure is given below:

(1) For given structural parameters of the spanned pipeline, the natural frequency (f_n) of the free span can first be evaluated with Equations (2)–(4a). The reduced velocity (Vr) is calculated with Equation (1) for a certain flow velocity (U).
(2) Once the value of Vr is obtained, the dimensionless vibration amplitude (A/D) can then be evaluated with reference to the recommended A/D–Vr curve (Figure 2). Subsequently, the maximum stress range $\Delta\sigma$ (= $2\sigma_{max}$) can be calculated with Equation (8a) or Equation (8b), where the maximum deflection count (w_{max}) is substituted by the vibration amplitude A.
(3) By referring to the recommended S-N curves (Figure 5), the fatigue life in terms of the vibration cycles (N) of the spanned pipeline can finally be obtained with Equation (9).

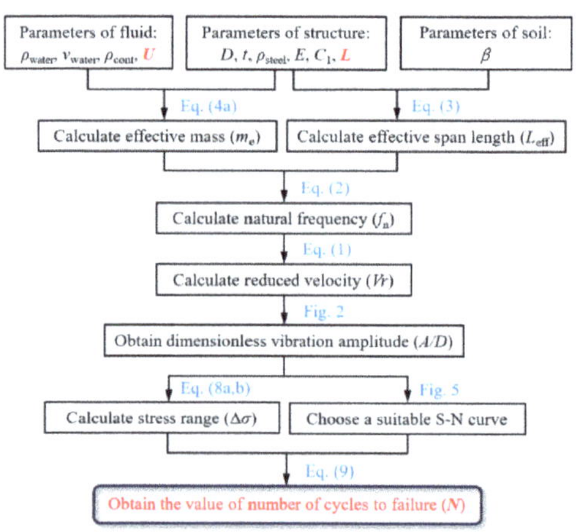

Figure 6. Flow chart for the fatigue life prediction of a free-spanning pipeline undergoing VIVs.

4.2. Parametric Study and Discussions

A subsea pipeline for gas transportation with an outer diameter of 20 inches (0.508 m) was considered for a parametric study. The parameters of the steel pipe, the seawater, the gas inside the pipe, and the soil stiffness are given in Table 1, and the values of the key parameters for the S-N curves are listed in Table 2. For the extreme scenario of large free spans, the span length values (L) were examined in a range from 40 m to 130 m. The correlations between the fatigue life (N) and the flow velocity (U) or span length (L) could be obtained by following the flow chart shown in Figure 6. Table 3 shows the calculation process for the free-spanning pipeline with various values for the span length under the pinned-pinned and the fixed-fixed end conditions. Note that a moderate flow velocity U (=0.57 m/s) was considered as an example.

Table 1. Input data for the parameters of the pipe spans.

Parameters	Values	Units
Outer diameter of the steel pipe (D)	0.508	m
Wall thickness of the steel pipe (t)	0.0379	m
Density of the steel (ρ_{steel})	7.870×10^3	kg/m^3
Density of the gas inside the pipe (ρ_{cont})	0.200×10^3	kg/m^3
Density of the seawater (ρ_{water})	1.024×10^3	kg/m^3
Kinematic viscosity of the seawater (υ_{water})	1.565×10^{-6}	m^2/s
Elastic modulus of the steel (E)	2.10×10^{11}	Pa
Moment of inertia of the steel pipe (I)	1.56×10^{-3}	m^4
Non-dimensional soil stiffness (β)	4.0	
Reference wall thickness of the pipe (t_{ref})	0.025	m

Table 2. Parameters for the S-N curves.

S-N Curve	$N \le 10^6$		$N > 10^6$		k
	m_1	$\mathrm{Log}_{10}\left(\bar{a}_1\right)$	m_2	$\mathrm{Log}_{10}\left(\bar{a}_2\right)$	
C1	3	12.049	5	16.081	0.10
F3	3	11.146	5	14.576	0.25

Figure 7a,b illustrates the correlations of the flow velocity (U) and the reduced velocity (Vr) for four selected span length values (i.e., L = 50 m, 70 m, 90 m, and 110 m) under the pinned-pinned and fixed-fixed end conditions, respectively. Note that the end condition of the free span had much influence on the natural frequency (f_n) of the free-spanning pipeline, which would further alter the values of Vr (see Equation (1)). As mentioned earlier, the four VIV branches are marked with various colors, which are related to various ranges of Vr (see Figure 2). A shorter free span L would increase the natural frequency f_n (Equation (5)), and thus a higher flow velocity U was required to keep the reduced velocity Vr unchanged (Equation (1)). As indicated in Figure 7, for a given flow velocity U (e.g., U = 0.57 m/s; see the horizontal arrow), the corresponding reduced velocity Vr increased with the increase in L.

Table 3. Calculation process for the fatigue life of a free-spanning pipeline with various span length values (U = 0.57 m/s).

L (m)	L_{eff} (m)	C1		f_n (Hz)		Vr		A/D		w (m)		σ_{max} (MPa)		N (Predicted with S-N Curve C1)		N (Predicted with S-N Curve F3)	
		P-P	F-F	P-P	F-F	P-P	F-F	P-P	F-F	P-P	F-F	P-P	F-F	P-P	F-F	P-P	F-F
40	51.78	1.57	3.56	0.407	0.916	2.76	1.23	0.019	0	0.010	0	1.82	0	1.52×10^{13}		3.49×10^{11}	
45	58.25	1.57	3.56	0.322	0.724	3.49	1.55	0.025	0	0.013	0	1.95	0	1.07×10^{13}		2.46×10^{11}	
50	64.72	1.57	3.56	0.261	0.586	4.31	1.91	0.410	0	0.208	0	25.44	0	2.87×10^{7}		7.78×10^{5}	
55	71.19	1.57	3.56	0.215	0.484	5.21	2.32	0.922	0.017	0.468	0.008	47.30	2.84	1.29×10^{6}	1.67×10^{12}	1.21×10^{5}	3.81×10^{10}
60	77.66	1.57	3.56	0.181	0.407	6.20	2.76	0.926	0.019	0.471	0.010	39.95	2.70	3.01×10^{6}	2.13×10^{12}	2.01×10^{5}	4.88×10^{10}
65	84.14	1.57	3.56	0.154	0.347	7.28	3.24	0.788	0.023	0.400	0.012	28.95	2.77	1.50×10^{7}	1.86×10^{12}	5.28×10^{5}	4.26×10^{10}
70	90.61	1.57	3.56	0.133	0.299	8.44	3.75	0.658	0.103	0.334	0.053	20.86	10.92	7.74×10^{7}	1.97×10^{9}	1.77×10^{6}	4.51×10^{7}
75	97.08	1.57	3.56	0.116	0.261	9.69	4.31	0.571	0.413	0.290	0.210	15.75	37.97	3.16×10^{8}	3.88×10^{6}	7.22×10^{6}	2.34×10^{5}
80	103.55	1.57	3.56	0.102	0.229	11.02	4.90	0.438	0.871	0.223	0.443	10.63	70.46	2.26×10^{9}	3.53×10^{5}	5.16×10^{7}	3.66×10^{4}
85	110.02	1.57	3.56	0.090	0.203	12.44	5.53	0.157	0.928	0.080	0.472	3.37	66.49	7.01×10^{11}	4.20×10^{5}	1.60×10^{10}	4.36×10^{4}
90	116.50	1.57	3.56	0.080	0.181	13.95	6.20	0.124	0.926	0.063	0.471	2.37	59.17	4.06×10^{12}	5.96×10^{5}	9.28×10^{10}	6.18×10^{4}
95	122.97	1.57	3.56	0.072	0.162	15.54	6.91	0.120	0.857	0.061	0.435	2.06	49.12	8.24×10^{12}	1.07×10^{6}	1.88×10^{11}	1.08×10^{5}
100	129.44	1.57	3.56	0.065	0.147	17.22	7.66	0.111	0.727	0.057	0.369	1.73	37.61	1.98×10^{13}	4.07×10^{6}	4.53×10^{11}	2.41×10^{5}
105	135.91	1.57	3.56	0.059	0.133	18.99	8.44		0.658		0.334		30.89		1.09×10^{7}		4.34×10^{5}
110	142.38	1.57	3.56	0.054	0.121	20.84	9.27		0.592		0.301		25.34		2.93×10^{7}		7.87×10^{5}
115	148.86	1.57	3.56	0.049	0.111	22.78	10.13		0.548		0.278		21.44		6.75×10^{7}		1.54×10^{6}
120	155.33	1.57	3.56	0.045	0.102	24.80	11.03		0.437		0.222		15.72		3.19×10^{8}		7.30×10^{6}
125	161.80	1.57	3.56	0.042	0.094	26.91	11.96		0.225		0.114		7.45		1.34×10^{10}		3.05×10^{8}
130	168.27	1.57	3.56	0.039	0.087	29.11	12.94		0.142		0.072		4.35		1.97×10^{11}		4.52×10^{9}

Note: "P-P" is the abbreviation for the "pinned-pinned" end condition, and "F-F" is the abbreviation for the "fixed-fixed" end condition.

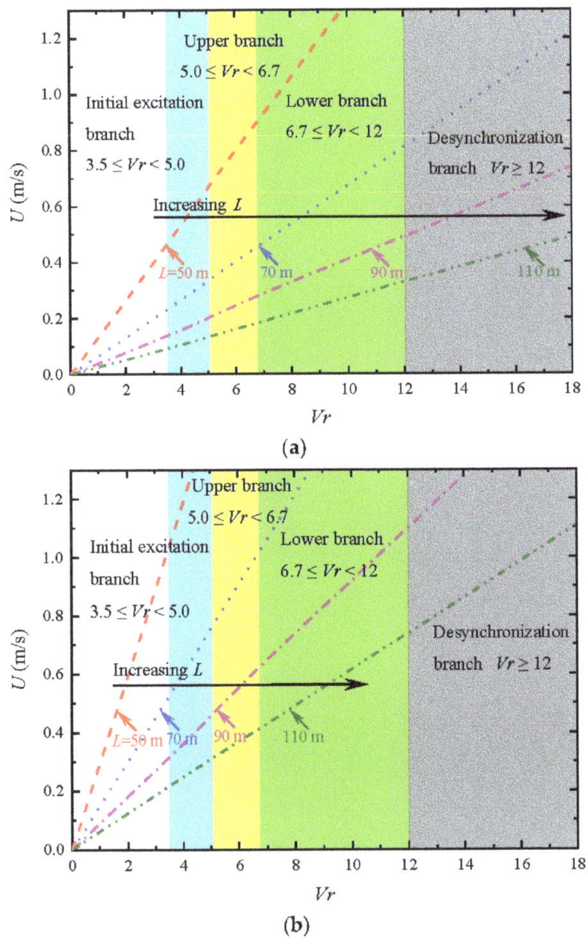

Figure 7. Illustration for VIV branch variations with flow velocities (U) for four selected span length values (L = 50 m, 70 m, 90 m, and 110 m) under (**a**) the pinned-pinned end condition and (**b**) the fixed-fixed end condition. (Input data for the parameters of the pipe spans are given in Table 1).

The variations in the predicted fatigue life (N) for two recommended S-N curves (see Figure 5) with a span length L at a given flow velocity U = 0.57 m/s are presented in Figure 8. Both the pinned-pinned and fixed-fixed end conditions were taken into account. It is indicated that the relationships between the fatigue life N and the span length L under a given flow velocity U were essentially nonlinear. As depicted in Figure 8, for $45 \leq L < 54$ (m) under the pinned-pinned end condition or $68 \leq L < 81$ (m) under the fixed-fixed end condition, the free span vibrated in the initial excitation branch (marked in blue; see Figure 8) such that the fatigue life N decreased with an increase in the span length L. For the span length $54 \leq L < 62$ (m) (pinned-pinned) or $81 \leq L < 94$ (m) (fixed-fixed), the free span vibrated in the upper branch (yellow) with larger amplitudes such that the corresponding fatigue life first became shorter to a minimum value and then increased with the increase in L. But for $62 \leq L < 83$ (m) (pinned-pinned) or $94 \leq L < 125$ (m) (fixed-fixed), the free span vibrated in the lower branch (green) (i.e., the vibration amplitudes were reduced). Therefore, the fatigue life became longer with an increasing L, and for $L \geq 83$ (m) (pinned-pinned) or $L \geq 125$ (m) (fixed-fixed), the free span vibrated in the desynchronization branch (gray) (i.e., the vibration amplitude continued to decrease with an increasing L) such that the fatigue life became much larger or even $N > 10^8$. From Figure 8a,b, it can be recognized that the critical values of the span length to differentiate the four VIV branches under the pinned-pinned end condition were much smaller than those under the fixed-fixed end condition. This was due to the fact that for a specified value of L, the natural frequency f_n for the pinned-pinned end condition was much smaller than that for the fixed-fixed end condition. As such, to keep the natural frequency f_n unchanged (or the reduced velocity Vr unchanged for a certain flow velocity U; see Equation (1)), a shorter span length L under the pinned-pinned end condition was required than that under the fixed-fixed end condition.

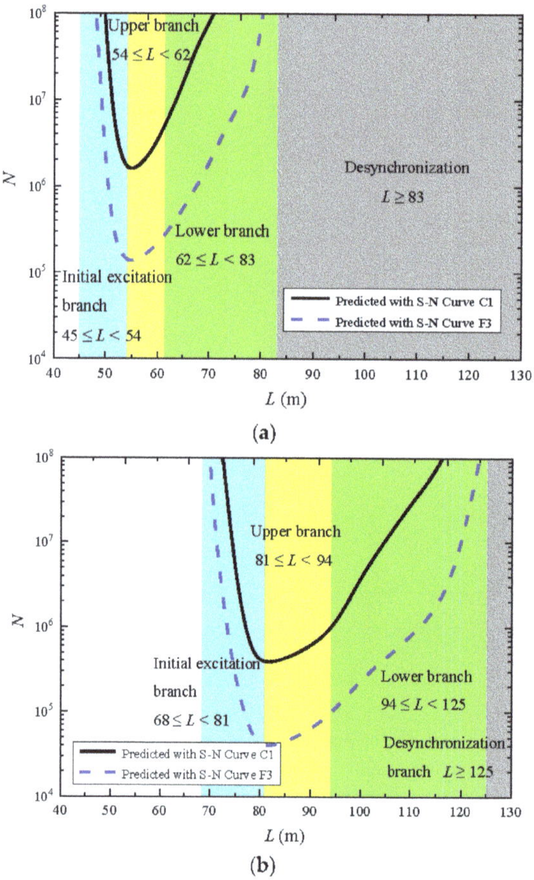

Figure 8. Variations in number of cycles to failure (N) with the span length (L) for the specified flow velocity value (U = 0.57 m/s): (**a**) pinned-pinned end condition and (**b**) fixed-fixed end condition.

The effect of the flow velocity (U) on the fatigue life (N) was further investigated. The variations in the fatigue life N with the flow velocity U for three span length values L = 70 m, 90 m, and 110 m are shown in Figure 9a–c, respectively. If the free span vibrated in the initial excitation branch ($3.5 \leq Vr < 5.0$) (i.e., $0.53 \leq U < 0.76$ (m/s) for L = 70 m (Figure 9a), $0.32 \leq U < 0.46$ (m/s) for L = 90 m (Figure 9b), and $0.22 \leq U < 0.31$ (m/s) for L = 110 m (Figure 9c)), then the fatigue life N decreased with the increase in the flow velocity U. If the free span vibrated in the upper branch with large amplitudes ($5.0 \leq Vr < 6.7$) (i.e., $0.76 \leq U < 1.02$ (m/s) for L = 70 m (Figure 9a), $0.46 \leq U < 0.62$ (m/s) for L = 90 m (Figure 9b), and $0.31 \leq U < 0.41$ (m/s) for L = 110 m (Figure 9c)), then the fatigue life N was first reduced slightly to a minimum value and then gradually increases with an increasing

U. In the lower branch of VIVs ($6.7 \leq Vr < 12$) (i.e., $1.02 \leq U < 1.82$ (m/s) for $L = 70$ m (Figure 9a), $0.62 \leq U < 1.10$ (m/s) for $L = 90$ m (Figure 9b), and $0.41 \leq U < 0.74$ (m/s) for $L = 110$ m (Figure 9c)), the free span vibrated with moderate amplitudes, and the corresponding fatigue life N became longer with the increase in U. Similarly, in the desynchronization branch ($Vr \geq 12$) (i.e., $U > 1.10$ (m/s) for $L = 90$ m (Figure 9b) and $U > 0.74$ (m/s) for $L = 110$ m (Figure 9c)), the free span vibrated with much smaller amplitudes, and the fatigue life N further became longer with the increase in U. It should be noted that for $L = 70$ m (Figure 9a), the desynchronization branch of VIVs was not involved in the examined range of $U < 1.3$ m/s.

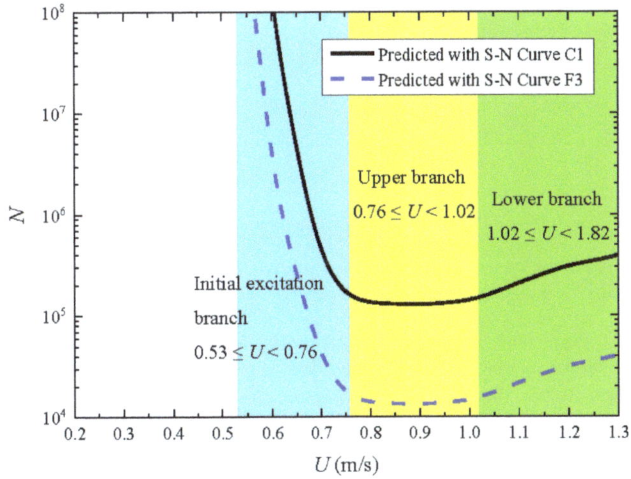

(**a**)

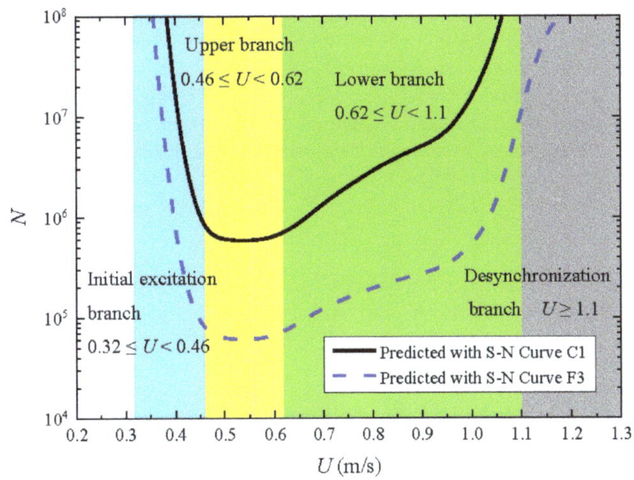

(**b**)

Figure 9. *Cont.*

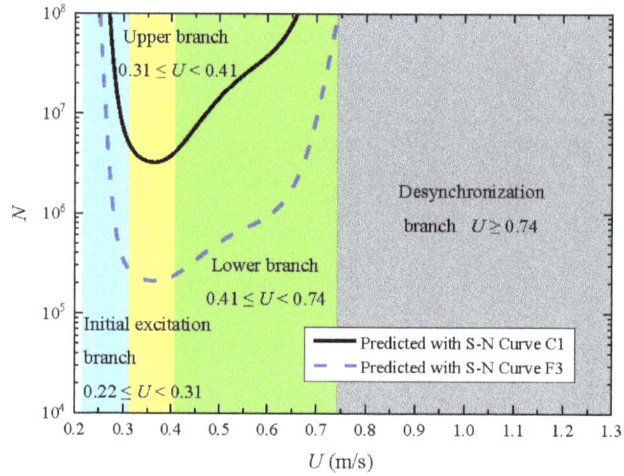

(**c**)

Figure 9. Variations in number of cycles to failure (N) with flow velocity (U) for three selected span length values: (**a**) L = 70 m; (**b**) L = 90 m; and (**c**) L = 110 m.

The above parametric study shows that the fatigue life (N) of the free spans of subsea pipelines is quite sensitive to the span length (L) and the flow velocity (U). By following the flow chart for fatigue life prediction (Section 4.1), the variations in N with U and L could be obtained, as shown in Figure 10. Note that the S-N curve C1 (see Figure 5) was adopted to predict the fatigue life N of the extreme scenario of large free spans. As illustrated by the two-dimensional contour map (Figure 10), for the span length range $70 \leq L \leq 110$ (m) and the flow velocity range $0.2 \leq U \leq 1.3$ (m/s), the minimum value of N ($\approx 1.28 \times 10^5$) emerged at $U \approx 0.9$ m/s for L = 70 m (see also Figure 9a). With a further decrease or increase in U for a specified span length value (L), the fatigue life N would be remarkedly enhanced, which could be within the regime of very high cycle fatigue (VHCF) (i.e., $N > \sim 10^7$; see Figure 10). As discussed above, such nonlinearity of the relationship between the fatigue life (N) and the flow velocity (U) and span length (L) is mainly attributed to the intrinsic characteristics of VIVs (see Figure 2) (i.e., the nonlinear variation in the dimensionless vibration amplitude with the reduced velocity).

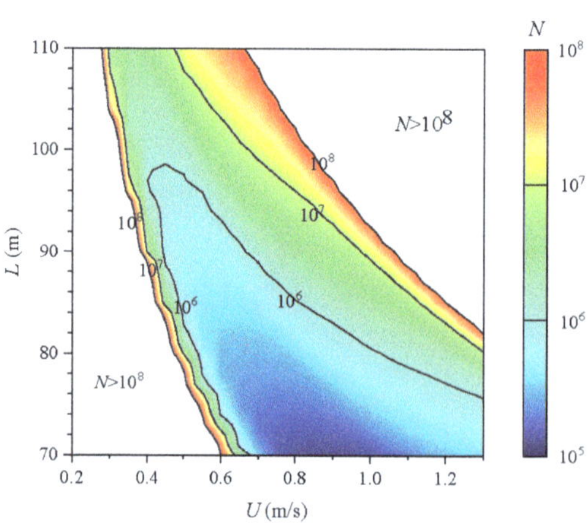

Figure 10. Variations in the number of cycles to failure (N) with flow velocity (U) and span length (L). (Input data for the related pipe span parameters are listed in Tables 1 and 2).

5. Concluding Remarks

In subsea environments, free-spanning subsea pipelines may suffer from vortex-induced vibrations (VIVs), which could further result in fatigue failure after a substantial period of service. In this study, the fatigue failure of free spans caused by VIVs was investigated, with a focus on very high cycle fatigue (VHCF) in the extreme scenario of large free spans under flows with moderate velocities. The following conclusions can be drawn:

(1) With reference to the benchmark flume observations of the VIV response amplitudes of a low mass damping system, the whole curve for the nonlinear variation in A/D with Vr (A/D–Vr curve) is recommended, in which four distinct branches of VIV responses can be identified (i.e., the initial excitation branch, the upper branch, the lower branch, and the desynchronization branch). Bilinear S-N curves were chosen to describe the relationship between the stress range and the fatigue life of the free spans in a complex seawater environment.

(2) On the basis of the recommended A/D–Vr curve and the S-N curve for VHCF of high-strength steel in seawater and with cathodic protections, a prediction method was proposed for the fatigue life of free spans undergoing VIVs. If the pipeline and environmental parameters are given, then the fatigue life in terms of the number of cycles to failure N can be evaluated by following the flow chart.

(3) The nonlinear relationships between the fatigue life N and the flow velocity U and span length L were obtained in a parametric study. For the examined range of span lengths and flow velocities, the minimum fatigue life emerged at a certain flow with a moderate velocity for a given span length L. With a further decrease or increase in U, the fatigue life N would be enhanced correspondingly, which could be within the regime of VHCF. Such nonlinearity of the relationship of N with U and L is mainly attributed to the intrinsic characteristics of VIVs (i.e., the nonlinear variation in the dimensionless vibration amplitude with a reduced velocity).

In the present investigation, a quantitative correlation was established between the VHCF life of a spanning pipeline undergoing VIVs and the key influential parameters (i.e., the span length and the flow velocity). The VHCF life of a spanning pipeline was examined under two extreme (F-F and P-P) end conditions by taking into account the effective span length. This study may provide an instructive reference or methodology for evaluating the service life of subsea-spanning pipelines. It should be noted that the above conclusions are predominantly limited to the engineering scenario of an isolated single span on a relatively flat seabed. Much uncertainty still exists for accurate prediction of the fatigue life of free spans (e.g., the complex pipe–soil interaction at the span shoulders), which needs to be further investigated.

Author Contributions: Conceptualization, F.G.; experiments and data analyses, Q.S. and J.L.; writing—original draft, Q.S.; validation, J.L. and F.G.; review and editing, F.G. and J.L. All authors have read and agreed to the published version of the manuscript.

Funding: This work was funded by the National Natural Science Foundation of China (Grant Nos. 12061160463 and 11825205).

Institutional Review Board Statement: Not applicable.

Informed Consent Statement: Not applicable.

Data Availability Statement: The data presented in this study are available upon request from the corresponding author.

Conflicts of Interest: The authors declare no conflicts of interest.

Notations

A	Vibration amplitude of the cylinder
$\bar{a}$	Coefficient in Equation (9)
$\bar{a}_1$	Parameter for the S-N curve in Table 2
$\bar{a}_2$	Parameter for the S-N curve in Table 2
C_1	Boundary condition coefficient in Equation (2)
C_3	Boundary condition coefficient in Equation (2)
C_A	Added mass coefficient
D	Outer diameter of the cylinder or the pipe
D_i	Inner diameter of the pipe
e	Gap between the pipe's bottom and the seabed surface
E	Elastic modulus of the pipe
$(EI)_{conc}$	Stiffness of concrete coating
F_{CS}	Concrete stiffness enhancement factor for the pipe
f_n	Natural frequency of the pipe or the cylinder
f_v	Frequency of vortex shedding from a fixed cylinder
G	Shear modulus of the soil
I	Moment of inertia
k	Exponent of thickness of the pipe
k_c	Empirical constant
K	Soil stiffness
K_s	Combined mass damping parameter
K_{vd}	Dynamic vertical stiffness of the soil
L	Free span length of the pipeline
L_{eff}	Effective span length of the pipeline
M	Bending moment of the pipeline
m_1	Parameter for the S-N curve in Table 2
m_2	Parameter for the S-N curve in Table 2
m_a	Added mass of the pipeline per meter
m_c	Mass of the content inside the pipeline per meter
m_e	Effective mass of the pipeline per meter
m_p	Mass of the steel pipeline per meter
N	Number of cycles to fatigue failure
P_{cr}	Critical buckling load
q	Uniform load on the pipeline
SCF	Stress concentration factor
S_{eff}	Effective axial force along the pipeline
t	Wall thickness of the pipe
t_{ref}	Reference wall thickness of the pipe
U	Flow velocity
Vr	Reduced velocity
Vr_{cr}	Critical reduced velocity
w	Deflection of the suspended pipeline
w_{max}	Maximum deflection of the suspended pipeline
β	Non-dimensional soil stiffness
δ	Static deflection of the pipeline
$\Delta\sigma$	Stress range
ρ_{steel}	Density of the steel
ρ_{water}	Density of the seawater
σ	Normal stress
σ_{max}	Maximum normal stress
υ_{water}	Kinematic viscosity of the seawater
ν	Poisson ratio of the soil

References

1. Drago, M.; Mattioli, M.; Bruschi, R.; Vitali, L. Insights on the design of free spanning pipelines. *Philos. Trans. R. Soc. A* **2015**, *373*, 20140111. [CrossRef]
2. Sumer, B.M.; Jensen, H.R.; Mao, Y.; Fredsøe, J. Effect of lee-wake on scour below pipelines in current. *J. Waterw. Port Coast. Ocean Eng.* **1988**, *114*, 599–614. [CrossRef]
3. Gao, F.P. Flow-pipe-soil coupling mechanisms and predictions for submarine pipeline instability. *J. Hydrodyn.* **2017**, *29*, 763–773. [CrossRef]
4. Blevins, R.D. *Flow-Induced Vibration*, 2nd ed.; Krieger Publishing Company: New York, NY, USA, 1990.
5. Williamson, C.H.K.; Govardhan, R. Vortex-induced vibrations. *Annu. Rev. Fluid Mech.* **2004**, *36*, 413–455. [CrossRef]
6. Liu, J.; Gao, F.P. Triggering mechanics for transverse vibrations of a circular cylinder in a shear flow: Wall-proximity effects. *J. Fluids Struct.* **2022**, *108*, 103423. [CrossRef]
7. Gao, F.P.; Yang, B.; Wu, Y.X.; Yan, S.M. Steady currents induced seabed scour around a vibrating pipeline. *Appl. Ocean Res.* **2006**, *28*, 291–298. [CrossRef]
8. Sharma, A.; Oh, M.C.; Ahn, B. Recent advances in very high cycle fatigue behavior of metals and alloys—A review. *Metals* **2020**, *10*, 1200. [CrossRef]
9. Murakami, Y.; Nomoto, T.; Ueda, T.; Murakami, Y. On the mechanism of fatigue failure in the superlong life regime ($N>10^7$ cycles). Part I: Influence of hydrogen trapped by inclusions. *Fatigue Fract. Eng. Mater. Struct.* **2000**, *23*, 893–902. [CrossRef]
10. Shiozawa, K.; Lu, L.; Ishihara, S. S-N curve characteristics and subsurface crack initiation behaviour in ultra-long life fatigue of a high carbon-chromium bearing steel. *Fatigue Fract. Eng. Mater. Struct.* **2001**, *24*, 781–790. [CrossRef]
11. Sakai, T.; Sato, Y.; Oguma, N. Characteristic S-N properties of high-carbon-chromium-bearing steel under axial loading in long-life fatigue. *Fatigue Fract. Eng. Mater. Struct.* **2002**, *25*, 765–773. [CrossRef]
12. Hong, Y.; Zhao, A.; Qian, G.; Zhou, C. Fatigue strength and crack initiation mechanism of very-high-cycle fatigue for low alloy steels. *Metall. Mater. Trans. A* **2012**, *43*, 2753–2762. [CrossRef]
13. Song, Q.Y.; Sun, C.Q. Mechanism of crack initiation and early growth of high strength steels in very high cycle fatigue regime. *Mater. Sci. Eng. A* **2020**, *771*, 138648. [CrossRef]
14. Det Norske Veritas and Germanischer Lloyd (DNV GL). *DNVGL-RP-C203: Fatigue Design of Offshore Steel Structures*; DNV: Oslo, Norway, 2016.
15. Det Norske Veritas and Germanischer Lloyd (DNV GL). *DNVGL-RP-F105: Free Spanning Pipelines*; DNV: Oslo, Norway, 2017.
16. Liu, J.; Gao, F.P. Evaluation for allowable span length of a submarine pipeline considering VIV hysteresis effect. *Int. J. Offshore Polar Eng.* **2021**, *31*, 325–332. [CrossRef]
17. He, Z.F.; Wei, Y.; Liu, S.L. Analysis of safe span length and fatigue life of submarine pipelines. *China Ocean Eng.* **2020**, *34*, 119–130. [CrossRef]
18. Det Norske Veritas and Germanischer Lloyd (DNV GL). *DNVGL-RP-F114: Pipe-Soil Interaction for Submarine Pipelines*; DNV: Oslo, Norway, 2017.
19. Guo, B.; Song, S.; Ghalambor, A.; Lin, T.R. *Offshore Pipelines: Design, Installation, and Maintenance*, 2nd ed.; Gulf Professional Publishing: Waltham, MA, USA, 2014.
20. Sarpkaya, T. A critical review of the intrinsic nature of vortex-induced vibrations. *J. Fluids Struct.* **2004**, *19*, 389–447. [CrossRef]
21. Bearman, P.W. Circular cylinder wakes and vortex-induced vibrations. *J. Fluids Struct.* **2011**, *27*, 648–658. [CrossRef]
22. Wu, X.; Ge, F.; Hong, Y. A review of recent studies on vortex-induced vibrations of long slender cylinders. *J. Fluids Struct.* **2012**, *28*, 292–308. [CrossRef]
23. Wang, X.K.; Hao, Z.; Tan, S.K. Vortex-induced vibrations of a neutrally buoyant circular cylinder near a plane wall. *J. Fluids Struct.* **2013**, *39*, 188–204. [CrossRef]
24. De Oliveira Barbosa, J.M.; Qu, Y.; Metrikine, A.V.; Lourens, E.M. Vortex-induced vibrations of a freely vibrating cylinder near a plane boundary: Experimental investigation and theoretical modelling. *J. Fluids Struct.* **2017**, *69*, 382–401. [CrossRef]
25. Daneshvar, S.; Morton, C. On the vortex-induced vibration of a low mass ratio circular cylinder near a planar boundary. *Ocean Eng.* **2020**, *201*, 107109. [CrossRef]
26. Khalak, A.; Williamson, C.H.K. Investigation of the relative effects of mass and damping in vortex-induced vibration of a circular cylinder. *J. Wind. Eng. Ind. Aerodyn.* **1997**, *69–71*, 341–350. [CrossRef]
27. Sun, X.F.; Fang, X.S.; Guan, L.T. *Mechanics of Materials*, 5th ed.; Higher Education Press: Beijing, China, 2009.
28. Sakai, T.; Oguma, N.; Morikawa, A. Microscopic and nanoscopic observations of metallurgical structures around inclusions at interior crack initiation site for a bearing steel in very high-cycle fatigue. *Fatigue Fract. Eng. Mater. Struct.* **2015**, *38*, 1305–1314. [CrossRef]
29. Hong, Y.; Liu, X.; Lei, Z.; Sun, C. The formation mechanism of characteristic region at crack initiation for very-high-cycle fatigue of high-strength steels. *Int. J. Fatigue* **2016**, *89*, 108–118. [CrossRef]
30. Avateffazeli, M.; Webster, G.; Tahmasbi, K.; Haghshenas, M. Very high cycle fatigue at elevated temperatures: A review on high temperature ultrasonic fatigue. *J. Space Saf. Eng.* **2022**, *9*, 488–512. [CrossRef]
31. Sakai, T.; Nakagawa, A.; Oguma, N.; Nakamura, Y.; Ueno, A.; Kikuchi, S.; Sakaida, A. A review on fatigue fracture modes of structural metallic materials in very high cycle regime. *Int. J. Fatigue* **2016**, *93*, 339–351. [CrossRef]

32. Nadot, Y. Fatigue from defect: Influence of size, type, position, morphology and loading. *Int. J. Fatigue* **2022**, *154*, 106531. [CrossRef]
33. Schmiedel, A.; Henkel, S.; Kirste, T.; Morgenstern, R.; Weidner, A.; Biermann, H. Ultrasonic fatigue testing of cast steel G42CrMo4 at elevated temperatures. *Fatigue Fract. Eng. Mater. Struct.* **2020**, *43*, 2455–2475. [CrossRef]
34. Bayraktar, E.; Mora, R.; Garcia, I.; Bathias, C. Heat treatment, surface roughness and corrosion effects on the damage mechanism of mechanical components in the very high cycle fatigue regime. *Int. J. Fatigue* **2009**, *31*, 1532–1540. [CrossRef]
35. Gorash, Y.; Comlekci, T.; Styger, G.; Kelly, J.; Brownlie, F. Investigation of S275JR+AR structural steel fatigue performance in very high cycle domain. *Procedia Struct. Integr.* **2022**, *38*, 490–496. [CrossRef]
36. Palin-Luc, T.; Pérez-Mora, R.; Bathias, C.; Domínguez, G.; Paris, P.; Arana, J. Fatigue crack initiation and growth on a steel in the very high cycle regime with sea water corrosion. *Eng. Fract. Mech.* **2010**, *77*, 1953–1962. [CrossRef]
37. Pérez-Mora, R.; Palin-Luc, T.; Bathias, C.; Paris, P. Very high cycle fatigue of a high strength steel under sea water corrosion: A strong corrosion and mechanical damage coupling. *Int. J. Fatigue* **2015**, *74*, 156–165. [CrossRef]
38. Behvar, A.; Haghshenas, M. A critical review on very high cycle corrosion fatigue: Mechanisms, methods, materials, and models. *J. Space Saf. Eng.* **2023**, *10*, 284–323. [CrossRef]
39. Song, Q.Y. Observation of Characteristic and Analysis of Mechanism in Very High Cycle Fatigue Regime of High-Strength Steel. Ph.D. Thesis, University of Chinese Academy of Sciences, Beijing, China, 2021.
40. Lemos, M.; Kwietniewski, C.; Clarke, T.; Joia, C.J.B.; Altenhofen, A. Evaluation of the fatigue life of high-strength low-alloy steel girth welds in aqueous saline environments with varying carbon dioxide partial pressures. *J. Mater. Eng. Perform.* **2012**, *21*, 1254–1259. [CrossRef]
41. Gao, Z.; Gong, B.; Xu, Q.; Wang, D.; Deng, C.; Yu, Y. High cycle fatigue behaviors of API X65 pipeline steel welded joints in air and H_2S solution environment. *Int. J. Hydrogen Energy* **2021**, *46*, 10423–10437. [CrossRef]
42. Yan, Y.; Zhong, S.; Chen, Z.; Sun, Y.; Xu, L.; Zhao, L.; Han, Y. Corrosion fatigue behavior of X65 pipeline steel welded joints prepared by CMT/GMAW baking process. *Corros. Sci.* **2023**, *225*, 111568. [CrossRef]

Journal of
Marine Science and Engineering

MDPI

Article

Visualized Experimental Study of Soil Temperature Distribution around Submarine Buried Offshore Pipeline Based on Transparent Soil

Hui Li [1,2], Yajing Meng [1,3], Yilong Sun [1,4,*] and Lin Guo [1,2]

1 College of Civil Engineering and Architecture, Wenzhou University, Wenzhou 325035, China; lihuice21@163.com (H.L.); mengyajing99@foxmail.com (Y.M.); lingpray@126.com (L.G.)
2 Key Laboratory of Engineering and Technology for Soft Soil Foundation and Tideland Reclamation of Zhejiang Province, Wenzhou 325035, China
3 Zhejiang Engineering Research Center of Disaster Prevention and Mitigation for Coastal Soft Soil Foundation, Wenzhou 325035, China
4 Collaborative Innovation Center of Tideland Reclamation and Ecological Protection, Wenzhou 325035, China
* Correspondence: sunyilong@wzu.edu.cn

Abstract: The temperature distribution around the offshore burial pipeline is an important factor affecting its safety design and economic operation. The traditional test method cannot obtain the continuous temperature distribution of soil owing to the constraints of placing measurement sensors in soil. The transparent soil model test is an alternative method to realize the visualization research of soil temperature. In this paper, a relationship between the temperature of transparent soil and pixel intensity was first established. Then, the transparent soil test and numerical simulation, considering the natural convection, were carried out to study the temperature distribution around the submarine pipeline during start-up and stable operation. The influence of buried depth and pipeline diameter was analyzed. The results suggest that the continuous temperature distribution can be obtained visually by using a transparent soil test, and the observed heating zone of influence extended to a radial distance of 2.6 pipe diameters. The numerical analysis results show that the influence zone of the temperature of pipeline is a distance of four pipeline diameters at a temperature difference of 45 °C. The buried depth and pipeline diameter have little influence on the influence zone. In addition, the contour curves of soil temperature around the pipeline with different diameter are similar in shape. With the decrease in the buried depth of pipeline, the temperature gradient of soil around the pipeline decreases, which is caused by the natural convection.

Keywords: offshore pipeline; transparent soil; thermal field; numerical simulation

1. Introduction

Offshore buried pipelines are often used in transportation engineering of marine crude oil with the advantages of low cost, large transportation volume and continuous transportation. Crude oil produced from offshore fields usually has a high wax content, resulting in poor fluidity at the marine environmental temperature. To ensure the smooth and safe transportation of crude oil, it is often raised to a very high temperature. However, the sea water temperature decreases with the increase in water depth. The temperature of sea water at 2000 m depth is about 3 °C [1]. There is a large temperature difference between the pipeline and the surrounding environment, which means that the heat of the pipeline transmits to the surrounding soil. Heat transmission causes the temperature of oil to drop, which, in turn, affects its transportation. In addition, it changes the temperature of the soil around the pipeline, which effects the interaction between the pipeline and soil. There is a huge impact on the safety and stability of the pipeline. It is very important to have a thorough understanding of the soil temperature distribution around the pipeline, which is an important basis for the safe design and economic operation of the pipeline. Therefore, it is necessary to study the soil temperature distribution around the pipeline.

Subsea burial is a common method to prevent pipeline damage and heat loss for submarine pipelines. The heat transfer of the buried pipeline is an important and complex problem. In the previous research on the temperature distribution of soil around the buried pipeline, the influence of the thermal conductivity of the insulation layer, properties and geometric properties of backfill materials, soil stratification, soil moisture content and void ratio, natural convection and other factors were studied by using numerical and experimental methods [2–6]. However, the experimental test usually uses temperature sensor embedded in the soil to measure the soil temperature. The temperature measurement has spatial discontinuity. The measurement is not accurate under large temperature gradient, such as media stratification, natural convection and local material phase change. In most numerical simulations, the soil is assumed to be solid material, ignoring the porous media characteristics of soil.

Transparent soil test technology can realize the visualization of the displacement of soil. The previous studies of transparent soil test mainly focus on the physical and mechanical properties of transparent soil materials [7–11] or use transparent soil tests to study the deformation of soil around the pile, tunnel, anchor or other structures [12–16]. The transparency of the transparent soil is expected to be improved for larger model tests. The changes in transparency caused by temperature, saturation and impurities are considered to be unfavorable factors in the transparent soil test; these factors need to be strictly controlled in the test. However, the change in the transparency of transparent soil can also be a favorable factor in the test [17,18]. Based on this understanding, the relationship between transparency of transparent soil and temperature was established through experiments, and the temperature field around the energy pile was studied [17]. It was proved that the continuous temperature field around the structure could be obtained based on the change in the transparency of transparent soil.

Therefore, aiming at studying the discontinuity of temperature monitoring in the traditional experimental test for the thermal fields of the submarine buried pipeline, a visual model test of the soil temperature distribution around the submarine buried pipeline by using transparent soil model test was carried out. The temperature distribution of soil around the submarine buried pipeline during operation was analyzed. Combined with the numerical simulation, the effect of the buried depth and pipe diameter on the soil temperature distribution around the submarine pipeline was studied.

2. Test Method and Process

2.1. Test Method for Transparent Soil

Transparent soil consists of transparent solid particles and a fluid with a matching refractive index. The matching degree of the refractive index between solid particles and pore fluid is one of the main factors affecting the transparency of transparent soil. The refractive index of a material is usually affected by the wavelength of light and temperature. The refractive index of solid particles is generally considered not to vary with temperature. However, the refractive index of fluid varies greatly with temperature. Kong et al. [19] measured the variation in the refractive index of fluid that may be used in the transparent soil model test with temperature. It found that the refractive index of those pore fluids mostly decreased with the increase in temperature. Therefore, with the variation in temperature, the matching degree of refractive index between solid particles and pore fluid decreases, resulting in the decrease in the transparency of transparent soil. The previously transparent medium becomes opaque or appear less clear with the change in temperature.

Individual pixels are the smallest unit that forms a digital image. The pixel intensity is used to describe the color and brightness of the pixels. The most commonly used pixel format is where the pixel number is stored as an 8-bit integer. The corresponding decimal values range from 0 (black) to 255 (white). When the pixel intensity is between the two extreme values, the pixel will appear gray, as shown in Figure 1. We prepared the transparent soil at the specified temperature (i.e., 20 °C) with the optimum refractive index match. We placed a black background behind the transparent soil and shot it through the soil. At this time, the black background was clearly visible as the soil was transparent. The pixel intensities of the background were sufficiently low due to the relative high transparency of the transparent soil. Conversely, when the temperature of the transparent soil increased (i.e., 50 °C), the transparency of transparent soil decreased due to the reduction in refractive index matching degree. The pixel intensity increased such that the soil appeared to be gray in color. Therefore, the visual measurement of transparent soil temperature could be achieved by establishing the relationship between pixel intensity and the temperature of transparent soil.

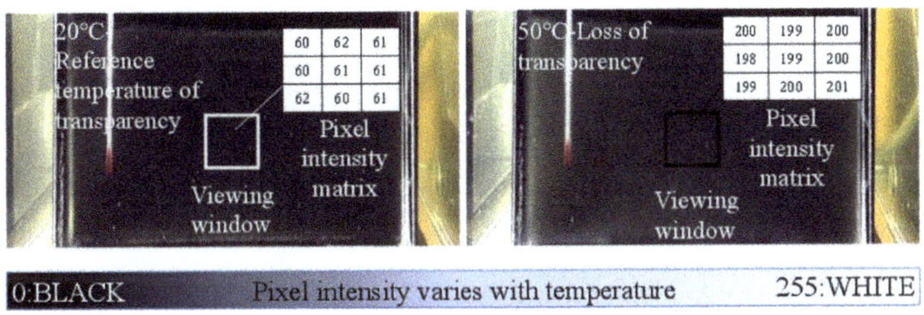

Figure 1. The measurement of temperature in transparent soil.

2.2. Test Materials

The transparent soil material used in this study was manufactured using fused quartz and mixed oil. Fused quartz is a commonly used solid particle material for transparent soil configuration due to its high purity, high transparency, high angular and similar chemical composition to natural sand [20]. The particle size distribution of the fused quartz is shown in Figure 2, which is close to that of natural marine sand [21–23]. The mixed oil was a blend of mineral oil (white oil) and paraffinic solvent (Norpar 12) with a mass ratio of 4:1. The refractive index of mixed oil is 1.4585 at 20 °C, which is consistent with that of fused silica. The fused quartz and mixed oil were thoroughly mixed by using a blender to form a uniform slurry. The sample was then placed in a vacuum to evacuate the air. A two-phase continuum was produced, which enabled the transparency of the transparent soil. The other specific preparation process of transparent soil is presented in the literature [24]. The physical and mechanical parameters of the transparent soil are shown in Table 1.

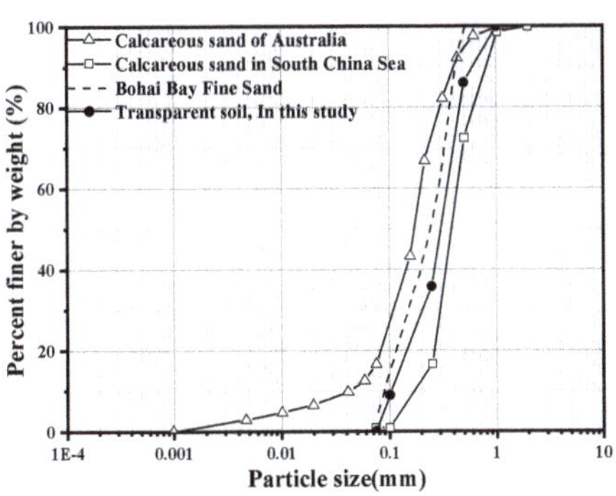

Figure 2. The particle distribution of fused quartz and natural marine sand including calcareous sand of Australia [21], calcareous sand in South China Sea [22], and Bohai Bay fine sand [23].

Table 1. Mechanical properties of the transparent soil.

Parameter	Unit	Value
Nonuniformity coefficient, C_u		3.51
Curvature coefficient, C_c		1.14
Saturated bulk density, $\gamma_{t,sat}$	kg/m^3	18.37
Thermal conductivity, K_t	W/(m·K)	0.785
Specific heat capacity, C_t	J/(kg·K)	2097.90

2.3. Test Apparatus for Transparent Soil

The visualization test apparatus for the submarine pipeline temperature based on transparent soil is shown in Figure 3. The test system mainly includes water bath, LED lighting device, temperature control device, model pipe, model box, image acquisition device and other parts. The water bath with dimensions of 400 mm (length) by 400 mm (width) by 450 mm (height) was mainly used to provide a constant temperature boundary for the model box filled with transparent soil. The water bath was filled with water, which was warmed by the temperature control device. The temperature control device includes a heating/cooling element and a circulating water pump to achieve a constant temperature of the water bath through the circulation of water. The model box for the transparent soil is 300 mm × 40 mm × 350 mm in size and constructed of 10 mm thick plexiglass panels. The model pipe was manufactured by using hollow copper pipe with black surface spraying. A water inlet and outlet were arranged in the inner space of the pipe. The inlet and outlet were connected with another temperature control device. The temperature of the pipe was controlled through the circulation of water inside the pipe. The specific dimensions and relative positions of the model box and model pipe are shown in Figure 4.

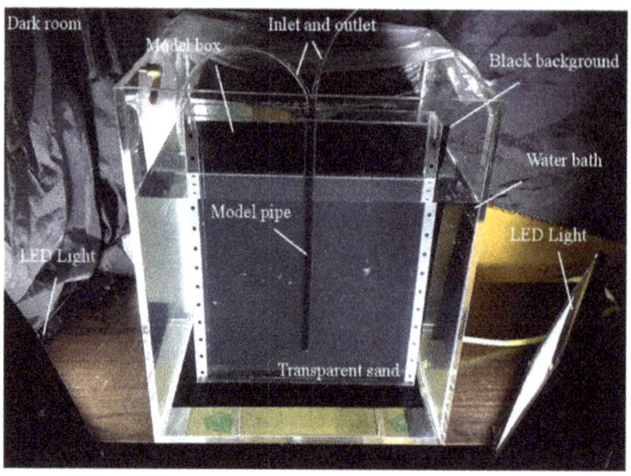

Figure 3. The physical figure of visual test system of offshore pipeline temperature.

Nonuniform lighting conditions are not conducive to producing consistent pixel intensity. In order to produce a consistent background, the test device was placed in a dark room. Two LED lights with constant power were used to provide uniform illumination of the water bath and model box. A CCD high-speed high-definition camera with fixed aperture, exposure time and focus was used to record the test image. During the test, the CCD camera was installed on a tripod at a distance of 0.5 m from the front of the water bath. The CCD camera was connected to the computer. The images of the test were captured without contact through control the software on the computer, which ensures long-term testing without operator intervention. The temperature field around the model pipe can be obtained through the image analysis software.

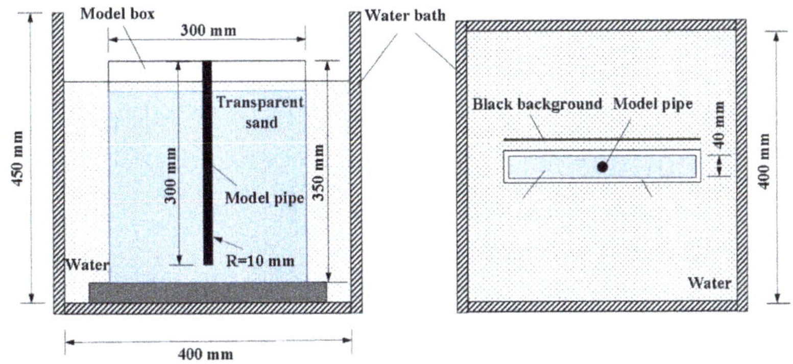

Figure 4. The schematic diagram of model tank in a transparent soil temperature test.

2.4. Numerical Simulation Method

(1) Mathematical model

As pipes are always longer than they are wide, a two-dimensional plane model of the buried submarine pipeline was used in this study, as shown in Figure 5. The length and depth of the rectangular influence range of the pipeline on the soil temperature distribution is $2a$ and b, respectively. The pipe section is circular with diameter of D and buried depth of H. The coordinate system is established with the corresponding center point of the pipeline on the seabed surface as the origin. The horizontal direction is the x-axis, and the vertical direction is the y-axis. The sea water temperature above the seabed surface is T_W. The initial temperature of soil is $T_{s,ini}$. The thermal conductivity of soil is K_s. There is thermal convection between the seabed surface and seawater, including natural convection.

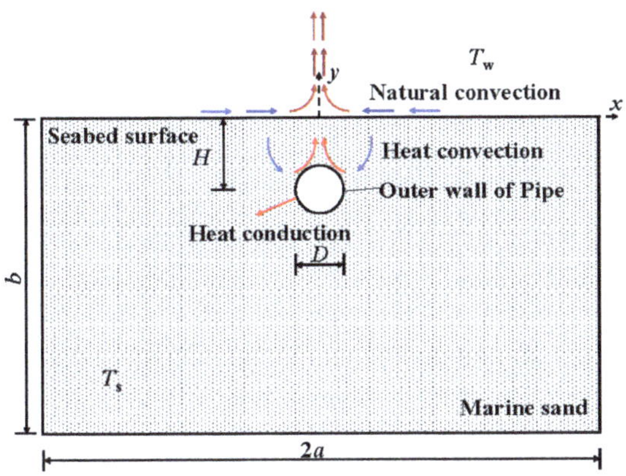

Figure 5. The two-dimensional plane model of heat transfers of offshore pipeline.

In order to simulate the physical process of heat transfer from the pipeline to the surrounding soil during start-up and operation, the following governing equations were used in the numerical model. Fourier's law expresses the conductive heat flux as a function of the temperature gradient as follows:

$$\mathbf{q} = -K\nabla T \tag{1}$$

The heat transfer in the soil, which is considered to be a porous medium, is governed by the heat transfer equation:

$$\rho C\frac{\partial T}{\partial t} + \nabla \cdot (-K\nabla T) = Q \tag{2}$$

The initial condition is that the temperature of soil is $T_{s,ini}$ for $t = 0$.

The thermal boundary conditions are that the temperature of the outer wall of the pipeline is constant as T_{wall}. The left, right and lower boundaries are adiabatic boundaries. The upper boundary is the heat flux boundary, and there is natural convection between the seabed surface and seawater. The convective heat flux is

$$q_0 = h(T_w - T) \tag{3}$$

The convective heat transfer coefficient h is determined by the following equation:

$$h = \begin{cases} \frac{K}{2a}0.54R_a^{1/4} & \text{if } T > T_w \text{ and } 10^4 \leq R_a \leq 10^7 \\ \frac{K}{2a}0.15R_a^{1/3} & \text{if } T > T_w \text{ and } 10^7 \leq R_a \leq 10^{11} \\ \frac{K}{2a}0.27R_a^{1/4} & \text{if } T \leq T_w \text{ and } 10^4 \leq R_a \leq 10^7 \end{cases} \tag{4}$$

It should be noticed that there is a sudden rise in the temperature of the outer wall of the pipeline at the beginning of the numerical simulation. In Table 2, all the employed parameters for the equations are reported.

Table 2. The employed parameters for the equations.

Symbol	Parameter	Unit
C	Specific heat capacity of material	J/(kg·K)
D, H	Diameter and buried depth of pipeline	m
h	Convective heat transfer coefficient	W/(m^2·K)
K	Thermal conductivity of material	W/(m·K)
ρ	Density of material	
Q	Heat	J
q_0	Heat flux	W/m^2
q	Heat flux vector	W/m^2
T	Temperature of material	K
R_a	Reynolds number	

(2) Numerical solution

The software of COMSOL Multiphysics 5.2 was employed to study the soil temperature distribution around the pipeline during stable operation, considering the natural convection.

To ensure the calculation of the model, the model was subject to the following assumptions: (1) the influence area of the pipeline on the surrounding soil was a rectangular area; (2) the soil was a continuous and uniform porous medium, and the soil was saturated and the pore water does not change phase; (3) the thermodynamic parameters of soil were kept constant and did not change with temperature; (4) the flow in soil was two-dimensional and followed Darcy's law; (5) there was no heat exchange between the soil skeleton and pore water; (6) the thermal conductivity of soil was independent of temperature; (7) the water in soil did not evaporate; (8) there was a natural convection between the seabed and seawater.

A two-dimensional pipe-soil model was used in this study, which means that the change in the pipeline temperature in the axial direction was not considered. Considering the boundary effect, the length of the model was set to 16 m, and the depth was set to 8 m. The section of pipeline was circular. The diameter D and buried depth H of the pipeline were set according to different test conditions. The soil material used in the numerical simulation was fine sand from the South China Sea, which was described as a porous medium saturated by water. The physical properties of the soil and seawater are shown in Table 3. To consider the natural convection between the soil and the seawater, the upper boundary was defined as the heat flux boundary. The left, right and lower boundaries were defined as adiabatic boundaries by setting the heat flux conditions as zero. At the beginning of the simulation, the soil temperature was assumed to be uniform and constant $T_{s,ini}$, which was same as the temperature of the seawater T_w. After heating, the soil temperature changed. In fact, we were interested in studying heat transfer from the outer surface of the pipe to the soil, rather than solving for the fluid flow inside the pipe. Therefore, the temperature of the outer wall of the pipe was set to the constant and uniform temperature T_{wall}. Due to the existence of temperature difference between the pipeline wall and soil, the heat of the pipeline continued to transfer to the soil, which meant that the soil was heated.

Table 3. Mechanical properties of the soil.

Soil Mass	Parameter	Unit	Value
Marine sand	Thermal conductivity, K_s	W/(m·K)	0.867
	Specific heat capacity, C_s	J/(kg·K)	3465
	Bulk density, γ_s	kg/m^3	18.0
	Porosity, n		0.6
Seawater	Thermal conductivity, K_w	W/(m·K)	0.65
	Specific heat capacity, C_w	J/(kg·K)	3850
	Bulk density, γ_w	kg/m^3	10.25

The specific mesh division used in the model is shown in Figure 6, where the mesh is finer near the pipe to ensure the convergence of numerical calculation and accuracy of the simulation. The triangular elements are used in the model, which are linear. The discretization of the domain has a total of 537 nodes and 979 elements.

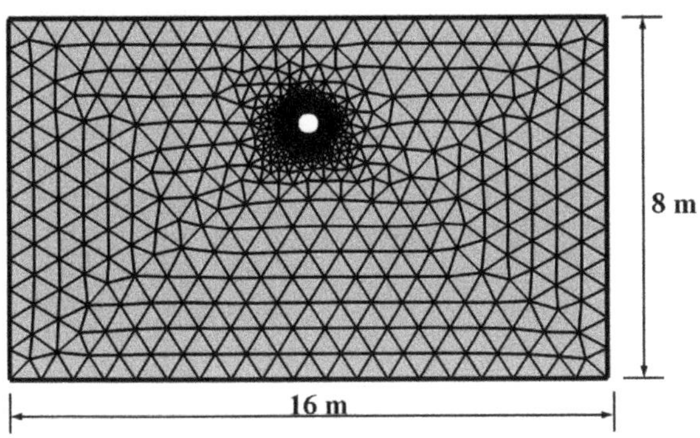

Figure 6. FEM mesh of numerical model.

2.5. Test Procedures and Progresses

In order to establish the relationship between the temperature of the transparent soil and the pixel intensity, the calibration test was carried out. Calibration of pixel intensity at different soil temperatures was achieved by placing the model box in the water bath and allowing the water circulation system to heat up. During the calibration test, the temperature of the water bath continuously increased from 20 °C to 50 °C, which was divided into 6 stages in 5 °C increments. At each heating stage, the temperature of the water bath was set to the desired value. The temperature of the soil was then precisely determined using a thermometer placed inside the model box. Images were captured during the heating process with an image acquisition frequency of 30 s. The image pixel intensities were obtained and recorded by using image processing software. Then, the relationship between pixel intensity and temperature were established.

To study the expansion of soil temperature around the pipeline along the horizontal direction, a transparent soil test was carried out to visualize the temperature. During the test, the temperatures of the outer wall of the model box filled with transparent soil and the model pipe were maintained at 20 °C and 50 °C, respectively, through the circulation of water warmed by the individual temperature control device. Due to the thin outer wall of the model pipe and the high thermal conductivity of copper, the pipe increased to the required temperature in a short time, which was not taken into account during the test. During the test, the images of the transparent soil were captured at a frequency of 5 min depending on the nature of the test. The measures ensured that sufficient changes in pixel intensity of each pixel were monitored. The variation in pixel intensity in the image was interpreted as changes in soil temperature according to the established relationship between pixel intensity and soil temperature. Therefore, the visualization of the temperature could be achieved.

Numerical simulation was carried out to study the soil temperature distribution around the pipeline under different buried depths and pipe diameters, considering natural convection. In order to verify the validity of the model, the soil parameters and initial conditions for the transparent soil test were first used by the model, including the thermal conductivity and initial temperature of the soil (20 °C) and the temperature of the model pile (50 °C). The results of the numerical simulation were compared with those of the transparent soil test. After the verification stage, the parameters of marine sand were substituted into the model for the numerical simulation. The specific calculation conditions of numerical simulation are shown in Table 4. Combined with reference [25], the calculation time of the model was 60 days. For heat transfer analysis during start-up, 4 temperature monitoring points (named as point 1, 2, 3, 4, respectively) were set in the center of the pipeline along the horizontal direction of the model, with the distances far away from the outer wall of the pipe being 0.5, 1, 1.5, and 2 pipe diameters, respectively, as shown in Figure 7.

Table 4. Modeling case in numerical simulation.

Temperature (°C)	Buried Depth (H/D)	Pipe Diameter (m)
$T_w = 5$ $T_{s,ini} = 5$ $T_{wall} = 50$	0 1.0 2.0 3.0	0.6 0.8 1.0 1.2

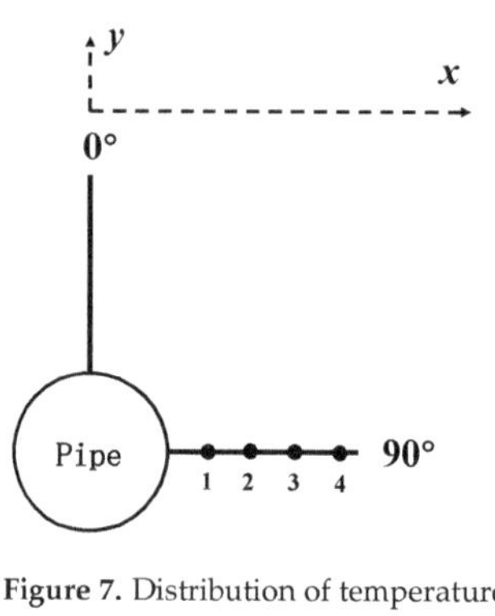

Figure 7. Distribution of temperature monitoring points for soil around the pipeline.

3. Experimental Results and Discussion

3.1. Calibration Test

The observation of transparent soil temperature is very sensitive to the change in image pixel intensity. In order to reduce the influence of abnormal pixel intensity on the test results, it is necessary to normalize the pixel intensity. The normalization method proposed by Siemens et al. [18] is chosen in this study:

$$I_{\mathrm{N}} = \frac{I - I_{\mathrm{cold}}}{I_{\mathrm{hot}} - I_{\mathrm{cold}}} \tag{5}$$

where I_{N} is the normalized pixel intensity, I is the pixel intensity at the specified temperature, I_{cold} is the reference pixel intensity at low temperature and I_{hot} is the reference pixel intensity at high temperature.

The relationship between temperature and normalized pixel intensity is shown in Figure 8. It can be seen that the normalized pixel intensity varies from 0 to 1 for the temperature of 20 °C to 50 °C. The temperature of transparent soil has a logarithmic relationship with the normalized pixel intensity. The specific equation is as follows:

$$T = 48.839 I_N^{0.163} \tag{6}$$

when the temperature is lower than 30 °C, the deviation between temperature data is apparent. It may be because the refractive of pore fluid is not sensitive to the variation in temperature at a lower temperature, resulting in low transparency changes. However, it is interesting to note that these results are in good agreement with the previous results [17,26]. The results of Siemens et al. [18] and Black et al. [17] are also listed in the figure. Siemens et al. [18] found a linear relationship between temperature and normalized pixel intensity in the test. Black et al. [17] found that the relationship curve between temperature and normalized pixel intensity is about a straight line when the temperature is greater than 30 °C, but it is nonlinear when the temperature is 20~30 °C. The variation in pixel intensity mainly depends on the change in the refractive index of pore fluid. Moreover, the research proposed by Kong et al. [19] showed that the refractive index of fluid was not uniformly sensitive to temperature. The difference of these test results may be caused by the difference in the pore fluid of transparent soil. Therefore, for the transparent soil prepared using different materials and processes, the relationship between temperature and normalized pixel intensity should be re-determined.

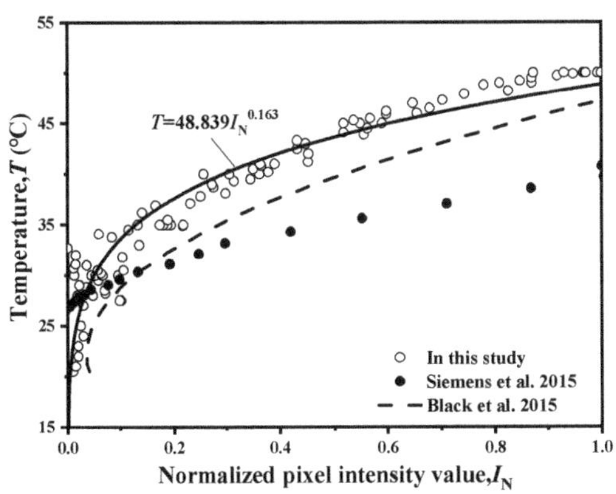

Figure 8. The relationship between temperature and normalized pixel intensity and the comparison with previous researches [17,18].

3.2. Horizontal Temperature Distribution of Soil around Pipeline

The test results of the horizontal temperature distribution of the submarine pipeline are shown in Figure 9. It can be seen from the figure that the temperature decreases with the increase in the distance from the pipe wall. After 5 min, the temperature expands to 0.5*D*. The influence range of pipeline temperature increases gradually with the increase in time. The influence range extends to 2.6*D* after 40 min, and tends to be stable.

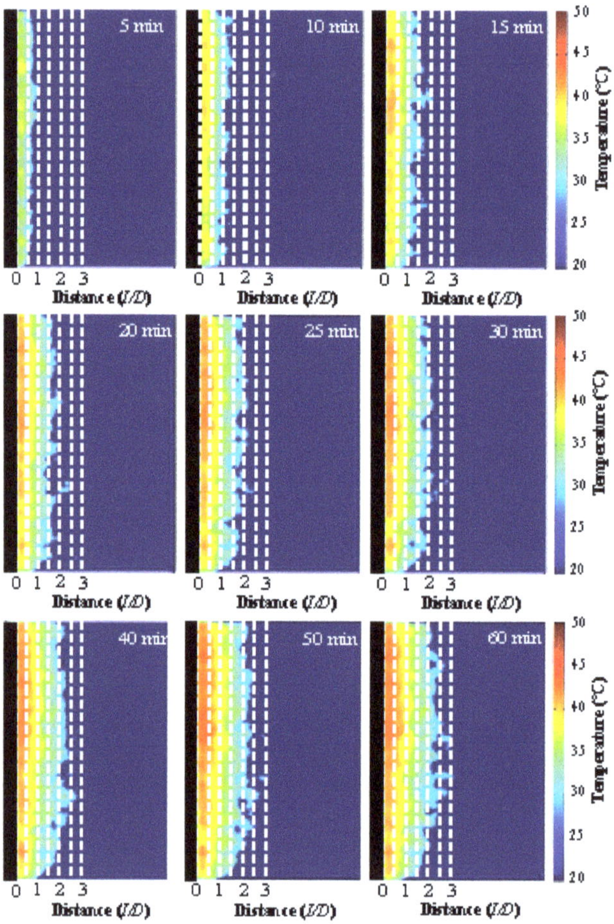

Figure 9. Temperature distribution around pipeline in transparent soil with different times.

When using transparent soil to study the temperature distribution, the test results may have errors due to light conditions, impurities in the transparent soil and other reasons. In order to reduce the test error, the average value of the temperature at the same distance from the pipe wall is taken as the temperature at that distance. The relationships between the temperature and heating time at the distances of 0.5*D*, 1.0*D*, 1.5*D* and 2*D* are shown in Figure 10. It can be seen that the temperature at different locations increases with heating time, and then tends to be stable. In addition, the temperature of the soil at a distance of 0.5*D* increases rapidly after the start of heating, while the temperature of the soil at distances of 1.0*D*, 1.5*D* and 2*D* begins to increase after 5 min, 10 min and 30 min, respectively. It indicates that the influence range of temperature gradually expands with the increase in time. The soil temperatures at different locations all tend to be stable after 60 min.

After the temperature is stable (60 min), the temperature distribution at different distances from the pipe wall is shown in Figure 11. It can be observed that the temperature gradually decreases with the increase in the distance from the pipe wall. The temperature influence range that can be observed is about 2.6*D*. At the distance far from 2.6*D*, it is difficult to use transparent soil to observe the temperature due to the small temperature change. Moreover, the test results of transparent soil at 0~0.3*D* decreased significantly, which may be related to the reflection of the model pipe wall. The light reflection of the pipe wall affects the test results.

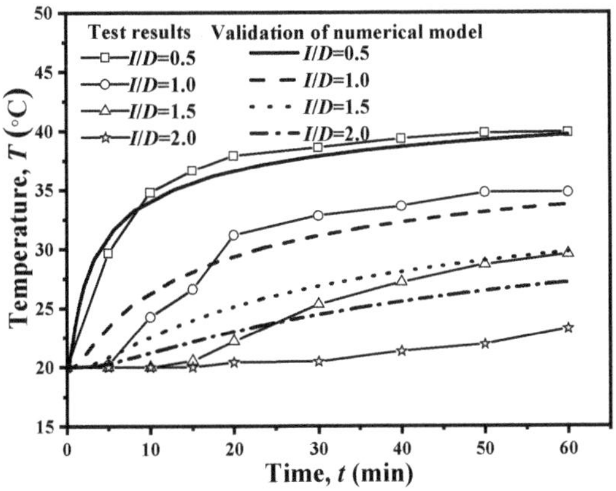

Figure 10. The temperature at different distances (I/D) far from the pipe wall varies with time.

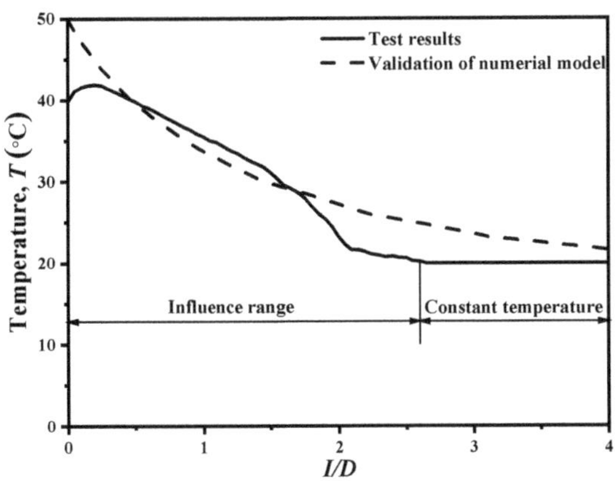

Figure 11. The influence range of pipeline temperature during stable operation in transparent soil.

3.3. Influence of Buried Depth and Pipe Diameter on Temperature Field

To verify the validity of the numerical model, the soil parameters and initial conditions for the transparent soil test were used via the numerical model. Their results are shown in Figures 10 and 11, respectively. It can be seen that numerical simulation results are very close to the results of transparent soil test, indicating that it is feasible to use the numerical model to study the distribution of temperature around the pipeline.

The soil temperature at different locations around the pipeline over time under different buried depths and pipeline diameters are shown in Figures 12 and 13. It can be seen that in the early stage of pipeline operation, the soil temperature increases rapidly. As the soil temperature continues to rise, the heat transferred by the pipeline to the soil is gradually reduced. The rise rate of the soil temperature decreases. The rate and magnitude of temperature rise at point 4 is much lower than that at point 1, which indicates that with the increase in the distance from the pipe wall, the influence of the pipe on the surrounding soil temperature gradually decreases.

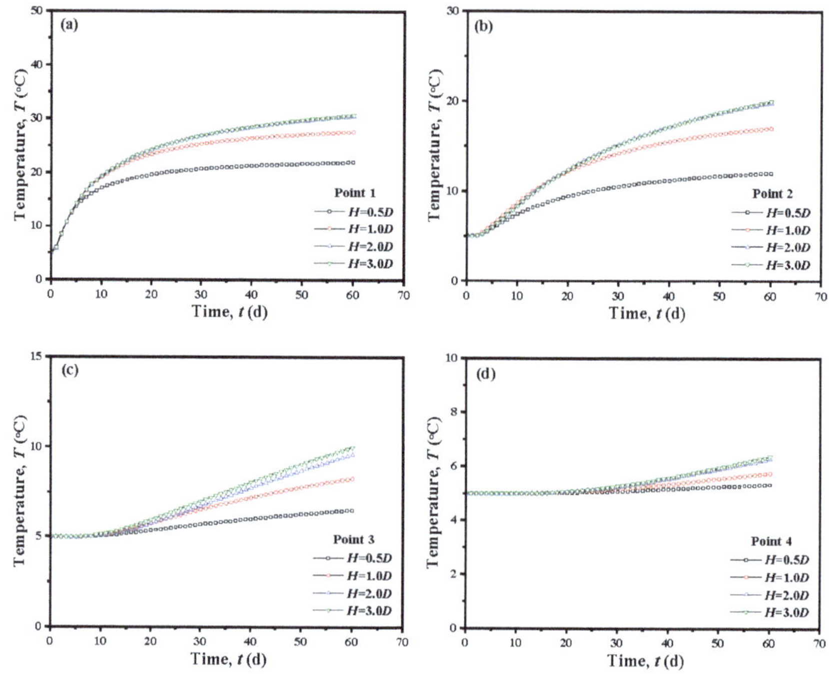

Figure 12. Temperature variation in soil with time at different locations with different buried depths: (**a**) point 1; (**b**) point 2; (**c**) point 3; (**d**) point 4.

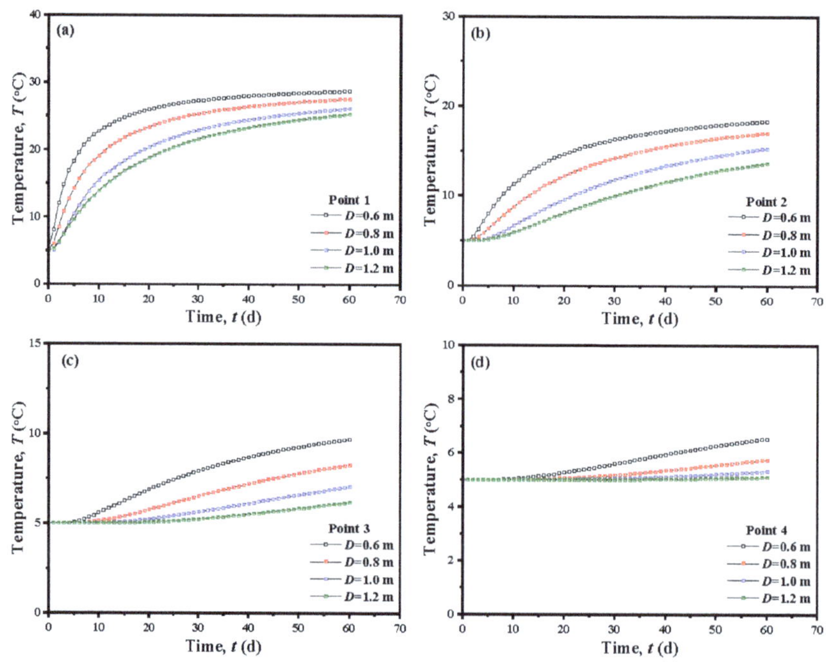

Figure 13. Temperature variation in soil with time at different locations with different diameters: (**a**) point 1; (**b**) point 2; (**c**) point 3; (**d**) point 4.

Figure 12 also shows that the soil temperature around the pipeline is closely related to the buried depth of the pipeline. As can be seen in Figure 12, the deeper the buried depth of the pipeline, the higher the temperature of the soil around the pipeline. This is influenced by the natural convection between seabed and seawater. The variation in soil temperature at different locations around the pipe with time under different pipe diameters is shown in Figure 13. The diameter of the pipeline affects the rise rate of the soil temperature around the pipeline. The larger the diameter, the slower the rise rate of soil temperature.

After 60 days of heating, the temperature distribution of soil around the pipeline tends to be stable. The temperature distribution around the pipeline during stable operation at different buried depths H is shown in Figure 14. It can be seen from the figure that the soil temperature gradient near the pipe wall is large, indicating that the soil is greatly affected by the pipe temperature. The soil temperature gradient far away from the pipe wall decreases, and the soil is less affected by the pipe temperature. The temperature gradient around the pipeline is relatively small at a small buried depth. With the increase in the buried depth, the temperature gradient tends to be stable, indicating that the influence of natural convection on the temperature field decreases with the increase in the buried depth.

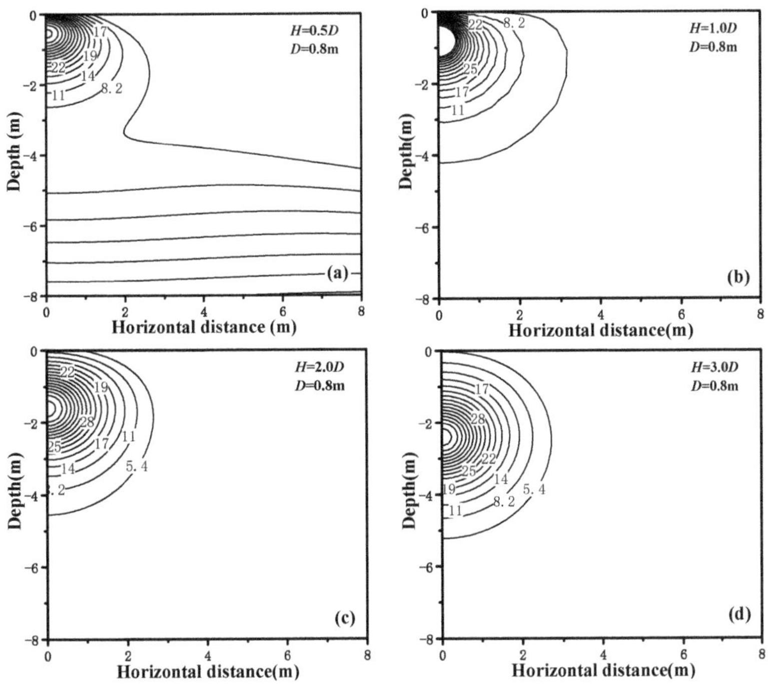

Figure 14. Temperature distribution of soil around pipeline at different buried depths: (**a**) $H = 0.5D$; (**b**) $H = 1.0D$; (**c**) $H = 2.0D$; (**d**) $H = 3.0D$.

The variation in soil temperature in the horizontal direction of the pipe center during stable operation at different buried depths is shown in Figure 15. It shows that the final influence range of heating on the soil temperature at different buried depths is about $4D$. However, the buried depth has little effect on the influence range of heating. At the same location, the smaller the buried depth, the lower the temperature, and the more obvious the influence of natural convection on the soil temperature.

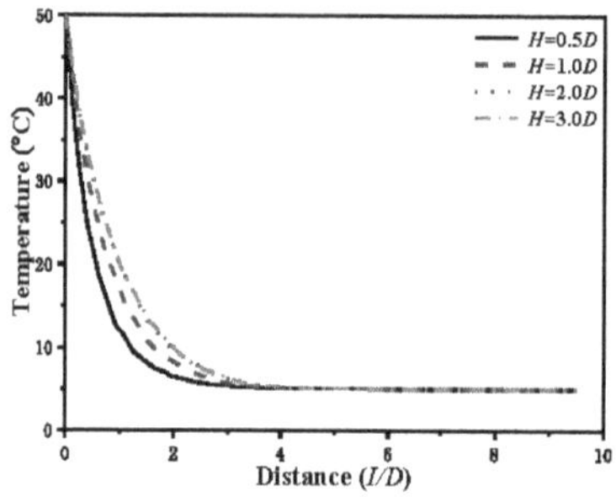

Figure 15. Temperature distribution of soil in horizontal direction of pipeline center at different buried depths.

The temperature distribution of soil around the pipeline with different pipe diameters during stable operation is shown in Figure 16. It can be seen that the contour shape of soil temperature around the pipeline under different pipe diameters is similar. But, the influence distance of heating will increase with the increased in the pipe diameter.

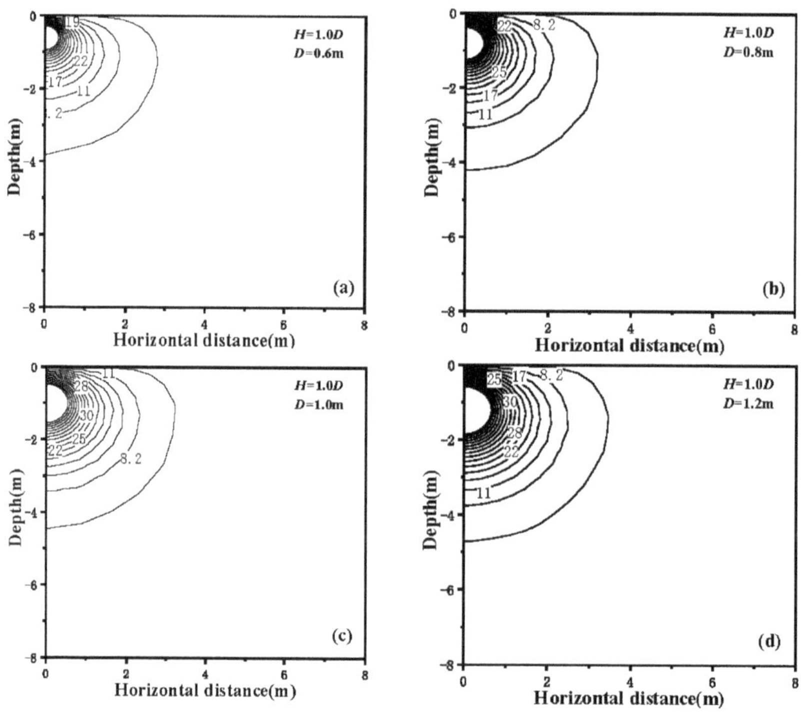

Figure 16. Temperature distribution of soil around pipeline at different diameters: (**a**) D = 0.6 m; (**b**) D = 0.8 m; (**c**) D = 1.0 m; (**d**) D = 1.2 m.

The variation in soil temperature in the horizontal direction of the pipe center during stable operation at different pipe diameters is shown in Figure 17. It shows that the influence range of the heating on the soil temperature under different pipe diameters is also about 4D. The pipe diameter has little influence on the temperature range of the pipe. At the same distance, the larger the pipe diameter, the lower the soil temperature.

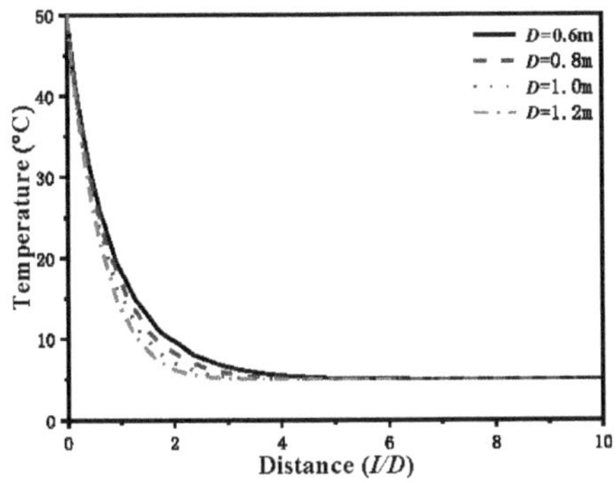

Figure 17. Temperature distribution of soil in horizontal direction of pipeline center at different pipeline diameters.

4. Conclusions

In this paper, an alternative experimental method to realize the visualization research on the temperature of soil around a submarine buried offshore pipeline is proposed based on transparent soil test technology. A set of test apparatuses was built, and the temperature distribution of the soil around the submarine pipeline during the operation was studied. Moreover, the influence of buried depth and pipe diameter on the soil temperature distribution was studied via numerical simulation, considering the natural convection. The following conclusions were obtained:

(1) The relationship between the soil temperature and normalized pixel intensity was established. The normalized pixel intensity varied from 0 to 1 for the temperature of 20 °C to 50 °C. The temperature of transparent soil had a logarithmic relationship with the normalized pixel intensity. It proved the feasibility of using transparent soil test technology to study the soil temperature distribution.

(2) The soil temperature at different locations increased with time, and then tended to be stable. The influence zone of pipeline temperature gradually expanded with the increase in time, and the final observable influence zone extended to a radial distance of 2.6*D*. At the distance far from 2.6*D*, it was difficult to use transparent soil to observe the temperature due to the small temperature change, which needed further researches to solve these problems.

(3) The shallower the buried depth of the pipeline, the smaller the soil temperature gradient around the pipeline and the more obvious the influence of natural convection of seawater. The contour shape of soil temperature around the pipeline was similar under different buried depths. The influence zone of pipeline temperature was a distance of 4*D* at a temperature difference of 45 °C, which was less effected by the buried depth and pipe diameter.

Author Contributions: Conceptualization, H.L. and Y.S.; methodology, Y.S. and L.G.; software, Y.M., Y.S. and H.L.; validation, L.G. and H.L.; formal analysis, L.G.; investigation, H.L. and Y.M.; resources, H.L., L.G. and Y.S.; data curation, H.L., Y.M., Y.S. and L.G.; writing—original draft preparation, H.L., Y.M. and Y.S.; writing—review and editing, Y.S. and L.G.; visualization, H.L. and Y.M.; supervision, L.G. All authors have read and agreed to the published version of the manuscript.

Funding: This study was funded by Youth Fund of the National Natural Science Foundation of China (52308371), the National Natural Science Foundation of China (51978532) and the Zhejiang Provincial Natural Science Foundation of China (LQ24E080013).

Institutional Review Board Statement: Not applicable.

Informed Consent Statement: Not applicable.

Data Availability Statement: The data presented in this study are available from the corresponding author by request.

Acknowledgments: The findings presented in this article were made possible through the diligent guidance of the authors' supervisor and the invaluable assistance of classmates, as well as the care and help of the teachers in the research group.

Conflicts of Interest: The authors declare no conflicts of interest.

References

1. Yang, X.; Shi, X.; Zhao, J.; Yu, C.; Gao, H.; Chen, A.; Lu, Y.-Z.; Chen, X.; Lin, W.; Zeng, X.; et al. Bottom Water Temperature Measurements in the South China Sea, Eastern Indian Ocean and Western Pacific Ocean. *J. Trop. Oceanogr.* **2018**, *37*, 86–97. [CrossRef]
2. Yu, J.; An, C.; Tang, Q.; Zhang, J.; Zhang, Y. Heat Transfer Characteristics of Subsea Long-Distance Pipeline Subject to Direct Electrical Heating. *Geoenergy Sci. Eng.* **2024**, *234*, 212679. [CrossRef]
3. Yuan, Q.; Luo, Y.; Shi, T.; Gao, Y.; Wei, J.; Yu, B.; Chen, Y. Investigation into the Heat Transfer Models for the Hot Crude Oil Transportation in a Long-Buried Pipeline. *Energy Sci. Eng.* **2023**, *11*, 2169–2184. [CrossRef]
4. Barletta, A.; Zanchini, E.; Lazzari, S.; Terenzi, A. Numerical Study of Heat Transfer from an Offshore Buried Pipeline under Steady-Periodic Thermal Boundary Conditions. *Appl. Therm. Eng.* **2008**, *28*, 1168–1176. [CrossRef]
5. Lu, T.; Wang, K. Numerical Analysis of the Heat Transfer Associated with Freezing/Solidifying Phase Changes for a Pipeline Filled with Crude Oil in Soil Saturated with Water during Pipeline Shutdown in Winter. *J. Pet. Sci. Eng.* **2008**, *62*, 52–58. [CrossRef]
6. Rossi di Schio, E.; Lazzari, S.; Abbati, A. Natural Convection Effects in the Heat Transfer from a Buried Pipeline. *Appl. Therm. Eng.* **2016**, *102*, 227–233. [CrossRef]
7. Li, Y.; Zhou, H.; Liu, H.; Ding, X.; Zhang, W. Geotechnical Properties of 3D-Printed Transparent Granular Soil. *Acta Geotech.* **2021**, *16*, 1789–1800. [CrossRef]
8. Iskander, M.; Lai, J.; Oswald, C.; Mannheimer, R. Development of a Transparent Material to Model the Geotechnical Properties of Soils. *Geotech. Test. J.* **1994**, *17*, 425–433. [CrossRef]
9. Zhao, H.; Ge, L. Investigation on the Shear Moduli and Damping Ratios of Silica Gel. *Granul. Matter* **2014**, *16*, 449–456. [CrossRef]
10. Liu, J.; Iskander, M.G. Modelling Capacity of Transparent Soil. *Can. Geotech. J.* **2010**, *47*, 451–460. [CrossRef]
11. Sun, Z.; Kong, G.; Zhou, Y.; Shen, Y.; Xiao, H. Thixotropy of a Transparent Clay Manufactured Using Carbopol to Simulate Marine Soil. *J. Mar. Sci. Eng.* **2021**, *9*, 738. [CrossRef]
12. Downie, H.; Holden, N.; Otten, W.; Spiers, A.J.; Valentine, T.A.; Dupuy, L.X. Transparent Soil for Imaging the Rhizosphere. *PLoS ONE* **2012**, *7*, e44276. [CrossRef] [PubMed]
13. Iskander, M.; Bathurst, R.J.; Omidvar, M. Past, Present, and Future of Transparent Soils. *Geotech. Test. J.* **2015**, *38*, 557–573. [CrossRef]
14. Ads, A.; Bless, S.; Iskander, M. Shape Effects on Penetration of Dynamically Installed Anchors in a Transparent Marine Clay Surrogate. *Acta Geotech.* **2023**, *18*, 3043–3059. [CrossRef]
15. Ads, A.; Shariful Islam, M.; Iskander, M. Longitudinal Settlements during Tunneling in Soft Clay, Using Transparent Soil Models. *Tunn. Undergr. Space Technol.* **2023**, *136*, 105042. [CrossRef]
16. Yu, S.; He, F.; Zhang, J. Experimental PIV Radial Splitting Study on Expansive Soil during the Drying Process. *Appl. Sci.* **2023**, *13*, 8050. [CrossRef]
17. Black, J.A.; Tatari, A. Transparent Soil to Model Thermal Processes: An Energy Pile Example. *Geotech. Test. J.* **2015**, *38*, 752–764. [CrossRef]
18. Siemens, G.A.; Mumford, K.G.; Kucharczuk, D. Characterization of Transparent Soil for Use in Heat Transport Experiments. *Geotech. Test. J.* **2015**, *38*, 620–630. [CrossRef]
19. Kong, G.Q.; Li, H.; Hu, Y.X.; Yu, Y.X.; Xu, W.B. New Suitable Pore Fluid to Manufacture Transparent Soil. *Geotech. Test. J.* **2017**, *40*, 658–672. [CrossRef]
20. Ezzein, F.M.; Bathurst, R.J. A New Approach to Evaluate Soil-Geosynthetic Interaction Using a Novel Pullout Test Apparatus and Transparent Granular Soil. *Geotext. Geomembr.* **2014**, *42*, 246–255. [CrossRef]
21. Zhang, J.; Stewart, D.P.; Randolph, M.F. Modeling of Shallowly Embedded Offshore Pipelines in Calcareous Sand. *J. Geotech. Geoenviron. Eng.* **2002**, *128*, 363–371. [CrossRef]
22. Zhang, J.M. Study on the Fundamental Mechanical Characteristics of Calcareous Sand and the Influence of Particle Breakage. Ph.D. Thesis, Chinese Academy of Sciences, Wuhan, China, 2007.
23. Liu, R.; Yan, S.W.; Wang, H.B.; Zhang, J.; Xu, Y. Model tests on soil restraint to pipelines buried in sand. *Chin. J. Geotech. Eng.* **2011**, *33*, 559–565.
24. Kong, G.Q.; Cao, Z.H.; Zhou, H.; Sun, X.J. Analysis of Piles Under Oblique Pullout Load Using Transparent-Soil Models. *Geotech. Test. J.* **2015**, *38*, 725–738. [CrossRef]

25. Bai, Y.; Niedzwecki, J.M.; Sanchez, M. *Numerical Investigation of Thermal Fields around Subsea Buried Pipelines*; American Society of Mechanical Engineers: New York, NY, USA, 2014.
26. Black, J.A.; Take, W.A. Quantification of Optical Clarity of Transparent Soil Using the Modulation Transfer Function. *Geotech. Test. J.* **2015**, *38*, 588–602. [CrossRef]

Journal of
Marine Science and Engineering

MDPI

Article

Stiffness Anisotropy and Micro-Mechanism of Calcareous Sand with Different Particle Breakage Ratios Subjected to Shearing Based on DEM Simulations

Yan Gao [1,2,*], Ketian Sun [1], Quan Yuan [3] and Tiangen Shi [1]

[1] School of Earth Sciences and Engineering, Sun Yat-sen University, Zhuhai 519080, China; sunkt@mail2.sysu.edu.cn (K.S.); shitg@mail2.sysu.edu.cn (T.S.)
[2] Southern Marine Science and Engineering Guangdong Laboratory, Zhuhai 519080, China
[3] Guangzhou Metro Design and Research Institute Co., Ltd., Guangzhou 510010, China; yuanquan@dtsjy.com
* Correspondence: gaoyan25@mail.sysu.edu.cn; Tel.: +86-159-1451-8020

Abstract: Stress-induced anisotropy in calcareous sand can cause an uneven displacement in island reef engineering. In this study, stiffness, as a quantitative indicator, is explored to reveal the stress-induced anisotropy in calcareous sand. Based on the discrete element method, the stiffness anisotropic characteristics of calcareous sand during shearing, as well as the impact of particle breakage, are investigated by numerical simulations. Both the macro and micro responses, i.e., the maximum shear modulus, contact normal, strong and weak contact normal force, and the direction of particle breakage, are explored for calcareous sand with different particle breakage ratios. The results show that calcareous sand exhibits notable anisotropy during shearing, with the maximum shear modulus in the vertical direction (deviatoric stress direction) being significantly greater than that in the horizontal direction. Moreover, the higher the particle breakage rate, the lower the stiffness and its anisotropy. The micro-mechanism results indicate that the primary particle breakage during the shearing process occurs in the vertical direction. That is, the particle breakage weakens the strong contact force in the vertical direction, leading to a redistribution of the strong contact forces from the vertical direction to other directions. This redistribution mainly manifests in a decrease in the anisotropy of contact normal and contact vector within the sample, as well as a decrease in the proportion of strong contact forces in the overall contacts. This, in turn, reduces the shear strength and stiffness of calcareous sand, particularly in the vertical direction, and results in a decrease in the maximum shear modulus and its anisotropy. The maximum reduction can be up to 50% of the original value. These insights can provide a certain theoretical support for the uneven displacement and long-term stability of calcareous sand for islands and reefs.

Keywords: calcareous sand; particle breakage; anisotropy; DEM

1. Introduction

In the South China Sea region, calcareous sand is widely distributed and typically chosen as the primary construction material for island and reef engineering projects [1–6]. Different from quartz sand, calcareous sand is an aggregate of insoluble carbonate minerals (mostly calcium carbonate), characterized by its irregular particle shape, high internal porosity, being easy to break, high angularity, and high internal friction angle [7–11]. Under conventional stress, particle breakage directly changes the internal structure of calcareous sand, including particle size distribution [12], packing density [13], friction angle [14], particle strength [15], permeability, and microstructure [16]. Currently, research on particle breakage is mainly based on laboratory experiments and field tests [17–20]. For example, Xiao et al. [13] found that the relationship between the particle strength and particle size of calcareous sand conforms to the Weibull distribution [21]. Coop et al. [22] discovered that under very large displacement conditions, calcareous sand reached a stable gradation that depends on the applied normal stress and initial gradation; at the same time, particle breakage occurred and caused volume continual compression until reaching a stable gradation. Liu et al. [19] proposed that particle breakage is related to particle hardness and stress state. Yu [18] suggested that a higher confining pressure would lead to more particle breakage. Particle breakage has a significant effect on the properties of calcareous sand. However, there are still shortcomings in the current research, such as its excessive cost, discrete results, and poor process repeatability. More importantly, due to the inherent dispersion and heterogeneity of calcareous sand, microscopic understandings of particle breakage are still insufficient.

Furthermore, extensive research has been conducted on calcareous sand to investigate its strength property, considering the influences of vertical stress, relative density, and particle size distribution. Gao et al. carried out a series of indoor experiments to investigate the shear characteristics [5], creep properties [6], and the relationship between particle breakage and plasticity work [4]. Zhang et al. [23] studied the results of triaxial shear tests on calcareous sand under different confining pressures and shear strains and analyzed the influencing factors and degree of particle breakage by the Hardin model. Yan et al. [24] conducted triaxial shear tests on calcareous sand with different particle sizes, relative densities, and confining pressures and introduced the stress-induced softening coefficient and shear dilatancy coefficient to quantitatively characterize the softening and dilatancy features. However, different stress paths on particle breakage and the anisotropy of shear strength were not considered. The anisotropy of shear strength in soil has long been considered a key issue and has been included in the analysis to improve the accuracy of soil strength prediction. In addition to the inherent anisotropy formed during the deposition process of soil particles, there is also an anisotropy caused by uneven external stresses. Toyota et al. [25] studied the variation of undrained shear strength of cohesive soil in different shear directions and stress conditions through hollow cylinder torsion tests. Hwang et al. [26] used finite element simulation to study the changes in stress state caused by excavation-induced slope sliding, resulting in the anisotropy of soil strength. Coccia and McCarteny [27] evaluated the effect of stress-induced anisotropy on saturated soil thermal volume change using a true triaxial test. Pan et al. [28] studied the undrained triaxial compression and tension behavior of saturated sand under different static pre-shearing conditions through experiments and analyzed the effects of pre-shearing on the anisotropy of sand elasticity, stiffness degradation, shear strength, and instability.

The different forces acting in the vertical and horizontal directions in island reef engineering, such as overlying soil pressure and building load pressure in the vertical direction and hydrostatic pressure in the horizontal direction, often lead to anisotropy of the foundation. However, for calcareous sand, research on macro- and micro-scale anisotropy with consideration of particle breakage is still in its initial stages. For example, He et al. [29] experimentally observed the pseudo-critical state, critical state, peak state, and phase transition state of calcareous sand under undrained and drained triaxial shear conditions. They studied the influence of different initial anisotropic stress states on the monotonic strength, deformation, and particle breakage characteristics of calcareous sand and found that the initial average effective stress and anisotropic stress ratio have a significant impact on the mechanical behavior of carbonate sand. However, the sieving method they used can only qualitatively evaluate the influence of particle breakage on the macroscopic properties and cannot reflect the micro-mechanism of particle breakage or reveal the influence of particle breakage on the macro/microstructure and deformation.

In recent years, the discrete element method [30] (DEM) has been proven to be a powerful tool for exploring the behavior of granular materials, such as shear strength, volume expansion, fracture evolution, contacts transmission, and energy distribution. Various DEM algorithms have been used to numerically simulate particle breakage and provide microscopic interpretations of experimental phenomena. These studies considered stress paths (e.g., confining pressure [31] and cyclic loading [32]), particle shape (e.g., particle roundness [33], sphericity [34], aspect ratio [35], particle size [36], boundary conditions [37], and intermediate principal stress [38]), etc. In this study, a series of triaxial shear numerical simulations is conducted based on the open-source 3D discrete element platform Yade [39], studying the shear strength, deformation characteristics, stiffness, microscopic behavior, and anisotropy of calcareous sand with different particle breakage rates.

2. Methods

2.1. Details of Numerical Simulations

2.1.1. Model Parameters

A cubic shape with dimensions of $10 \times 10 \times 10$ mm^3 was adopted into the DEM numerical model for the sample of calcareous sand. All generated particles with particle radii ranging from 0.24 to 0.48 mm are spherical in shape to avoid the anisotropy induced by internal structure. The particle gradation satisfies a uniform distribution with the average radius of 0.36 mm. The particle–particle contacts are modeled using the Hertz–Mindlin nonlinear contact model. The values of radius distribution, Young's modulus, Poisson's ratio, and density were determined based on the results of laboratory experiments [40], which are listed in Table 1.

Table 1. The values of the main parameters in simulations.

<table>
<tr><th></th><th>Parameter</th><th colspan="4">Value</th></tr>
<tr><td rowspan="9">Particles</td><td>Particle number</td><td colspan="4">3000</td></tr>
<tr><td>Radius</td><td colspan="4">0.24–0.48 mm</td></tr>
<tr><td>Density</td><td colspan="4">2810 kg/m^3</td></tr>
<tr><td>Young's modulus</td><td colspan="4">800 Mpa</td></tr>
<tr><td>Poisson's ratio</td><td colspan="4">0.3</td></tr>
<tr><td>Internal friction angle</td><td colspan="4">Sample preparation/compression: 0°
Shearing: 30°</td></tr>
<tr><td>Degree of particle breakage</td><td>None</td><td>Low</td><td>Mid</td><td>High</td></tr>
<tr><td>Uniaxial tensile strength (Mpa)</td><td>∞</td><td>4</td><td>2</td><td>1</td></tr>
<tr><td>Uniaxial compressive strength (Mpa)</td><td>∞</td><td>6</td><td>6</td><td>3</td></tr>
<tr><td rowspan="3">Walls</td><td>Young's modulus</td><td colspan="4">800 Mpa</td></tr>
<tr><td>Poisson's ratio</td><td colspan="4">0.05</td></tr>
<tr><td>Internal friction angle</td><td colspan="4">Sample preparation/compression: 0°
Shearing: 30°</td></tr>
</table>

To examine the influence of the degree of particle breakage, four levels of particle breakage, termed as high, medium, low, and none, were used, corresponding to approximately 10%, 5%, 1%, and 0% of the particle breakage rate, which is a ratio of the number of broken particles to the total number of particles. These four terms, "high, medium, low and none", were also consistently used in the following figures and discussion.

During the shearing process, an initial confining pressure of 400 kPa was applied, followed by maintaining a vertical strain rate of 2%/s until the vertical strain (axial strain) reached 18%, at which point the specimen was considered to be failed, and the shearing process was thereby terminated. Throughout the simulation, it was ensured that the unbalanced force rate does not exceed 0.01 to avoid unstable configurations.

2.1.2. Particle Breaking Mode

In the numerical simulations, the post-positioning method was used to simulate the particle breakage of calcareous sand. When the inter-particle contact force meets certain conditions, the particle breaks, and new particles are generated and replace the original particle. The strength criterion for determining whether the particle breaks is the modified Mohr–Coulomb–Weibull strength criterion proposed by Gladkyy [41], which uses the Mohr–Coulomb law with tensile and compressive cut-off and also considers the influence of particle size on the strength of a single particle. The relationship between particle strength and particle size follows the Weibull distribution. The particle replacement criterion is modified from the algorithm proposed by Brzeziński [42], and the specific replacement method is illustrated in Figure 1. When a particle is broken according to the Mohr–Coulomb–Weibull strength criterion, it is replaced by several small particles with equal radii. In this study, guided by previous studies [43–45], each particle is replaced with 9 small particles with equal radii when it breaks, and the basic parameters of the new particles (e.g., density, Young's modulus, etc.) are consistent with the original crushed particles. To ensure mass conservation before and after the breakage, the filling area was extended to the surrounding area. When the surrounding space is insufficient for successful replacement, the replacement is abandoned directly. This study optimizes the behavior when replacement fails and introduces a "delayed failure" mechanism guided by the experimental results of calcareous sand [40]. That is, when the space is limited, it is assumed to have been fractured, but the fractured particles are densely packed without affecting the surrounding particles, maintaining their original shape without replacement; when there is sufficient space later, the fractured particles are released and affect the surrounding particles, and regardless of whether they still meet the failure criteria, they are replaced.

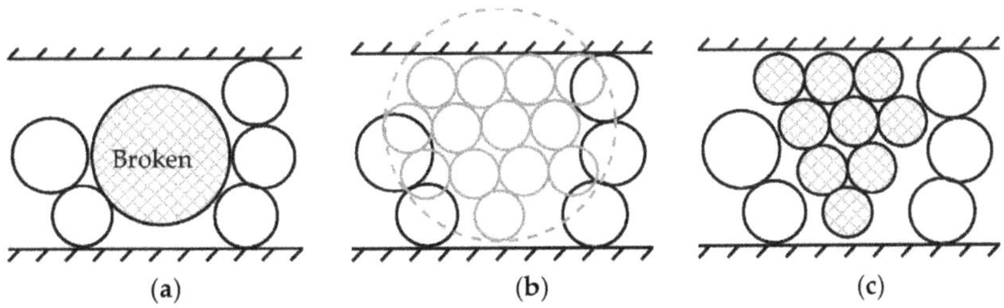

Figure 1. Spherical particle breakage mode process: (**a**) identify the particles to be crushed; (**b**) search for the potential locations of daughter particles; (**c**) fill when the daughter particles can be packed to a predetermined amount (modified from Gladkyy [41]).

2.1.3. The Maximum Shear Modulus, G_{max}

The anisotropy of soil properties is primarily attributed to two main factors: internal and external causes. The former arises from the irregular shape of the particles, while the latter results from uneven external loads. In this study, spherical particles were used; thus, the anisotropy generated by the particles themselves can be neglected, and the anisotropy generated during the shearing process is mainly caused by external loads. To characterize the anisotropy of calcareous sand during shearing, the shear modulus was used as an indicator. It is well known that the shear modulus itself decreases with increasing shearing strain, while it is almost constant when the shearing strain is very small ($<10^{-6}$ for sand), which corresponds to the maximum shear modulus G_{max}. Hence, the maximum shear modulus G_{max} is used to describe the anisotropic behavior of calcareous sand. In order to obtain G_{max} values at different shearing levels, the sample is first sheared to the target state by the triaxial test simulation (corresponding to different axial strains); then, another simple shearing mode is conducted, as shown in Figure 2. Since the shearing strain during the simple shearing process is very small ($<10^{-6}$) and the internal structure cannot be changed, the obtained G_{max} can represent the level of stiffness at different axial strains. As shown in Figure 2, the G_{max} of three orthogonal directions is denoted as G_{xz}, G_{yz}, and G_{xy}, respectively. Herein, x, y, and z represent three mutually perpendicular directions. By using G_{xz} as an example, the maximum shear modulus on the z–x plane can be determined by Equation (1):

$$G_{zx} = \frac{\tau_{zx}}{(\delta_{zx}/d_{zx})} \tag{1}$$

where τ_{zx} represents the shear stress on the z–x plane; δ_{zx} represents the lateral displacement between two walls' z-axes; and d_{zx} represents the sample dimension. This calculation method has been verified in previous studies [46].

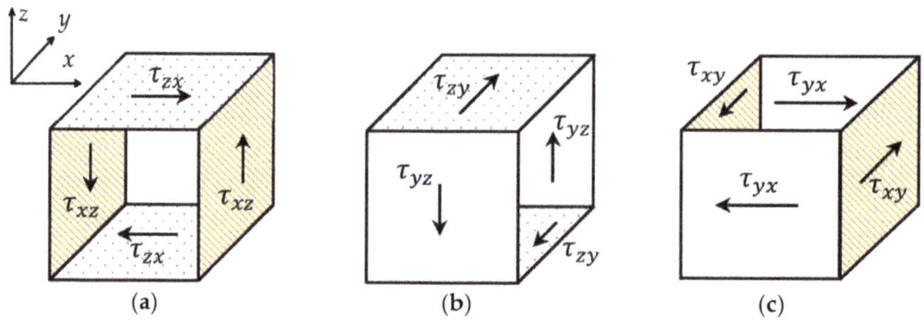

Figure 2. Shear modulus in different directions: (**a**) shear mode 1 G_{xz}; (**b**) shear mode 2 G_{yz}; and (**c**) shear mode 3 G_{xy}.

According to previous studies [47–51], in the elastic condition, the maximum shear modulus G_{max} is only related to the stress in the corresponding plane, i.e., G_{xy} is only controlled by the stress in the xy plane and is independent of the stress in the z direction.

2.2. Methodology for Micro-Scale Parameters

Guided by Radjai et al. [52], the contacts between particles are classified into two parts: strong contact normal force chain, where the contact normal force is greater than the average force, and weak contact normal force chain. These two types of contacts form two complementary networks with different geometric and mechanical properties. The strong contact normal force chain bears the entire additional stress generated by triaxial shearing, while the weak contact normal force chain only contributes to maintaining the average stress. To explore the intrinsic mechanisms of the anisotropic characteristics during the shearing of calcareous sand, the evolutions of microscopic parameters, such as the contact normal, contact vectors and their directions, the magnitude of contact normal forces, and the direction of particle breakage, were analyzed.

2.2.1. Anisotropy of Contact Normal

The contact normal force direction between two adjacent particles, known as the contact normal, refers to the direction perpendicular to the contact surface. To quantitatively characterize the anisotropy degree of contact normal, the fabric tensor of microstructures [53] was introduced as follows:

$$N = \frac{1}{N_c} \sum_{c=1}^{N_c} \left(n^c \bigotimes n^c \right) \tag{2}$$

where n^c is the unit vector of contact normal and N_c is the number of contacts between particles. The eigenvalues and eigenvectors of the fabric tensor can reflect the concentration of anisotropy of the contact normal and principal directions. The coefficient of anisotropy is given as follows:

$$\Delta\Phi = \sqrt{\frac{\left(e_1^N - e_2^N\right)^2 + \left(e_2^N - e_3^N\right)^2 + \left(e_1^N - e_3^N\right)^2}{2}} \tag{3}$$

where e_i^N (i = 1, 2, or 3) represents the three eigenvalues of the fabric tensor **N**. At the same time, selecting the eigenvector corresponding to the maximum eigenvalue represents the dominant direction of anisotropy.

2.2.2. Anisotropy of Contact Vector

The contact vector is the vector connecting the centroids of two contacting particles. The branch tensor can reflect the average length of the line connecting the centroids in different directions in space, equivalent to the average radius of the contacting particles. The higher the contact vector, the larger the related particles. In order to distinguish the anisotropy of the contact vector from the anisotropy of the contact direction, the formula proposed by Sitharam et al. [54] was used to calculate the contact vector tensor $\boldsymbol{B}^n$:

$$B_{ij}^n = \frac{1}{N_c}\sum_{c=1}^{N_c} \frac{l_c^n n_i^c n_j^c}{E(n^c)_{ij}} \tag{4}$$

where n^c represents the contact normal; l_c^n is the length of the contact vector, defined as the distance between the centroids of two contacted particles; and $E(n)$ is the probability density function of the contact normal. For any direction vector n:

$$\boldsymbol{E}(\boldsymbol{n}) = \frac{1}{4\pi}\left(1 + \boldsymbol{D} \bigodot (n \bigotimes n)\right) \tag{5}$$

in which $\boldsymbol{D}$ is the skew tensor of the fabric tensor $\boldsymbol{N}$ shown in Equation (2). On this basis, to conveniently compare the anisotropy degree under different axial strains, the normalized contact vector tensor $\boldsymbol{B}^{n'} = \boldsymbol{B}^n \bigodot \frac{1}{tr(\boldsymbol{B}^n)}$ was used to calculate the level of anisotropy. The calculation method for the coefficient of anisotropy for the contact vector $\Delta\Phi_{cv}$ is consistent with the form of $\Delta\Phi$ for the contact normal (Equation (3)), where e_i (i = 1, 2, or 3) represents the three eigenvalues of the contact vector tensor. The dominant direction is also represented by the related eigenvector corresponding to the maximum eigenvalue.

2.2.3. Anisotropy of Contact Normal Force

The external load is mainly borne by the contact normal force between particles, which is the component of the contact force between two particles along the normal direction. The formula defined by Sitharam [54] was also used for this calculation. The normalized degree of anisotropy $\Delta\Phi_{cf}$ was also used to facilitate the comparison of the anisotropy of contact normal force magnitudes under different axial strains, and the related eigenvector corresponding to the maximum eigenvalue also represents the dominant direction. For all contact normal force vectors, the contact normal force is demoted as f_c^n. The specific formula for calculating the contact normal force tensor is as follows:

$$\begin{cases} F_{ij}^n = \frac{1}{N_c}\sum\limits_{c=1}^{N_c} \frac{f_c^n n_i^c n_j^c}{E(n^c)_{ij}} \\ \boldsymbol{F}^{n'} = \boldsymbol{F}^n \bigodot \frac{1}{tr(\boldsymbol{F}^n)} \end{cases} \tag{6}$$

where $\boldsymbol{E}(\boldsymbol{n})$ is the probability density function of the contact normal.

2.2.4. Quantification of Particle Breakage Direction Anisotropy

To further explore the influence of particle breakage on the anisotropy of inter-particle contacts, the degree of anisotropy and the principal direction of particle breakage were studied. The direction of maximum contact normal force during particle breakage was taken as the direction of particle breakage. Given the particle breakage direction vector n^b and the related contact number N_b, the particle breakage tensor $\boldsymbol{N}^b$ was determined as follows:

$$\boldsymbol{N}^b = \frac{1}{N_b}\sum_{b=1}^{N_b}\left(n^b \bigotimes n^b\right) \tag{7}$$

The calculation of the degree of anisotropy and the principal direction were similar to that for the contact normal detailed in Section 2.2.1. The degree of anisotropy $\Delta\Phi_{pb}$

was calculated using the eigenvalues of N^b substituted into Equation (3). The principal direction of anisotropy was determined using the eigenvector corresponding to the maximum eigenvalue.

3. Results

3.1. Shear Characteristics and Stiffness Variation of Calcareous Sand with Different Particle Breakage Rates

3.1.1. Stress–Strain Relationship

As shown in Figure 3, the shear stress–strain response of calcareous sand increases rapidly in the initial stage (axial strain less than 3%), and the curves are basically overlapped for different particle breakage ratios. This is because the shear deformation in this stage is mainly controlled by the adjustment of particle positions caused by particle rotation and sliding and the filling of internal large voids, accompanied by insignificant particle crushing (Figure 4). As the deviatoric stress continues to increase, the stress–strain relationships of calcareous sand with different particle breakage ratios show differences: the deviatoric stress of calcareous sand with a higher particle breakage ratio is obviously lower than that with a lower particle breakage ratio at the same axial strain. More stress drops occur for calcareous sand with a higher particle breakage ratio during shearing, which is caused by the stress release induced by particle crushing.

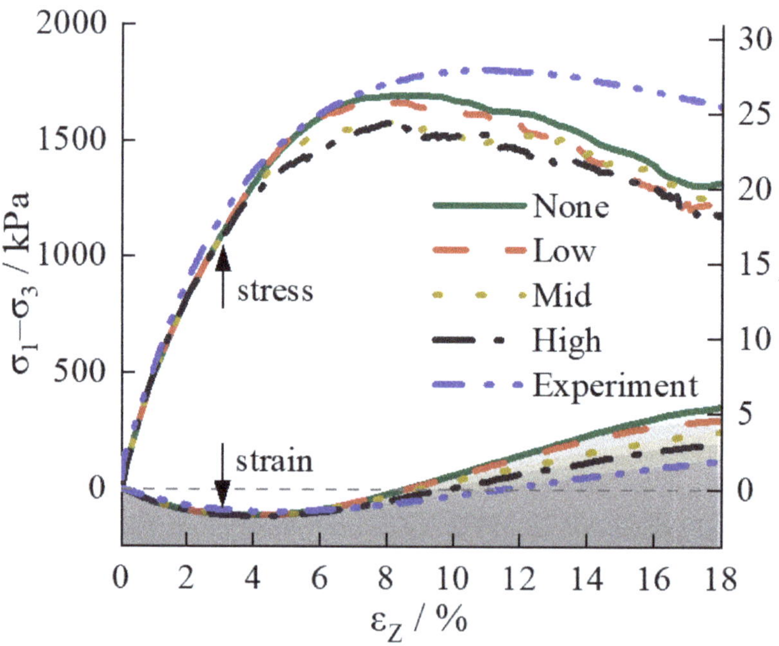

Figure 3. Deviatoric stress–axial strain and volume strain–axial strain curves for calcareous sand with different particle breakage rates.

The volume change during the shearing is the first contraction, followed by dilation. When the axial strain is less than 6%, the volumetric strain–axial strain curves of the calcareous sand with different particle breakage ratios are basically overlapped because, before this point, the overall particle breakage is at a relatively low level and changes little (Figure 4). With the increase in axial strain, the level of particle breakage increases. As shown in Figure 4, for the calcareous sand with a higher degree of particle breakage, particle breakage occurs earlier, and the rate of particle breakage growth is also faster. The continual increase in particle breakage makes the dilation tendency of the calcareous sand decrease even further.

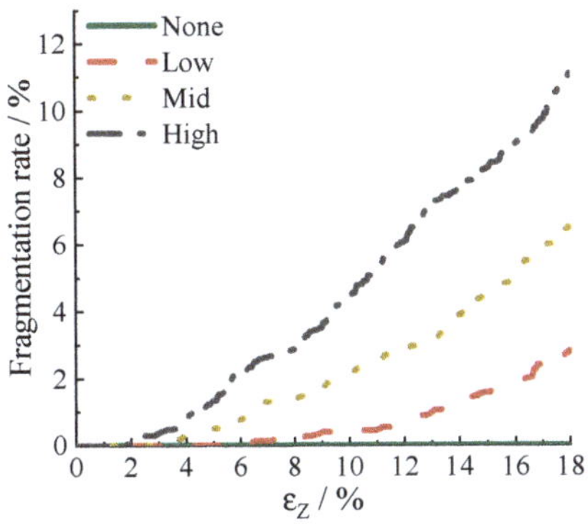

Figure 4. The variation in the particle breakage rate with axial strain for different calcareous sands.

These simulation results are consistent with previous studies (e.g., reference [55]) and laboratory experiments. There are discrepancies between the current simulation results and the experimental results at large strains, specifically manifested as a significant overestimation of volume dilation and an underestimation of peak stress, as shown in Figure 3. The possible reason is the lack of consideration of the actual particle shape, surface roughness, intra-particle voids, and so on. However, it is believed that the tendency of the current results is accurate for studying the causes of stress-induced anisotropy.

3.1.2. Macroscopic Stiffness Anisotropy

During shearing, as the axial strain increases, the changes in G_{xz}, G_{yz}, and G_{xy} are shown in Figure 5. For different particle breakage rates, with the increase in axial strain, both G_{xz} and G_{yz} first increase to a peak until the axial strain reaches 4–5% and then decrease. Meanwhile, G_{xy} remains basically unchanged until reaching the 2% axial strain and then decreases. This is because with the increase in axial strain, the stress in the z direction increases, while the stresses in the x and y directions remain unchanged. The maximum shear modulus in the elastic condition is only related to the stress in the shear plane. However, after reaching a certain level of strain, the specimen begins to exhibit plastic deformation, and the original structure of the calcareous sand is destroyed. Therefore, the maximum shear moduli in all directions decrease. Based on Figure 5, it can also be observed that at low axial strains (<2%), the maximum shear modulus G_{max} for different particle breakage rates basically overlaps as before this point, the degree of particle breakage is low and the influence of particle breakage on the macroscopic stiffness of the specimen is not significant. As the axial strain increases, with the particle breakage increasing, the level of increase in G_{max} decreases more for samples with higher particle breakage. At large axial strains (>6%), the higher the particle breakage rate, the smaller the G_{max} value. In other words, particle breakage weakens the stiffness of calcareous sand, and the reduction can be up to 50% compared with that without particle breakage, which is crucial for the stability and safety assessments of the foundations of calcareous sand.

The maximum shear moduli in different directions have different magnitudes and trends, showing significant anisotropy. The changes in G_{xz}/G_{yz} and G_{xz}/G_{xy} with the axial strain are shown in Figure 6 and Table 2. Throughout the entire shearing process, G_{xz}/G_{yz} is close to one, and G_{xz}/G_{xy} continually increases. There is a significant difference in the maximum shear moduli between the horizontal and vertical planes; that is, the stiffness in the vertical direction is greater than that in the horizontal direction during shearing. With the increase in axial strain, the degree of anisotropy increases. As the particle breakage rate increases, the ratio of G_{xz}/G_{yz} under the same axial stress does not change significantly at small strains, but at higher strains, it is much lower under higher particle breakage rates. That is, particle breakage can induce the macroscopic stiffness and related degree of anisotropy of the calcareous sand decrease, and the degree of decrease is related to the particle breakage rate; the larger the rate, the greater the decrease.

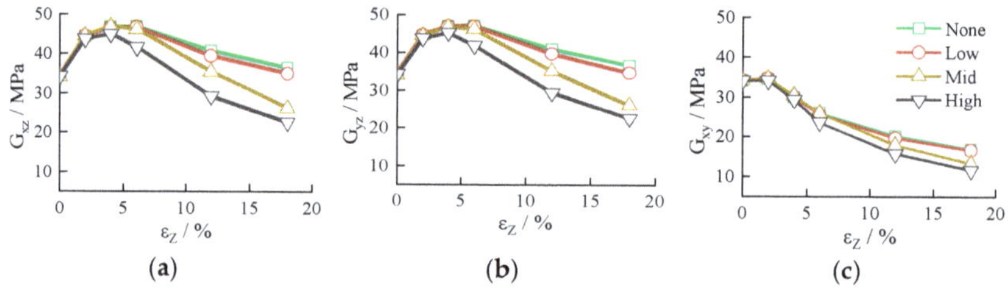

Figure 5. The evolution of small-strain shear moduli in different directions as the axial strain increases during the shearing process: (**a**) G_{xz}; (**b**) G_{yz}; and (**c**) G_{xy}.

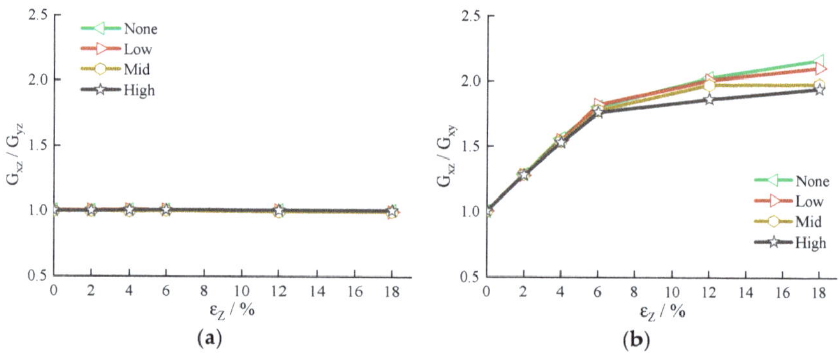

Figure 6. The difference in small-strain shear moduli in different directions during shearing with increasing axial strain: (**a**) G_{xz}/G_{yz} and (**b**) G_{xz}/G_{xy}.

Table 2. Variation in G_{xz}/G_{xy} for calcareous sand with different particle breakages.

Strain (%)	None	Low	Mid	High
0	1.004	1.003	1.001	1.004
2	1.286	1.281	1.285	1.282
4	1.562	1.550	1.539	1.529
6	1.805	1.821	1.775	1.761
12	2.025	2.009	1.974	1.860
18	2.163	2.103	1.979	1.944

3.2. Anisotropy of Internal Contact Forces

During shearing, the external stress causes different responses in the contact normal forces between the calcareous sand particles. As shown in Figure 7, taking the situation under the 18% axial strain as an example, the magnitudes of the contact forces between the particles in the horizontal direction (XOY plane) and the vertical direction (XOZ plane) are significantly different. In the XOY plane, the magnitude of the contact forces does not have a clear dominant direction. It is consistent in all directions, forming a circular distribution of force magnitudes. In the XOZ plane, the magnitude of the contact forces forms an elliptical distribution, with the contact forces in the Z direction being significantly greater than those in the X direction, consistent with the larger external stress in the vertical direction. The internal microstructural features can reflect macroscopic phenomena.

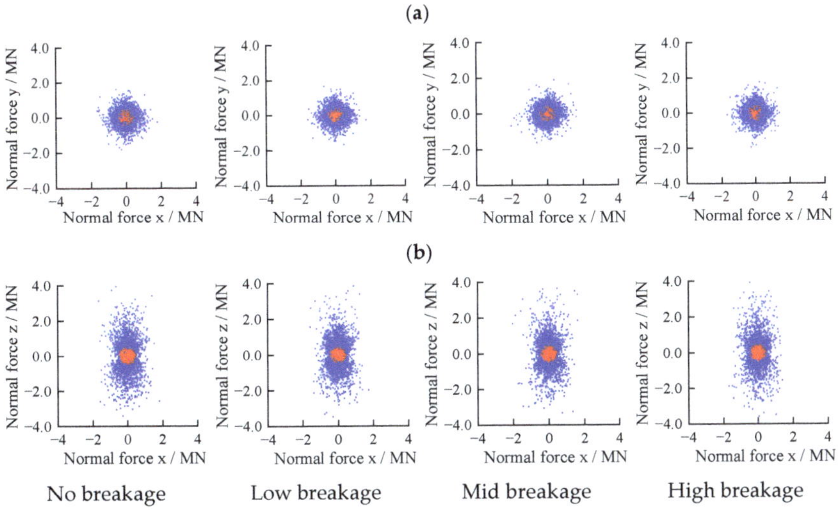

Figure 7. Projection of inter-particle contact forces in calcareous sand at different planes under an 18% axial strain: (**a**) XOY plane and (**b**) XOZ plane. In this figure, the blue points represent strong contact forces, and the red points represent weak contact forces.

4. Discussion

Macroscopic behavior can be reflected and related to the internal structure's evolution. Based on the relationship between the stress and strain for calcareous sand during shearing (Figure 3), two representative cases of particle breakage rate, including no particle breakage and high particle breakage, and three axial strains at 2% (before the peak deviatoric stress), 6% (around the peak deviatoric stress), and 18% (after the peak deviatoric stress) are selected for comparison. The indicators of internal structure for calcareous sand, i.e., contact normal, contact vector, distribution of contact normal forces, and the direction of particle breakage, are discussed.

4.1. Contact Normal

4.1.1. Distribution of Contact Normal

The projections of contact normal in the XOY and XOZ planes are shown by the density proportional rose diagrams in Figures 8 and 9, respectively. In the XOY plane, there is no significant preferential direction for both strong and weak contacts, and the distribution of contact normal is circular, with a significantly larger number of weak contacts than strong contacts. In the XOZ plane, there is a clear preferential direction for strong and weak contacts, mainly distributed in the vertical direction (Z direction) for strong contacts and in the horizontal direction (X direction) for weak contacts. Furthermore, with the increase in axial strain, the number of weak contacts in the horizontal direction significantly decreases.

The direction of maximum contact normal force for particle breakage is served as the direction of particle breakage [44,56], which is also shown in Figure 9. There is no significantly preferred direction for particle breakage in the XOY plane, but in the XOZ plane, particle breakage mainly occurs in the vertical direction (Z direction), indicating the existence of anisotropy in particle breakage. Under the same axial strain, the comparison of the samples with high particle breakage and without particle breakage is shown in Figure 10 (i.e., Figures 8 and 9). It is clearly seen that the particle breakage level reduces the strong contact, especially in the Z direction, while the weak contact increases in the Z direction. In other words, particle breakage effectively reduces the concentration of strong contact forces in the Z direction, thereby reducing the degree of anisotropy.

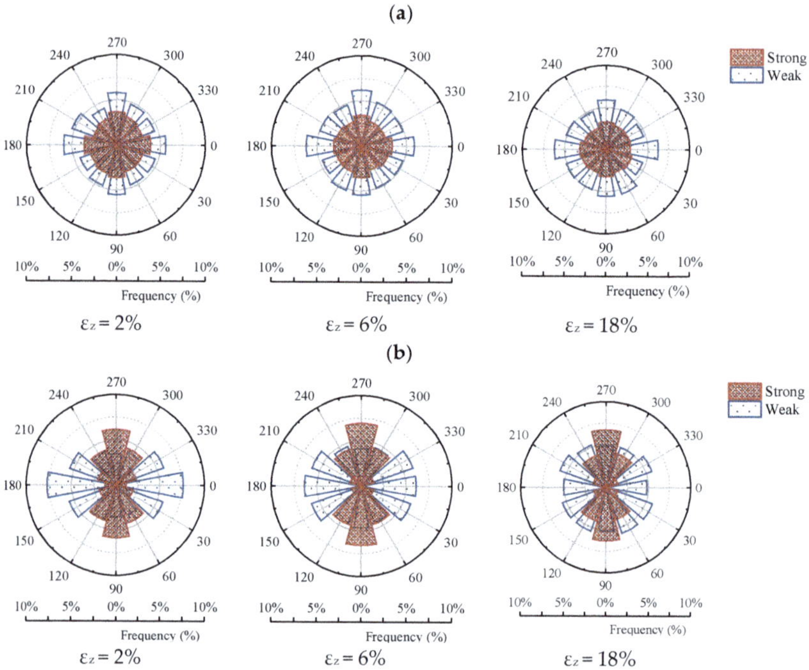

Figure 8. Distribution of inter-particle contact normal under different axial strains without particle breakage: (**a**) XOY plane and (**b**) XOZ plane.

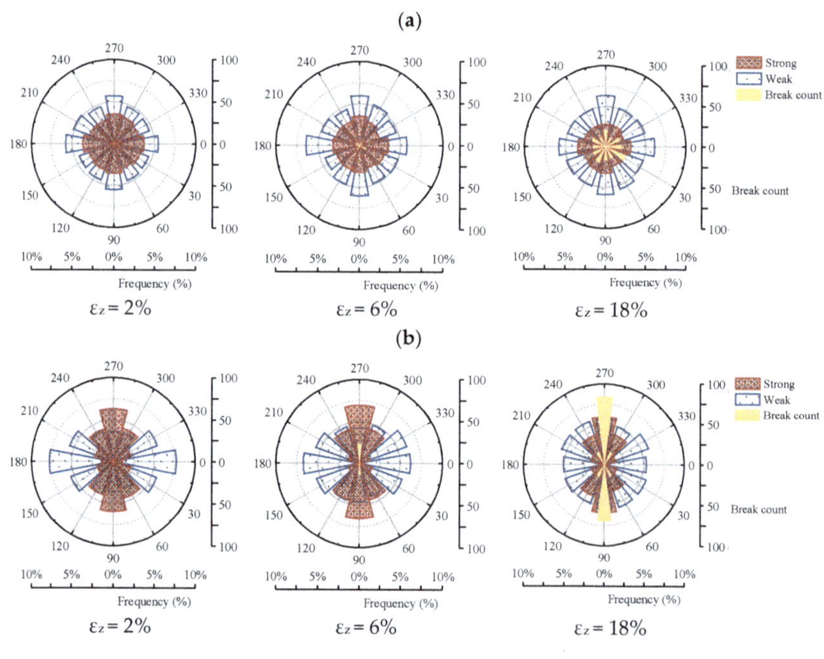

Figure 9. Distribution of inter-particle contact normal under different axial strains at high particle breakage rates: (**a**) XOY plane and (**b**) XOZ plane.

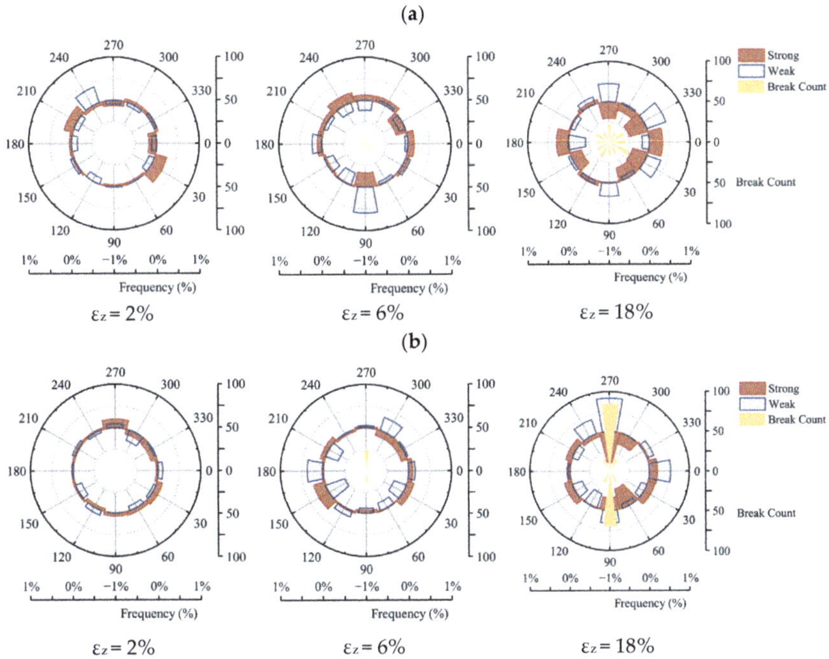

Figure 10. Influence of particle breakage on the distribution of inter-particle contact normal at different axial strains: (**a**) XOY plane and (**b**) XOZ plan.

4.1.2. Contact Normal Anisotropy

Figure 11 shows the anisotropy degree of the orientation of strong and weak contacts (represented by the fabric tensor) during the shearing process as a function of axial strain, where N, L, M, and H represent no, low, medium, and high particle breakage, and S and W denote strong and weak contacts, respectively. These symbols have the same meanings as in the following figures. Throughout the entire shearing process, the dominant direction of the strong contacts is consistently in the vertical direction (~90°; Z direction in Figure 11a), while the dominant direction of the weak contacts is in the horizontal direction (0°; X or Y direction in Figure 11a), with occasional fluctuations affected by particle breakage.

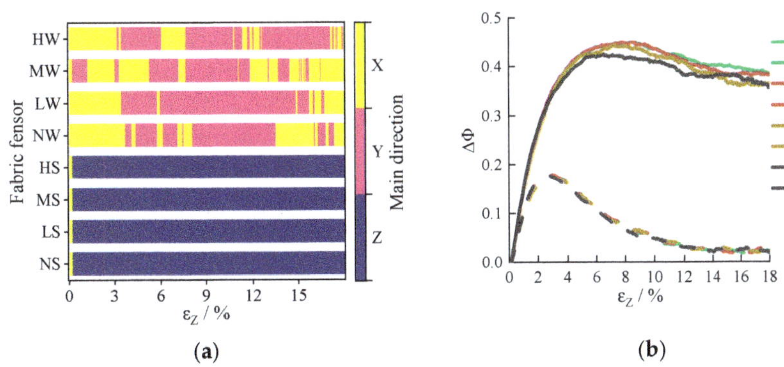

Figure 11. Anisotropy of particle contact orientation (fabric tensor) for different particle breakage ratios: (**a**) dominant direction and (**b**) anisotropy degree.

For weak contacts, with the increase in axial strain, the degree of anisotropy first increases, reaching its maximum value at around a 3% axial strain, and then significantly decreases to negligible levels. This variation reflects the transition of deformation in the calcareous sand from elastic to plastic deformation during shearing. The degree of anisotropy coefficients at different particle breakage rates are similar and much smaller than the anisotropy of strong contacts. This indicates that particle breakage has almost no effect on the anisotropy of weakly bonded particles.

For strong contacts, the maximum level of anisotropy appears at around 6–12% axial strain and then decreases slightly. For different particle breakage rates, there is no significant difference in the initial shear stage. The degree of anisotropy for the higher particle breakage rate of calcareous sand is significantly smaller when the axial strain exceeds 3%. It proves that particle breakage can affect the anisotropy of the strong contacts. The higher the particle breakage rate, the lower the degree of anisotropy. Due to the strong contacts controlling the overall strength and stability of calcareous sand [57,58], one of the intrinsic reasons for the reduction in the macroscopic anisotropy degree (e.g., the maximum shear modulus, as shown in Figure 4) is the reduction in the anisotropy of the strong contacts between particles.

4.2. Contact Vector Anisotropy

During shearing, the anisotropy degree of the strong contact vector with respect to the axial strain development is shown in Figure 12. Similar to the results of the contact normal, the dominant direction of the strong contact vector is always in the vertical direction (~90°), and the dominant direction of the weak contact vector is mainly in the horizontal direction (0°), but it fluctuates occasionally due to particle breakage. Particle breakage has almost no effect on the anisotropy magnitude of the weak contact vector. For strong contacts, after the 6% axial strain, with the increase in axial strain and particle breakage rate, the degree of anisotropy decreases, but the difference is not significant. That is, the higher the particle breakage rate, the lower the anisotropy degree of the contact vector, but the impact is weak. It is also indicated that the strong contacts in the Z direction is mainly borne by large particles. This is because the preferred direction of the contact vector for strong contacts is along the Z direction, and the corresponding anisotropy of the contact vector tensor $\boldsymbol{B}^{n'}$ is higher. This suggests that the average radius of the particles bearing strong contacts in the Z direction is larger. As the particle breakage rate increases, there is a slight decrease in the $\Delta\Phi_{cv}$ for strong contacts, whereas the $\Delta\Phi_{cv}$ for weak contacts remains relatively unchanged, which indicates that particle breakage is likely to occur more frequently in the larger particles that bear strong contacts. The reduced anisotropy of the contact vector of the strong contacts means that the size of the particles bearing the strong contacts along the Z direction decreases.

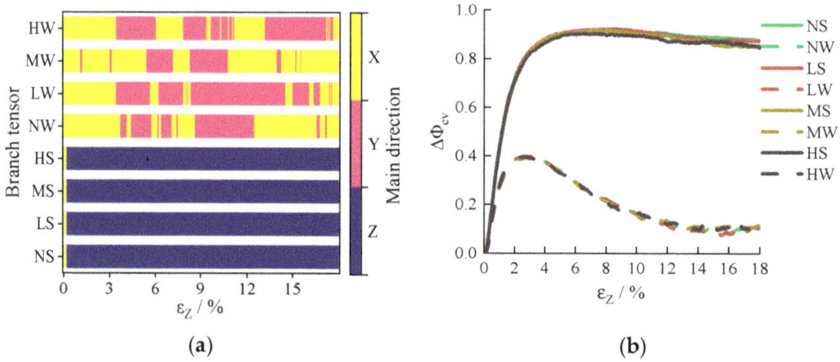

Figure 12. Anisotropy of contact branch vectors at different particle breakage rates: (**a**) dominant direction and (**b**) anisotropy degree.

4.3. Contact Normal Force

4.3.1. The Anisotropy of Contact Normal Forces

The normalized anisotropy degree and principal direction of the contact normal forces are shown in Figure 13. It is readily seen that no significant changes are obtained under different particle breakage rates. That is, particle breakage does not have a considerable influence on the contact normal forces in terms of the anisotropy degree and principal direction.

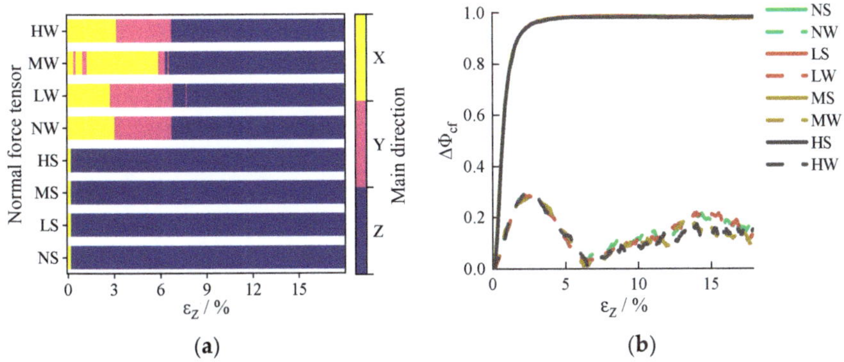

Figure 13. Anisotropy of normal contact forces in the strong and weak contacts at different particle breakage ratios: (**a**) dominant orientation and (**b**) anisotropy degree.

4.3.2. The Distribution of Contact Normal Forces

Taking the representatives at the axial strains of 2%, 6%, and 18% of calcareous sand with no particle breakage and high particle breakage, the contact normal force distribution histograms are shown in Figure 14. Under a small axial strain, the contact normal force distributions are almost identical due to the insignificant particle breakage for calcareous sand with different particle breakage rates. As the shearing progresses, the number of strong contact normal forces of calcareous sand with high particle breakage is significantly smaller than that with no particle breakage, while the number of weak contact normal forces is significantly larger. The evolution of the percentage of weak contacts during shearing in Figure 15 also indicates a decrease in the proportion of strong contact normal forces and an increase in the proportion of weak contact normal forces. At the same time, the increase in weak contact normal forces in calcareous sand with high particle breakage is significantly greater than in calcareous sand with no particle breakage. In other words, particle breakage can enlarge the destruction of strong contacts and the formation of weak contacts between particles in calcareous sand.

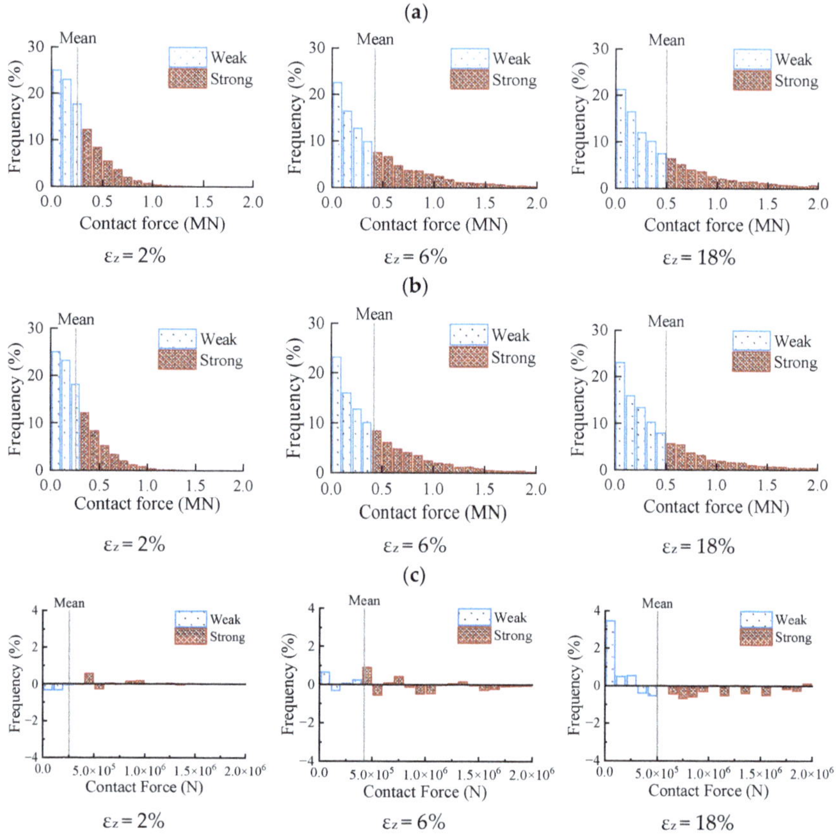

Figure 14. Influence of particle breakage on contact normal force distributions at different axial strains for calcareous sand with different particle breakage rates: (**a**) no particle breakage; (**b**) high particle breakage; and (**c**) variation (high breakage–no breakage).

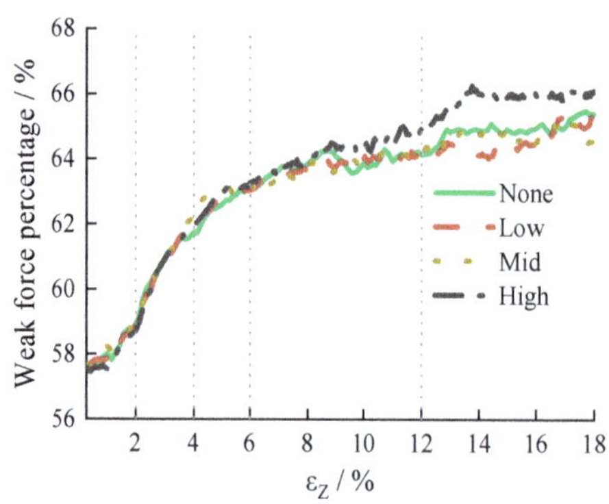

Figure 15. The variation in the proportion of weak contact normal forces with axial strain.

4.4. Anisotropy of Particle Breakage Direction

Figure 16 shows the degree of anisotropy of particle breakage and the principal direction. The principal direction of particle breakage of calcareous sand with different degrees of particle breakage is always the vertical direction, consistent with the external deviatoric stress direction. During shearing, with the increase in axial strain, especially after the peak stress (corresponding to a 6% axial strain), the degree of anisotropy of particle breakage decreases. This is because at lower strains, only strong contacts that are nearly parallel to the Z-axis can satisfy the failure criteria. Therefore, it is considered that the direction of particle breakage is more concentrated in the direction parallel to the Z-axis at lower strains. As the strain progresses, a large number of strong contacts parallel to the Z-axis have already been extensively damaged by particle breakage. Therefore, contacts with a certain angle to the Z-axis can also satisfy the failure criteria. As a result, the level of anisotropy in the direction of particle breakage decreases. Furthermore, the greater the degree of particle breakage, the greater the reduction in the degree of anisotropy. The reduction in the degree of anisotropy of particle breakage will also lead to a decrease in the macroscopic degree of anisotropy of calcareous sand.

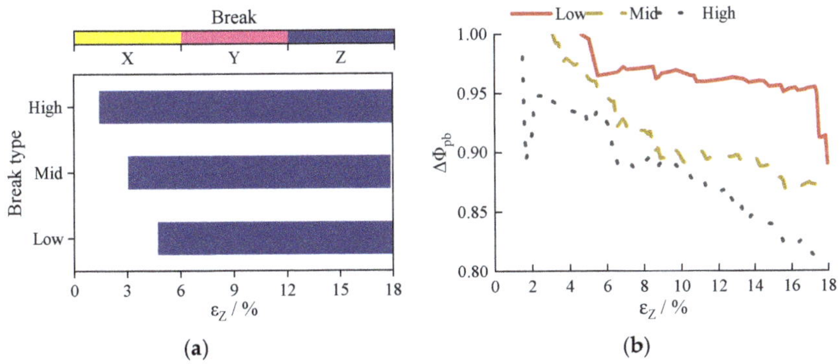

Figure 16. Anisotropy of particle breakage: (**a**) dominant direction and (**b**) anisotropy degree.

The findings of this study emphasize the importance of understanding the influence of particle breakage on the stiffness anisotropy of calcareous sand. It should be noted that the DEM method has certain limitations, such as the demand for computational efficiency and resources, along with the influence of size effects, and the increase in particle quantity due to particle breakage further slows down its computational speed. The results of this study can be used as a reference and guide for future engineering endeavors. In the future, further numerical simulations under conditions more akin to actual engineering environments could be required, considering the dynamic wave loads, the true shapes of calcareous sand particles, and more realistic forms of particle breakage, as well as variations in Young's modulus, Poisson's ratio, and so on. These simulations should be combined with field testing or laboratory experiments, which can enhance the reliability and application of research outcomes in practical engineering.

5. Conclusions

This study simulates the particle breakage behavior using the post-positioning method and the "delayed replacement" mechanism. The anisotropic characteristics and micro-mechanisms of calcareous sand with different particle breakage rates subjected to shearing are investigated using the discrete element numerical simulations. The main conclusions are as follows:

(1) During shearing, the macroscopic stiffness G_{max} of the calcareous sand is greater in the vertical direction than that in the horizontal direction, and the degree of stiffness anisotropy increases with the increase in axial strain. A distinct anisotropy is found. Particle breakage can reduce the stiffness in different directions, mainly in the direction of deviatoric stress, which, in turn, weakens the anisotropy characteristics of the calcareous sand induced

by external stress. As the particle breakage rate increases, the peak and residual shear stress and shear dilatancy of the calcareous sand decrease.

(2) The micro-anisotropy of the calcareous sand during shearing is mainly manifested in the distribution of contact normal directions, contact vectors, and contact normal force values between particles. During the shearing process, the strong contacts and related contact normal forces in the vertical direction are significantly greater than those in the horizontal direction, while the weak contacts and related contact normal forces in the horizontal direction are significantly greater than those in the vertical direction. Thus, particle breakages mainly occur in the vertical direction. The degree of anisotropy of particle breakage decreases with an increase in the particle breakage rate. Although particle breakage has no significant impact on the anisotropy of the contact normal force mean value and the principal direction, it mainly affects the distribution characteristics of contact normal forces. The strong contact forces decrease while the weak contact forces increase, which leads to a tendency of contact force redistribution and uniformity. The reducing strong contact normal forces results in a significant reduction in G_{max}. Additionally, the degree of anisotropy of the strong contact direction also decreases with the increase in particle breakage, thus reducing the degree of stiffness anisotropy.

Author Contributions: Conceptualization, K.S.; Methodology, Y.G.; Software, K.S.; Validation, Y.G. and Q.Y.; Formal analysis, K.S.; Investigation, Y.G.; Resources, Y.G.; Data curation, K.S.; Writing—original draft, K.S.; Writing—review & editing, Y.G., Q.Y. and T.S.; Supervision, Y.G.; Project administration, Y.G.; Funding acquisition, Y.G. All authors have read and agreed to the published version of the manuscript.

Funding: This research was supported by the National Natural Science Foundation of China (42072295), the Guangdong project (2017ZT07Z066), and the National Key R&D Program of China (2022YFC3005203). The third author, Quan Yuan, also acknowledges the support from the Guangdong Provincial Key Laboratory of New Construction Technology for Urban Rail Transit Engineering (2017B030302009).

Institutional Review Board Statement: Not applicable.

Informed Consent Statement: Not applicable.

Data Availability Statement: Data are contained within the article.

Conflicts of Interest: Author Quan Yuan was employed by the company Guangzhou Metro Design and Research Institute Co., Ltd. The remaining authors declare that the research was conducted in the absence of any commercial or financial relationships that could be construed as a potential conflict of interest.

References

1. Da, B.; Yu, H.; Ma, H.; Tan, Y.; Mi, R.; Dou, X. Experimental Investigation of Whole Stress-Strain Curves of Coral Concrete. *Constr. Build. Mater.* **2016**, *122*, 81–89. [CrossRef]
2. Kuang, D.; Long, Z.; Guo, R.; Zhao, T.; Wu, K. Experimental and Numerical Study on the Fragmentation Mechanism of a Single Calcareous Sand Particle under Normal Compression. *Bull. Eng. Geol. Environ.* **2021**, *80*, 2875–2888. [CrossRef]
3. Xiong, H.; Qiu, Y.; Liu, J.; Yin, Z.-Y.; Chen, X. Macro–Microscopic Mechanism of Suffusion in Calcareous Sand under Tidal Fluctuations by Coupled CFD-DEM. *Comput. Geotech.* **2023**, *162*, 105676. [CrossRef]
4. Gao, R.; Ye, J. Mechanical Behaviors of Coral Sand and Relationship between Particle Breakage and Plastic Work. *Eng. Geol.* **2023**, *316*, 107063. [CrossRef]
5. Gao, R.; Ye, J. A Novel Relationship between Elastic Modulus and Void Ratio Associated with Principal Stress for Coral Calcareous Sand. *J. Rock Mech. Geotech. Eng.* **2024**, *16*, 1033–1048. [CrossRef]
6. Ye, J.; Haiyilati, Y.; Cao, M.; Zuo, D.; Chai, X. Creep Characteristics of Calcareous Coral Sand in the South China Sea. *ACTA Geotech.* **2022**, *17*, 5133–5155. [CrossRef]
7. Chen, H.Y.; Wang, R.; Li, J.G.; Zhang, J.M. Grain Shape Analysis of Calcareous Soil. *Rock Soil Mech.* **2005**, *26*, 1389–1392. [CrossRef]
8. Ma, C.H.; Zhu, C.Q.; Liu, H.F.; Cui, X.; Wang, T.M.; Jiang, K.F.; Yi, M.X. State-of-the-Art Review of Research on the Particle Shape of Soil. *Rock Soil Mech.* **2021**, *42*, 2041–2058. [CrossRef]
9. Wang, X.Z.; Wang, X.; Liu, H.F.; Meng, Q.S.; Zhu, C.Q. Field Test Study of Engineering Behaviors of Coral Reef Foundation. *Rock Soil Mech.* **2017**, *38*, 2065–2070+2079. [CrossRef]

10. Qing, Y.; Yao, T.; Wang, R.; Zhu, C.Q.; Meng, Q.S. Particle Breakage-Based Analysis of Deformation Law of Calcareous Sediments under High-Pressure Consolidation. *Rock Soil Mech.* **2014**, *35*, 3123–3128. [CrossRef]
11. Wang, X.Z.; Wang, X.; Weng, Y.L.; Lv, S.Z.; Yan, K.; Zhu, C.Q. Characteristics of Dry Density of Calcareous Sand and Its Testing Methods. *Rock Soil Mech.* **2016**, *37*, 316–322. [CrossRef]
12. Varadarajan, A.; Sharma, K.G.; Venkatachalam, K.; Gupta, A.K. Testing and Modeling Two Rockfill Materials. *J. Geotech. Geoenviron. Eng.* **2003**, *129*, 206–218. [CrossRef]
13. Xiao, Y.; Liu, H.L.; Ding, X.M.; Chen, Y.M.; Jiang, J.S.; Zhang, W.G. Influence of Particle Breakage on Critical State Line of Rockfill Material. *Int. J. Geomech.* **2016**, *16*, 04015031. [CrossRef]
14. Konrad, J.-M.; Salami, Y. Particle Breakage in Granular Materials—A Conceptual Framework. *Can. Geotech. J.* **2018**, *55*, 710–719. [CrossRef]
15. Lade, P.V.; Yamamuro, J.A.; Bopp, P.A. Significance of Particle Crushing in Granular Materials. *J. Geotech. Eng.* **1996**, *122*, 309–316. [CrossRef]
16. Wang, Y.; Santosh, M.; Luo, Z.H.; Hao, J.H. Large Igneous Provinces Linked to Supercontinent Assembly. *J. Geodyn.* **2015**, *85*, 1–10. [CrossRef]
17. Jia, Y.F.; Wang, B.S.; Chi, S.C. Particle breakage of rockfill during triaxial tests. *Chin. J. Geotech. Eng.* **2015**, *37*, 1692–1697. [CrossRef]
18. Yu, F.W. Characteristics of Particle Breakage of Sand in Triaxial Shear. *Powder Technol.* **2017**, *320*, 656–667. [CrossRef]
19. Liu, S.H.; Mao, H.Y.; Wang, Y.S.; Weng, L.P. Experimental Study on Crushable Coarse Granular Materials during Monotonic Simple Shear Tests. *Geomech. Eng.* **2018**, *15*, 687–694. [CrossRef]
20. Wu, Y.Q.; Tahmasebi, P.; Lin, C.Y.; Zahid, M.A.; Dong, C.M.; Golab, A.N.; Ren, L.H. A Comprehensive Study on Geometric, Topological and Fractal Characterizations of Pore Systems in Low-Permeability Reservoirs Based on SEM, MICP, NMR, and X-ray CT Experiments. *Mar. Pet. Geol.* **2019**, *103*, 12–28. [CrossRef]
21. Weibull, W. A Statistical Distribution Function of Wide Applicability. *J. Appl. Mech.* **1951**, *18*, 293–297. [CrossRef]
22. Coop, M.R.; Sorensen, K.K.; Freitas, T.B.; Georgoutsos, G. Particle Breakage during Shearing of a Carbonate Sand. *Geotechnique* **2004**, *54*, 157–163. [CrossRef]
23. Zhang, J.M.; Zhang, L.; Jiang, G.S.; Wang, R. Research on Particle Crushing of Calcareous Sands under Triaxial Shear. *Rock Soil Mech.* **2008**, *29*, 2789–2793. [CrossRef]
24. Yan, C.P.; Long, Z.L.; Zhou, Y.C.; Kuang, D.M.; Chen, J.M. Investigation on the Effects of Confining Pressure and Particle Size of Shear Characteristics of Calcareous Sand. *Rock Soil Mech.* **2020**, *41*, 581–591+634. [CrossRef]
25. Toyota, H.; Susami, A.; Takada, S. Anisotropy of Undrained Shear Strength Induced by K0 Consolidation and Swelling in Cohesive Soils. *Int. J. Geomech.* **2014**, *14*, 04014019. [CrossRef]
26. Hwang, J.; Dewoolkar, M.; Ko, H.-Y. Stability Analysis of Two-Dimensional Excavated Slopes Considering Strength Anisotropy. *Can. Geotech. J.* **2002**, *39*, 1026–1038. [CrossRef]
27. Coccia, C.R.; McCartney, J. A Thermo-Hydro-Mechanical True Triaxial Cell for Evaluation of the Impact of Anisotropy on Thermally Induced Volume Changes in Soils. *Geotech. Test. J.* **2012**, *35*, 227–237. [CrossRef]
28. Pan, K.; Yang, Z.X.; Xu, T.T. Impact of Static Preshearing on Undrained Anisotropy and Shear Characteristics of Sand. *Int. J. Geomech.* **2018**, *18*, 04018162. [CrossRef]
29. He, S.H.; Goudarzy, M.; Ding, Z.; Sun, Y.F. Strength, Deformation, and Particle Breakage Behavior of Calcareous Sand: Role of Anisotropic Consolidation. *J. Geotech. Geoenvironmental Eng.* **2023**, *149*, 04023002. [CrossRef]
30. Cundall, P.A.; Strack, O.D.L. A Discrete Numerical Model for Granular Assemblies. *Géotechnique* **1979**, *29*, 47–65. [CrossRef]
31. Zhang, T.; Wu, J. Discriminative Frequency Filter Banks Learning with Neural Networks. *Eurasip J. Audio Speech Music Process.* **2019**, *2019*, 1. [CrossRef]
32. Dahal, B.; Mishra, D. Discrete Element Modeling of Permanent Deformation Accumulation in Railroad Ballast Considering Particle Breakage. *Front. Built Environ.* **2020**, *5*, 145. [CrossRef]
33. Afshar, T.; Disfani, M.M.; Arulrajah, A.; Narsilio, G.A.; Emam, S. Impact of Particle Shape on Breakage of Recycled Construction and Demolition Aggregates. *Powder Technol.* **2017**, *308*, 1–12. [CrossRef]
34. Zhang, T.; Zhang, C.; Zou, J.; Wang, B.; Song, F.; Yang, W. DEM Exploration of the Effect of Particle Shape on Particle Breakage in Granular Assemblies. *Comput. Geotech.* **2020**, *122*, 103542. [CrossRef]
35. Ueda, T.; Matsushima, T.; Yamada, Y. DEM Simulation on the One-Dimensional Compression Behavior of Various Shaped Crushable Granular Materials. *Granul. Matter* **2013**, *15*, 675–684. [CrossRef]
36. McDowell, G.R.; Harireche, O. Discrete Element Modelling of Soil Particle Fracture. *Geotechnique* **2002**, *52*, 131–135. [CrossRef]
37. Qu, T.; Feng, Y.T.; Wang, Y.; Wang, M. Discrete Element Modelling of Flexible Membrane Boundaries for Triaxial Tests. *Comput. Geotech.* **2019**, *115*, 103154. [CrossRef]
38. Liu, Y.; Liu, H.; Mao, H. DEM Investigation of the Effect of Intermediate Principle Stress on Particle Breakage of Granular Materials. *Comput. Geotech.* **2017**, *84*, 58–67. [CrossRef]
39. Smilauer, V.; Angelidakis, V.; Catalano, E.; Caulk, R.; Chareyre, B.; Chevremont, W.; Dorofeenko, S.; Duriez, J.; Dyck, N.; Elias, J.; et al. Yade Documentation. *arXiv* **2023**, arXiv:2301.00611.
40. Gao, Y.; Shi, T.; Yuan, Q.; Sun, K. The Creep Characteristics and Related Evolution of Particle Morphology for Calcareous Sand. *Powder Technol.* **2024**, *431*, 119077. [CrossRef]

41. Gladkyy, A.; Kuna, M. DEM Simulation of Polyhedral Particle Cracking Using a Combined Mohr–Coulomb–Weibull Failure Criterion. *Granul. Matter* **2017**, *19*, 41. [CrossRef]
42. Brzeziński, K.; Gladky, A. Clump Breakage Algorithm for DEM Simulation of Crushable Aggregates. *Tribol. Int.* **2022**, *173*, 107661. [CrossRef]
43. Ciantia, M.O.; Arroyo, M.; Calvetti, F.; Gens, A. An Approach to Enhance Efficiency of DEM Modelling of Soils with Crushable Grains. *Geotechnique* **2015**, *65*, 91–110. [CrossRef]
44. Zhang, K.; Zhang, S.; Teng, J.; Sheng, D. 3D Numerical Simulation of Particle Breakage Using Discrete Element Method. *Rock Soil Mech.* **2017**, *38*, 2119–2127. [CrossRef]
45. Zhao, F.; Chi, S. A Simulation Method for Particle Breakage with Random Fragment Size Based on Discrete Element Method. *J. Northeast. Univ. Nat. Sci.* **2023**, *44*, 408–414.
46. Benjamin, M. An Investigation of Strain Localization in Cemented Sands and Mechanisms of Stiffness Anisotropy Using the Dem Simulations—Rare & Special e-Zone. Master's Thesis, The Hong Kong University of Science and Technology, Hong Kong, China, 2007.
47. Stokoe, K.; Lee, S.; Knox, D.P. Shear Moduli Measurements under True Triaxial Stresses. In *Advances in the Art of Testing Soils Under Cyclic Conditions*; ASCE: Reston, VA, USA, 1985.
48. Stokoe, K.H.I.; Hwang, S.K.; Lee, J.N.K.; Andrus, R.D. Effects of Various Parameters on the Stiffness and Damping of Soils at Small to Medium Strains. In *Pre-Failure Deformation of Geomaterials, Proceedings of the International Symposium on Pre-Failure Deformation Characteristics of Geomaterials, Sapporo, Japan, 12 September 1994*; CRC Press: Boca Raton, FL, USA, 1994; pp. 785–816.
49. Stokoe, K.H.I.; Lee, J.N.K.; Lee, S.H.H. Characterization of Soil in Calibration Chambers with Seismic Waves. In *Calibration Chamber Testing, Proceedings of the First International Symposium on Calibration Chamber Testing/ISOCCT1, Potsdam, NY, USA, 28–29 June 1991*; Elsevier: Amsterdam, The Netherlands, 1991; pp. 363–376.
50. Ismail, M.A.; Sharma, S.S.; Fahey, M. A Small True Triaxial Apparatus with Wave Velocity Measurement. *Geotech. Test. J.* **2005**, *28*, 113–122. [CrossRef]
51. Chaney, R.; Demars, K.; Fioravante, V.; Capoferri, R. On the Use of Multi-Directional Piezoelectric Transducers in Triaxial Testing. *Geotech. Test. J.* **2001**, *24*, 243. [CrossRef]
52. Radjai, F.; Wolf, D.E.; Jean, M.; Moreau, J.-J. Bimodal Character of Stress Transmission in Granular Packings. *Phys. Rev. Lett.* **1998**, *80*, 61–64. [CrossRef]
53. Satake, M. Fabric Tensor in Granular Materials. In *Deformation and Failure of Granular Materials, Proceedings of the IUTAM-Conference on Deformation and Failure of Granular Materials, Delft, The Netherlands, 31 August–3 September 1982*; Elsevier: Amsterdam, The Netherlands, 1982; pp. 63–68.
54. Sitharam, T.G.; Vinod, J.S.; Ravishankar, B.V. Post-Liquefaction Undrained Monotonic Behaviour of Sands: Experiments and DEM Simulations. *Géotechnique* **2009**, *59*, 739–749. [CrossRef]
55. Cai, Z.; Chen, Y.; Zhu, X.; Tang, Y. Influences of Gradation on Particle Breakage and Deformation Characteristics of Coral Sand. *Chin. J. Geotech. Eng.* **2023**, *45*, 661–670.
56. Russell, A.R.; Muir Wood, D. Point Load Tests and Strength Measurements for Brittle Spheres. *Int. J. Rock Mech. Min. Sci.* **2009**, *46*, 272–280. [CrossRef]
57. Gao, Y.; Yu, J.Y.; Wang, Y.B.; Li, W.L.; Shi, T.G. Breakage Effect of Calcareous Sand on Pile Tip Resistance and the Surrounding Soil Stress. *Energy Rep.* **2022**, *8*, 183–190. [CrossRef]
58. Chen, Q.; Gao, Y.; Yuan, Q.; Yu, J.Y. Microstructural Response during Shearing Process of Dense Sand. *Water Esources Hydropower Eng.* **2021**, *52*, 178–187. [CrossRef]

Journal of
Marine Science and Engineering

MDPI

Article

The Influence of Structural Design on the Hydrodynamics of Floating Offshore Wind Turbine Platforms

Nilotpal Dhar [1,*], Charlie J. Lloyd [1], John Walker [2] and Robert M. Dorrell [3]

[1] Energy and Environment Institute, University of Hull, Hull HU6 7RX, UK; c.j.lloyd@hull.ac.uk
[2] Offshore Renewable Energy Catapult, Inovo, 121 George St, Glasgow G1 1RD, UK; john.walker@ore.catapult.org.uk
[3] School of Architecture, Building and Civil Engineering, Loughborough University, Loughborough LE11 3TU, UK; r.m.dorrell@lboro.ac.uk
* Correspondence: n.dhar-2021@hull.ac.uk

Abstract: Floating offshore wind turbine (FOWT) platforms are subject to a wide range of hydrodynamic loading and dynamic movement, making hydrodynamic force evaluation difficult. Amongst various floating platforms, submersible platforms are structurally complex, with multiple members held together by cross-braces. The influence of these members on hydrodynamic loading is poorly understood. An investigation of the effect of these members on loads is essential to optimise the design of FOWT platforms, mooring systems, and protective coatings, leading to a reduction in construction and maintenance costs. This paper numerically investigates the effect of structural members on the forces acting on a static semi-submersible platform in a unidirectional current flow of Reynolds number (*Re*) ranging from 2000 to 200,000, based on structural diameter and tidal velocity. The OC4 semi-submersible is chosen as the baseline platform. For each *Re*, this study is divided into three stages, such that in each stage, the number of members increased. These stages are as follows: (1) a finite cylinder (FC), (2) a finite cylinder with a heave plate (FCHP), (3) three cylinders with heave plates (TCHP) in an equilateral triangle arrangement, and (4) the OC4 semi-sub. The drag coefficient ($\overline{C}_d$) increases with increasing structural members and weakly varies with increasing *Re*. However, the viscous drag coefficient ($\overline{C}_f$) decreases with increasing *Re*, and a reverse trend is seen in the case of the pressure drag coefficient ($\overline{C}_p$), with pressure drag dominating over friction drag. Further, the contribution of individual members is observed to vary with *Re*. The contribution of cylinders towards $\overline{C}_d$ is higher than heave plates, showing that contributions directly depend on the aspect ratio of members. In the case of TCHP and OC4, the contribution of the rear members is higher than that of the leading members due to the strong wake effect of the former. Also, the braces and pontoons of OC4 have contributed substantially towards total $\overline{C}_d$, unlike the central cylinder, which has experienced low drag due to the wake effect of the front cylinder and heave plate. Also, flow visualisation has shown vortex cores, and recirculating flows in the near wake of the cylinders and under the heave plates. Recirculation zones under the heave plates lead to vertical pressure on the structures. This vertical pressure increases with the number of structural members and the vertical pressure coefficient ($\overline{C}_v$), varying with *Re* due to three-dimensionality in the wake. Further, this pressure varies across the bottom surfaces of structures. Analyses of the streamwise pressure coefficient have shown it is highest on the front surfaces of cylinders. The highest friction is on the top and sides of the heave plates, and there is considerable friction on the sides of the cylinder.

Keywords: offshore wind energy; floating offshore wind turbine platform; hydrodynamic loading; pressure coefficient; friction coefficient

1. Introduction

Renewable energy sources, including offshore wind energy, are essential for reducing fossil fuel consumption and achieving decarbonisation. To this end, many countries across the world have set targets to achieve NetZero. Globally, offshore wind capacity is forecast to nearly double over the next five years, from 72.5 GW in 2023 to 138 GW in 2028 [1]. So far, offshore wind farm developers have preferred installing wind turbines at near-shore sites due to easy access and associated reduced construction, operation, and maintenance costs [2,3]. However, this also means that suitable near-shore sites available for development are limited. Therefore, new wind farms are expected to be developed at far-offshore, deep water locations ($\geq$60 m) where wind resources are stronger and more consistent. This move towards deeper waters requires the deployment of floating offshore wind turbines (FOWT), as opposed to the current fixed-bottom technology. Floating platforms are extensively used in the oil and gas (O and G) industry for fuel extraction from deep-sea sites [4]. In fact, the initial conceptualisation of FOWT platforms is based on those employed in the O and G industry [5]. However, the scale of FOWT foundations, location, and loading on them are vastly different than O and G floating platforms. This necessitates the design and structural optimisation of the FOWT foundations to meet the requirements of the sector. As a result, different types of FOWT platforms have been developed. Amongst them, the most widely adopted are the spar-buoy [6], semi-submersible [7], and tension-leg platforms [8] (Figure 1). Spar-buoys are cylindrical structures with a heave ballast attached at the lower end to stabilise the platform. Semi-submersible platforms have multiple short columns held together by cross-braces/struts and are water-plane stabilised. Both these platforms are held at the station using mooring systems. The tension-leg platform resembles a truncated cylinder with horizontal members extending outwards from the lower end. The foundation is stabilised by tendons connecting the horizontal arms to the sea bed. Further, the draft of the platform should be such that the buoyancy force acting on the structure keeps the mooring system tensioned [9]. Hywind Scotland is the world's first floating wind farm where a spar-buoy type foundation has been used, whereas, in the Kincardine offshore wind farm, the wind turbines are supported by semi-submersible platforms [6,10]. Compared to onshore wind turbines, offshore wind turbines are subjected to additional hydrodynamic loading due to waves and currents, including loading due to wave-induced diffraction and platform-induced radiation [11]. Therefore, accounting for this wide range of loads, especially under extreme weather conditions, is essential for the design optimisation of the entire FOWT system and cost reduction [12]. However, the baseline understanding of loading on floating structures is poor. Therefore, at the onset, load evaluation on FOWT platforms under simplified conditions is essential.

Unlike spar-buoy and TLP, which structurally resemble a single cylinder supporting the wind turbine with smaller plan areas, semi-submersible platforms usually have three cylinders with heave plates spread apart in a triangular arrangement. Heave plates are usually thinner but wider than cylinders and help to reduce the heave motion of structures [13]. The wind turbine is held on the top of one of the cylinders or on a third cylinder at the centre (Figure 1). All these members are held together by multiple braces/struts and pontoons. These additional members make semi-submersible platforms much more structurally complex than other FOWT platforms. Further, these members are bound to affect the flow field around the floating foundation, the loading, and its distribution on platforms. However, the influence of these members on hydrodynamic forces is poorly understood. Also, the forces are expected to vary with increasing Reynolds number, which needs to be considered.

Figure 1. The three different floating platforms for offshore wind turbines. From left to right, spar-buoy, semi-submersible, and tension-leg platforms. All three foundations are tethered to the sea bed through an anchoring–mooring system. Image by Joshua Bauer, NREL [14].

Tran and Kim [15] successfully studied the free-decay and regular wave motions of the OC4 semi-submersible platform through two different CFD methods. The results obtained with the shear stress model and Spalart–Almaras model agreed well with experiments compared with the $k - \epsilon$ model. A second CFD study by Tran and Kim [16] investigating the motion response amplitude operators (RAOs) in regular waves showed good agreement with experiments and OpenFast. However, there was a large difference in tension acting on mooring systems compared to experimental results [17,18]. Wang et al. [19] numerically studied the pitch decay of the OC4 semi-submersible platform and compared the numerical solution to the experimental study. Although very small discretisation errors were observed in platform maxima motion and period, large differences in linear and quadratic damping coefficients were also observed. A complex catenary morning line, drag of the mooring system, power cables, aerodynamic damping due to wind turbine and tower, and three-degrees-of-freedom adopted in simulations were some aspects that influenced the results [20].

Further, in a separate study, Wang et al. [21] showed that the incorporation of dynamic modelling of the catenary system significantly improved CFD predictions of pitch period and linear pitch damping coefficient. However, the quadratic pitch damping coefficient was still underpredicted. Li and Bachynski-Polic [22] simulated the low-frequency radiation characteristics of OC4 under free-decay motions and forced oscillations in heave, surge, and pitch and compared them with experiments [23,24]. Although pitch and heave decay numerical predictions matched with experimental results, there was a substantial difference in surge decay. This excessive damping was attributed to the mechanical mooring system in the experimental setup, which was not considered in simulations. Burmester et al. [25] studied surge decay motions through CFD simulations and formally quantified discretisation errors through time-step and grid convergence studies. The influence of discretisation schemes, free-surface, scaling, domain size, catenary mooring models, wave-absorption, turbulence models, etc., was also studied [26]. The addition of a wave-absorption zone was found to have a limited influence on hydrodynamic damping. Also, parameters, such as line weight and drag coefficient, associated with catenary modelling strongly influenced the structure's motion. Wang et al. [27] simulated platform motion in regular waves and found good agreement of surge RAO with experiments. However, heave and pitch RAOs were underpredicted due to unsatisfactory non-linear modelling of the catenary system.

Liu et al. [28] investigated the vortex-induced motion (VIM) of the OC4 submersible platform due to unidirectional current. Three different incidence angles (α) and a range of flow velocities were considered. The flow velocities were non-dimensionalised by the natural frequency of the platform and the diameter of the cylinder, which are called reduced velocities (V_r). Although the streamwise motion for all α was limited, large amplitude transverse motions were observed from $V_r = 6$ to 30 for $\alpha = 0°$ and 90°. However, the transverse motion at $\alpha = 180°$ was comparatively low due to upstream vortices interacting with downstream cylinders. In a separate CFD study, Liu et al. [29] compared the VIM of cylinders with and without a heave plate. The heave plate was found to expand the V_r range over which VIM takes place. Further, the large surface area of the heave plate increased viscous damping, reducing the VIM response. Benitz et al. [30] compared the hydrodynamic coefficient on a stationary semi-submersible platform obtained from CFD solver, OpenFOAM, against results from Fast HydroDyn. They showed that the presence of a free surface, free ends, and multiple members reduced drag. However, these factors were not considered in HydroDyn. Ma et al. [31] studied the influence of drag on low-frequency surge motion of the semi-submersible platforms and concluded that under moderate sea conditions, drift motion was directly related to the drag coefficient and could significantly affect motion under extreme sea conditions. Also, the choice of the drag coefficient, depending on the sea state, was essential to simulate the drift and damping forces [32]. Existing state-of-the-art engineering tools were found to underpredict hydrodynamic loading and corresponding motions, especially at low frequencies, which could lead to resonance and extreme loading [17]. The drag forces on heave plates and cylinders were crucial to understanding low-frequency motions. In addition, the heave and pitch motion of the platform increased with increasing the drag coefficient of the heave plates [20]. Biofouling, which is the settlement and growth of marine organisms on offshore structures, could lead to significant changes in the drag coefficient and force response of floating platforms [33,34] and fatigue failure [35]. Also, this could also make maintenance of the structures difficult and expensive [36].

So far, these studies have included aero-hydrodynamic forcing on semi-submersible platforms under dynamic conditions, motions, and decay of the platforms. The presence of multiple members in a semi-submersible platform is bound to influence the hydrodynamic loading on the entire structure. However, the contribution of different components, such as cross-braces and pontoons, towards the drag coefficient is not yet understood. Further, very little attention has been paid towards the scaling of hydrodynamic forces with Reynolds number (Re). Therefore, understanding the contribution of structural components towards drag and scaling of drag with respect to Re can help understand the influence of the drag coefficient towards the dynamic motion of the platform. Also, this can be fed into high-fidelity numerical tools for evaluating FOWT systems and optimising the design of floating platforms and mooring systems. This can also inform decision-making processes for construction material selection, operation, and maintenance and cost reduction. This paper numerically investigates the influence of structural components of a static semi-submersible platform on hydrodynamic loading under unidirectional current flow. This study has been carried out without considering the influence of waves. Drag coefficients ($\overline{C}_d$) acting on the four structures and contributions of structural components of the structures towards $\overline{C}_d$ are presented for Re = 2000 to 200,000. Further, flow visualisation, pressure, friction coefficient, etc.; results are presented for Re = 2900, 43,000, and 200,000. The NREL OC4 semi-submersible platform developed as part of the DeepCWind project is used as the baseline semi-submersible platform for this study [37]. The investigation is divided into three stages for each Re. In the first stage, a finite cylinder (FC) is considered. In the second stage, a finite cylinder with a heave plate (FCHP) is investigated. In the next stage, three cylinders with heave plates (TCHP) in an equilateral arrangement are considered, and a central column and braces are added to the TCHP to form a 1:50 model of the OC4 semi-sub. The paper is organised as follows: in Section 2, we introduce the CFD methodology, numerical set-up, and validation of flow past a finite cylinder. This is followed by the results in Section 3, where $\overline{C}_d$ for Re = 2000 to 200,000 for all structures and contributions of their components, flow visualisation, pressure, and friction coefficient results are presented. Finally, the discussion is presented in Section 4, and our findings are summarised in Section 5.

2. Methodology

The Reynolds averaged Navier–Stokes (RANS) method, which is based on the incompressible Navier–Stokes equation, is adopted. Here, the variables are decomposed into an ensemble-averaged mean component and a fluctuating component:

$$\phi = \overline{\phi} + \phi', \tag{1}$$

where ϕ is an instantaneous field, $\overline{\phi}$ is the ensemble averaged field, and ϕ' is the fluctuating field. The Reynolds averaged Navier–Stokes equations are given as follows:

$$\frac{D\overline{U}_j}{Dt} = -\frac{\partial \overline{P}}{\partial x_j} + \nu \nabla^2 \overline{U}_j - \frac{\partial \overline{u'_i u'_j}}{\partial x_i}, \tag{2}$$

and

$$\frac{\partial \overline{U}_i}{\partial x_i} = 0, \tag{3}$$

where $\overline{U_i}$ is the ensemble averaged velocity, $\overline{P}$ is the ensemble averaged kinematic pressure, and ν is the kinematic molecular velocity of the fluid. The instantaneous velocity fluctuations are given as $u'_i = u_i - \overline{U_i}$, the flow is assumed to be incompressible, and density, ρ, is accounted through ν and $\overline{P}$. Using the Boussinesq approximation, the Reynolds stresses, $-\overline{u_i u_j}$, are closed by

$$-\overline{u'_i u'_j} = 2\nu_t S_{ij} - \frac{2}{3} k \delta_{ij}, \tag{4}$$

and

$$S_{ij} = \frac{1}{2}\left(\frac{\partial \overline{U}_i}{\partial x_j} + \frac{\partial \overline{U}_j}{\partial x_i}\right), \tag{5}$$

where S_{ij} is the symmetric velocity gradient tensor, $k = \frac{1}{2}\overline{u'_i u'_i}$ is the turbulent kinetic energy, and ν_t is the kinematic turbulent viscosity, the calculation of which depends on the turbulence closure model adopted. In this work, the model adopted is Menter $k - \omega$ SST, which has shown excellent performance in external fluid dynamics problems [38]. This model solves transport equations for turbulent kinetic energy, k, and specific dissipation rate, ω. These equations are as follows:

$$\frac{Dk}{Dt} = \tilde{P} - \beta^* \omega k + \frac{\partial}{\partial x_j}\left[\left(\nu + \frac{\nu_t}{\sigma_k}\right)\frac{\partial k}{\partial x_j}\right], \tag{6}$$

and

$$\frac{D\omega}{Dt} = \gamma P - \beta\omega^2 + \frac{\partial}{\partial x_j}\left[\left(\nu + \frac{\nu_t}{\sigma_\omega}\right)\frac{\partial k}{\partial x_j}\right] + (1 - F_1)2\sigma_{\omega 2}\frac{1}{\omega}\frac{\partial k}{\partial x_j}\frac{\partial \omega}{\partial x_j}, \quad (7)$$

with

$$\tilde{P} = min(P; c_l\epsilon). \quad (8)$$

The model coefficients are functions of F_1:$\phi = F_1\phi_1 + (1 - F_1)\phi_2$, where ϕ_1, ϕ_2 are the coefficients of $k - \omega$ and $k - \epsilon$ models, respectively. The blending function F_1 is calculated as follows:

$$F_1 = tanh(arg_1^4) \qquad arg_1 = min\left(max\left(\frac{\sqrt{k}}{\beta^*\omega y}; \frac{500\nu}{y^2\omega}\right)\frac{4\rho\sigma_{w2}k}{CD_{k\omega}y^2}\right), \quad (9)$$

$$CD_{k\omega} = max\left(2\rho\sigma_{\omega 2}\frac{1}{\omega}\frac{\partial k}{\partial x_j}\frac{\partial \omega}{\partial x_j}; 1.0e^{-10}\right). \quad (10)$$

The model coefficients are $\sigma_{k1} = 0.85$, $\sigma_{k2} = 1.0$, $\sigma_{\omega 1} = 0.5$, $\sigma_{\omega 2} = 0.856$, $\beta_1 = 0.075$, $\beta_2 = 0.0828$, $\beta^* = 0.09$, $\alpha_1 = 5/9$, $\alpha_2 = 0.44$, $\alpha_1 = 0.31$, and $c_l = 10.0$. The validation of the adopted eddy viscosity model is presented in Appendix A.1.

2.1. Numerical Set-Up

The OC4 semi-submersible platform was developed to support the NREL offshore 5 MW wind turbine to build aero-hydro-servo-elastic models as part of the DeepCwind project by NREL (Figure 2). The platform consists of three circular columns with a diameter of 12 m and a height of 26 m, arranged in an equilateral triangle. Heave plates of diameter of 24 m and a thickness of 6 m are attached at the bottom of each cylinder. The wind turbine is fixed on the top of a fourth cylinder of diameter 6.5 m and a height of 30 m. This cylinder is connected to three outer cylinders and heave plates by a number of cross-braces and pontoons. The length of these members ranges from 38 to 13.62 m with a diameter of 0.0175 m. A detailed description of the OC4 semi-submersible platform and its structural details can be found in [37].

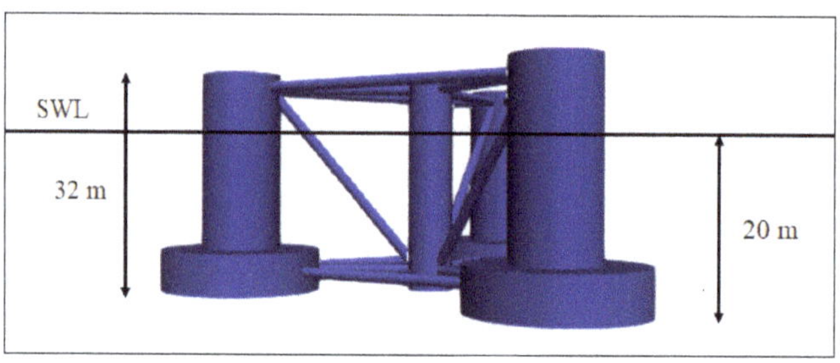

Figure 2. The OC4 semi-submersible platform with the 5 MW baseline wind turbine, as conceptualised by NREL for developing numerical models [37].

This study is divided into four stages for each Re. At Stage 1, a finite cylinder (FC) with a diameter of 0.24 m and a draft of 0.48 m is considered. Stage 2 considers a finite cylinder with a heave plate (FCHP) with a diameter of 0.48 m and a thickness of 0.12 m. In Stage 3, there are three cylinders with heave plates (TCHP) in an equilateral arrangement, and in Stage 4, the OC4 semi-submersible platform (OC4) is considered (Figure 3). The diameter and draft of cylinders of TCHP and OC4 are kept at 0.24 m and 0.48 m, respectively. This ensures that the aspect ratio (AR = draft of cylinder (L)/diameter of cylinder (D)) is fixed at 2 throughout the study for all structures, and the OC4 is scaled as 1:50 of the original model, as in the case of Goupee et al. [39]. Computer-aided drawing (CAD) models of FC, FCHP, TCHP, and OC4 are prepared using SALOME, an open-source software for CAD, meshing, visualisation, etc. These CAD drawings are then imported to OpenFOAM V10 for simulation [40]. The domain size in streamwise, cross-stream, and vertical directions for all structures is summarised in Table 1. To ensure that side wall effects are minimal, the length of the domain is kept slightly larger than the one used by Alkishriwi et al. [41]. A larger domain in the streamwise direction is used for TCHP and OC4 as these structures have multiple members spread apart, covering a larger plan area. The meshes for the simulations are generated using OpenFOAM's blockMesh, refineMesh, and snappyHexMesh facilities. To capture the flow separation in the wake of the structures, the mesh is refined locally, and the aspect ratio is kept close to 1. Figures 4–7 show the plan views from the top profile view at $y/D = 0$ of the computational meshes for the four structures. A refined area has been added around the structure to capture the wake.

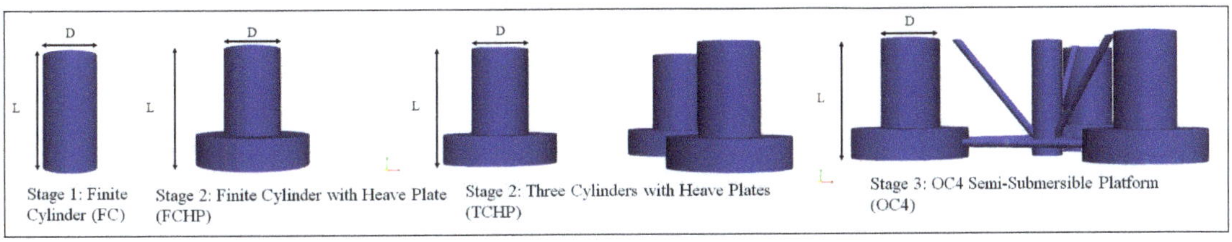

Figure 3. The study is carried out in four stages for each Re. These stages are stage 1: finite cylinder (FC), stage 2: finite cylinder with heave plate (FCHP), stage 3: three cylinders with heave plates (TCHP), and stage 4: OC4 semi-submersible platform (OC4). The diameter (D) and drafts (L) of cylinders are 0.24 m and 0.48 m, respectively.

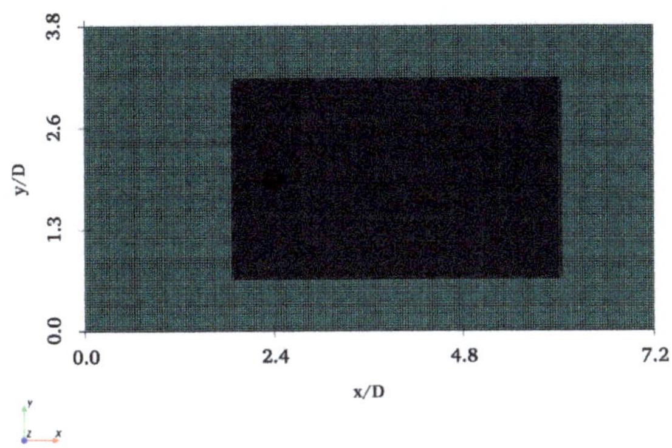

(a)

Figure 4. *Cont.*

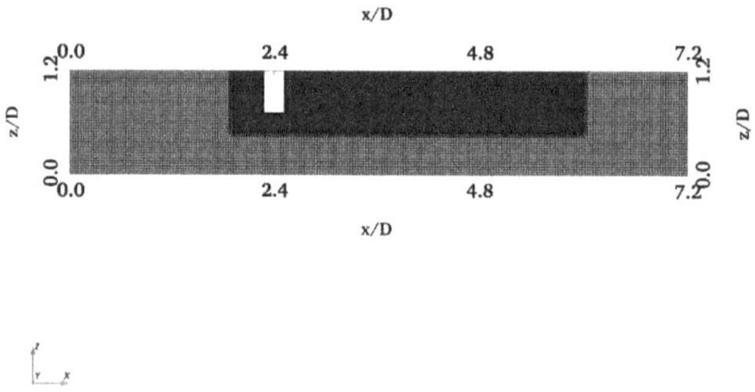

(**b**)

Figure 4. Computation mesh adopted for simulation past FC: (**a**) plan view of the mesh from the top showing the refined area; (**b**) profile view of the mesh through an x–z plane at $y/D = 0$.

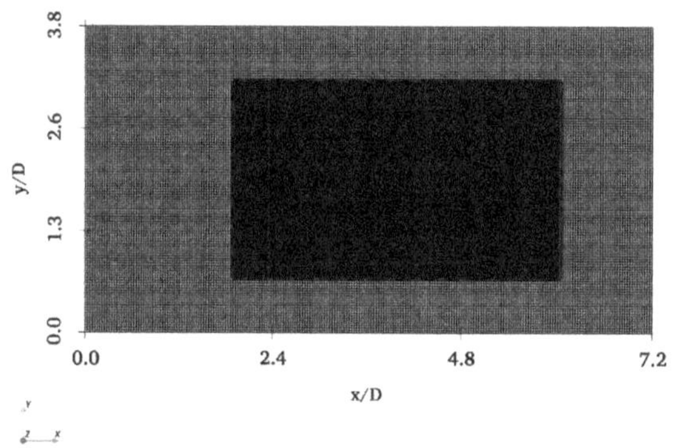

(a)

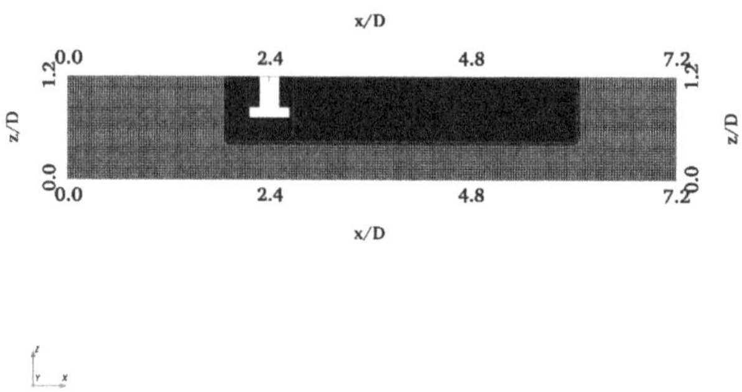

(**b**)

Figure 5. Computation mesh similar to that of FC is adopted for FCHP: (**a**) plan view of the mesh from the top; (**b**) profile view through a slice in the streamwise direction.

Table 1. Dimensions of the domains in streamwise, cross-stream, and vertical directions used for simulations past FC, FCHP, TCHP, and OC4. Here, D refers to the diameter of the cylinder, which is 0.24 m.

Structure	Streamwise Direction	Cross-Stream Direction	Vertical Direction
FC, FCHP	$-10 \leq x/D \leq 14$	$-8 \leq x/D \leq 8$	$-3 \leq x/D \leq 1.2$
TCHP, OC4	$-14 \leq x/D \leq 26$	$-12 \leq x/D \leq 12$	$-3 \leq x/D \leq 1.2$

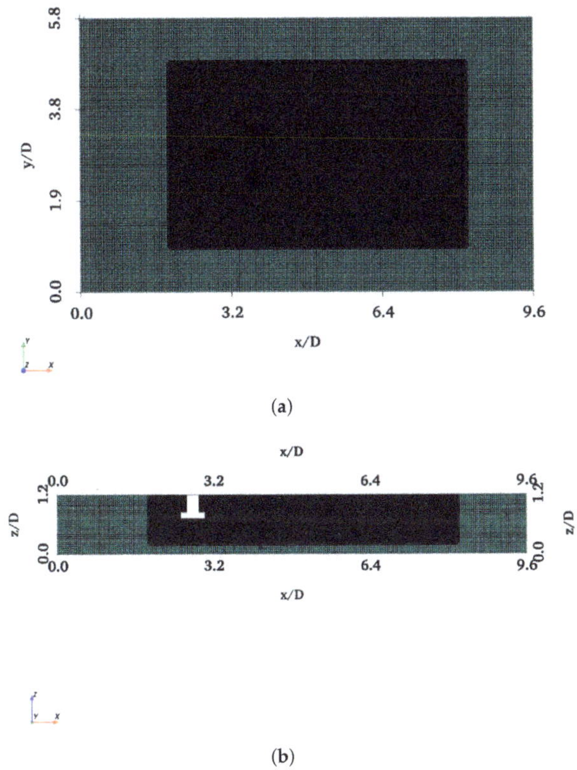

Figure 6. A typical computational mesh used for simulations past TCHP: (**a**) plan view of the mesh showing the refined zone in the mesh; (**b**) profile view of a slice through the domain in the streamwise direction.

For boundary conditions, a uniform streamwise velocity is defined at the inlet, and a no-slip boundary condition is defined at the surfaces of structures. Pressure is defined as a zero gradient at the inlet, whereas at the outlet, a fixed value of zero is specified. In addition, the boundary conditions for k and ω are specified based on the following equations:

$$k = \frac{3}{2}(IU_\infty)^2, \qquad \omega = \frac{\rho k}{\mu}\left(\frac{\mu_t}{\mu}\right)^{-1}, \tag{11}$$

where I is the turbulent intensity, U_∞ is the free-stream velocity, ρ is the density of the fluid, and μ_t/μ is the ratio between the turbulent and molecular viscosities. To specify the initial conditions of k and ω, I of 1.0 % and μ_t/μ of 10^{-3} are adopted. A symmetry boundary condition is defined at all remaining boundaries (Table 2). This gives a zero gradient to the scalar and tangential components of vector variables and a zero fixed value of the normal component to the plane [42]. The air–water interface is approximated with a shear-free (symmetry) boundary condition. The influence of a free surface on hydrodynamics is reserved for future work. The simulations are run in OpenFOAM [40]. A second-order bounded scheme is adopted for time-step discretisation. The advective terms and transport equations for k and ω are discretised by first/second-order accurate total variation diminishing (TVD) schemes, whereas a second-order linear scheme is defined for the diffusion terms. For pressure–velocity coupling, the PISO algorithm (pressure-implicit with splitting of operators) is used [43]. The Generalized Geometric–Algebraic MultiGrid (GAMG) solver with a GaussSeidel smoother is used to solve the pressure equation. For velocity and turbulence quantities, smoothSolver with a symGaussSeidel smoother is adopted. For all the variables, a tolerance of 10^{-8} is specified throughout the simulation. The simulations are run at a non-dimensional time step ($\Delta T^* = \Delta TU/D$) = 0.00417 up to a duration of ΔT^* = 420. The simulations are run from uniform initial conditions until a statistically steady state is reached before data collection is commenced. The $\overline{C}_d$ is monitored throughout the simulations to ensure a steady state is reached.

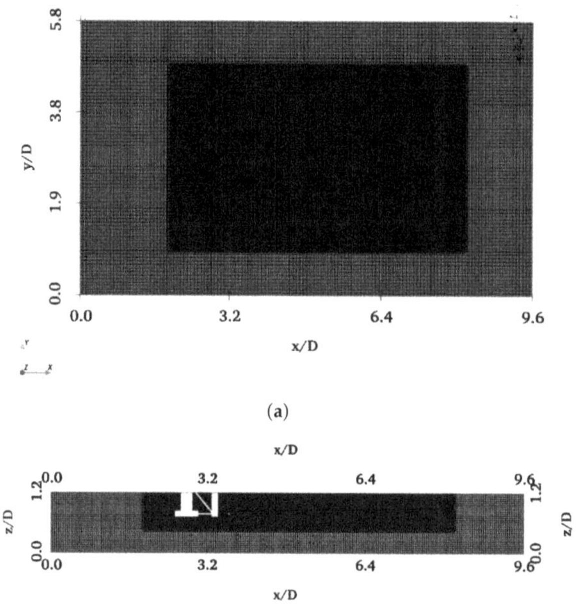

(b)

Figure 7. A mesh similar to that of TCHP is used for simulations past OC4: (**a**) plan view of the mesh from top with the refined region; (**b**) profile view of the mesh in the streamwise direction.

Table 2. Summary of the boundary conditions specified at various sections of the domain and wall of the structures across all the three structures and *Re* cases. The initial condition for *k* and ω at the inlet is based on Equation (11).

Boundary	U_∞ (m/s)	p (m^2/s^2)	k (m^2/s^2)	ω (s^{-1})
Inlet	(1, 0, 0)	zeroGradient	1.5×10^{-4}	150
Outlet	zeroGradient	0	zeroGradient	zeroGradient
Wall	noSlip	zeroGradient	1×10^{-9}	1×10^{10}
Top and Bottom	Symmtery	Symmetry	Symmetry	Symmetry
Sides	Symmtery	Symmetry	Symmetry	Symmetry

2.2. Validation of Flow Past a Finite Cylinder

The flow past a surface-piercing finite cylinder with a submerged free end has received far less attention than the flow past an infinite or finite cylinder resting on the bed. There are only a few works which have investigated the flow past a finite cylinder with a free end [44]. Benitz et al. [44] studied the flow past finite cylinders of *AR* ranging from 1 to

19 at $Re = 2900$ to understand the effect of AR change on force coefficients and vortex shedding. Similarly, Rosetti et al. [45] and He et al. [46] investigated the flow past a finite cylinder of $AR = 2$ at Re = 43,000. The former validated the numerical simulations against experiments, whereas the latter investigated the vortex-induced vibration of a finite cylinder. So far, there are no studies investigating the flow past a finite cylinder at high Re. However, inferences can be made from flow past a bed-resting finite cylinder of $AR = 2$. Frederich et al. [47] compared two different numerical methodologies for flow past a finite cylinder of $AR = 2$ resting on the bed at Re = 200,000. Therefore, these works are chosen to validate flow past FC of $AR = 2$ at Re = 2900, 43,000, and 200,000. In this work, the coefficient of drag ($\overline{C}_d$) of the simulations is compared to $\overline{C}_d$ reported in the literature. The $\overline{C}_d$ is given as follows:

$$\overline{C}_d = \frac{\overline{F}_d}{\frac{1}{2}\rho U_\infty^2 A_{ref}}, \tag{12}$$

where $\overline{F}_d$ is the ensemble-averaged drag force, ρ is the fluid density, U_∞ is the free-stream velocity, and A_{ref} is the projected area of the structure. A mesh independence study is carried out for flow past an FC at each Re, and $\overline{C}_d$ is reported in Table 3, along with $\overline{C}_d$ reported in the literature. Although the $\overline{C}_d$ of the simulations at Re = 2900 and 43,000 agree well with those reported in the literature, there is a difference at Re = 200,000. This is attributed to differences in boundary conditions. In the literature, a bed-resting finite cylinder is considered, whereas here, a surface piercing cylinder is adopted.

Table 3. Mesh independence study: meshes with their cell count, $\overline{C}_d$ for each mesh at Re = 2900, 43,000, and 200,000, and results from the literature for each Re.

Mesh	Cells ($\times 10^6$)	$\overline{C}_d$ $Re = 2900$	$\overline{C}_d$ Re = 43,000	$\overline{C}_d$ Re = 200,000
M1	1.8	0.716	0.644	0.693
M2	4.6	0.718	0.680	0.695
M3	10.0	0.720	0.710	0.699
M4	16.2	0.724	0.715	0.703
M5	16.3	0.724	0.715	0.703
Benitz et al. [44]	-	0.725	-	-
He et al. [46]	-	-	0.71	-
Frederich el al. [48]	-	-	-	0.8

3. Results

In this section, results of numerical simulations of flow past FC, FCHP, TCHP, and OC4 at Re = 2900, 43,000, and 200,000 are presented. At the onset, the $\overline{C}_d$ of all four structures for Re = 2000 to 200,000 is presented in Figure 8 to understand the variation in $\overline{C}_d$ with respect to Re. It is clear from the plots that adding structural members increases the $\overline{C}_d$ of the structures; however, there is no significant change with increasing Re. Gonçalves et al. [49] showed that for a finite cylinder of given AR, $\overline{C}_d$ is independent of Re, except for $AR = 0.2$. The projected area of TCHP is thrice that of FCHP. Therefore, it is expected that it will experience a drag force thrice that of FCHP. In other words, the $\overline{C}_d$ of TCHP and FCHP should be the same. However, the differences in $\overline{C}_d$ of TCHP and FCHP at some Re can be attributed to the flow interference of upstream structures on downstream members. Figure 9 shows the viscous ($\overline{C}_f$) and pressure drag ($\overline{C}_p$) coefficients acting on all the structures for Re = 2000 to 200,000. Although the variation of total $\overline{C}_d$ with Re is insignificant, the $\overline{C}_f$ decreases with increasing Re and a reverse trend for $\overline{C}_p$. This shows that pressure drag dominates over friction drag as Re increases. The heave plate of FCHP significantly increases $\overline{C}_f$ compared to FC. This is attributed to higher friction on the heave plate, as presented in the streamwise friction results later in this section. However, adding more cylinders with plates, as in the case of TCHP, did not increase $\overline{C}_f$ compared with FCHP.

This is attributed to the rear members experiencing lower friction due to the wake effect of the leading member. Also, adding pontoons and braces increases viscous drag acting on the OC4. Interestingly, the variation of $\overline{C}_f$ with Re for FC resembles the variation in $\overline{C}_d$ with Re for an infinite cylinder, as can be seen in [50]. Apart from the total $\overline{C}_d$ and $\overline{C}_f$, an investigation on comparison of individual contributions of each member of the structures is also necessary. Also, in cases of TCHP and OC4, where the flow is symmetrical around the structure, the difference between $\overline{C}_d$ acting on the two rear members is of the order $\leq 10^{-7}$. Figure 10 shows the drag coefficient ($\overline{C}_{dm}$) contributions of individual structural members of structures normalised by the total $\overline{C}_d$ of the structure. Across all three different structures, the contribution of the individual members varies with Re. The contribution of the cylinders is higher than that of heave plates across all structures. Gonçalves et al. [49] observed that $\overline{C}_d$ is directly related to AR. This explains the higher $\overline{C}_{dm}/\overline{C}_d$ of finite cylinders compared to heave plates, which are shorter and thicker. Further, variations in $\overline{C}_{dm}/\overline{C}_d$ with respect to Re of cylinders are similar to the variation of $\overline{C}_d$ with Re for an infinite cylinder, suggesting that the flow regime is similar for finite and infinite cylinders [50]. The $\overline{C}_{dm}/\overline{C}_d$ of heave plates increases until $Re \approx 30{,}000$, after which the variation is insignificant. A similar variation of $\overline{C}_d$ is observed for finite cylinders of $AR = 0.2$ [49]. This variation is attributed to three-dimensionality in the wake due to the formation of recirculating flow under the heave plates ($AR = 0.35$), as has been presented in the flow visualisation results later in this section. Although the contributions of central cylinders and braces and pontoons vary from a low Re until $Re \approx 10{,}000$, these are mostly independent of Re as Re increased (Figure 10c). This is due to a reduction in the wake effect from upstream members. In the case of TCHP and OC4, the contribution of leading members is lower than that of rear members (Figure 10b,c). This is also observed in the flow past infinite circular cylinders in an equilateral triangle arrangement with various spacing to diameter ($\delta = d/D$) ratios. When the cylinders are closely placed ($\delta \leq 1.5$), the $\overline{C}_d$ acting on the rear cylinders is lower than on an infinite cylinder due to the wake effect of the front cylinder. However, as the spacing ratio between cylinders increases ($\delta \geq 2.5$), this wake effect decreases, and the vortex shed by each cylinder is independent of each other, thereby increasing drag on rear members. Also, wake suppression by rear cylinders reduces drag on the front cylinder [51–54]. It should be noted that δ of TCHP and OC4 are $\sim$4.1; this explains the higher contribution of rear members than leading members towards total drag. The contribution of the central cylinder is the lowest, whereas the contribution of braces and pontoons is substantially high, showing that the central cylinder experiences low drag due to the wake effect of the front cylinder.

Further, adding more structural members is expected to influence the flow behaviour in the vicinity of the structures. Streamlines of time-averaged streamwise velocity are presented to understand the flow physics around floating structures. Figure 11 shows the mean velocity streamlines in the wake of FC at $Re = 2900$. Figure 11a shows two symmetric vortex cores formed in the near wake of the FC, whereas Figure 11b shows a recirculating flow in the near wake. Also, a recirculation zone is formed behind the cylinders due to upward flow. The fluid flows down along the sides of the cylinder and then under the free end in the upstream direction. Similarly, Figure 12 shows velocity streamlines in the wake of FCHP. Although symmetric vortex cores are formed in the near wake (Figure 12a), these are not as distinct as in the case of FC. However, the recirculation zones in the near wake and under the heave plate are much more distinct compared with FC (Figure 12b). The effect of multiple heave plates of TCHP and central cylinder and braces of OC4 on the flow behaviour is presented in Figures 13 and 14, respectively, for $Re = 2900$. Similar to the FCHP, the vortex cores formed behind the front cylinder of the FCHP are not as distinct as in the case of FC, showing the effect of the heave plate (Figure 13a). Further, vortex cores behind the two back cylinders are not symmetric due to the influence on the flow field from the upstream structure. Similar to the FCHP, two distinct recirculation zones are formed behind the front cylinder and under the heave plate (Figure 13b). In the case of OC4, the central cylinder slightly inhibits the formation of vortex cores behind the front cylinder. Also, similar to TCHP, the effect of the upstream members on the vortex cores behind the two back cylinders can be observed (Figure 14a). However, the inclined cross-brace between the front and the central cylinder restricts the recirculation region behind the front cylinder. Also, recirculating flows can be observed under the heave plate (Figure 14b). Similar recirculating flow and vortices are also seen at Re = 43,000 and 200,000, and flow visualisation plots are presented in Appendix A.1.

The vortices and recirculating flow around the structures influence the distribution of pressure and friction acting on the structures. Therefore, analyses of pressure and friction coefficients on structures are essential. The recirculating flows under the heave plates are the negative pressure pockets that exert considerable pressure on the structures in a downward direction. FOWT platforms are subjected to dynamic movement in response to hydrodynamic loading. Therefore, investigation of downward force is essential. Figure 15 shows the vertical force coefficient ($\overline{C}_V$) normalised by the density (ρ), free-stream velocity (U_∞), and projected area of structures in the vertical direction (A_b) and vertical force ($\overline{F}_V$), respectively. Adding the heave plate and braces significantly increases the downward vertical force. This is further evident from the $\overline{C}_V$ plot. Although the FC experiences significantly less $\overline{F}_V$ than other structures, a lower A_b leads to a higher $\overline{C}_V$. Further, the $\overline{C}_V$ increases with Re for $Re \geq$ 40,000. This increase in $\overline{C}_V$ acting on FC can be attributed to the change in the recirculating flow under the free end of the FC, as can be seen in Figure 16. Here, the recirculation zone under the free end of the FC can be observed at Re = 2900, 10,000, 100,000, and 200,000. The zone is largest at Re = 200,000 and lowest at Re = 2900, and the size increases with increasing Re. This increase in size is due to the increase in flow going downwards along the back of FC and then under the free end.

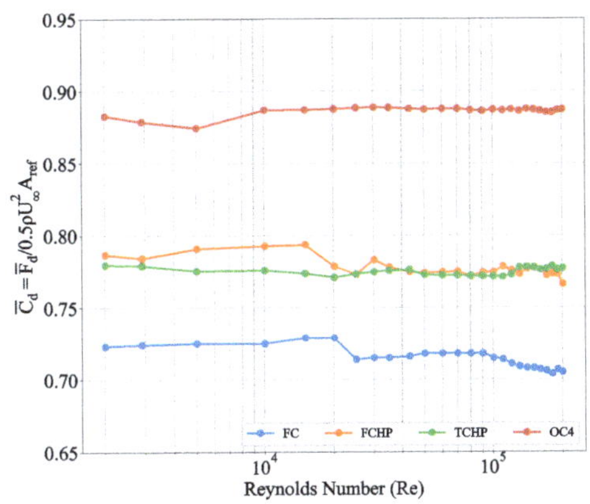

Figure 8. Plot of time-averaged drag coefficient ($\overline{C}_d$) for all structures at Re = 2000 to 200,000. Adding more structural members increases the $\overline{C}_d$ across all Re. However, no significant change with respect to Re is observed.

Further, it is expected that the vertical pressure on the bottom surfaces of structures will vary due to recirculating flow. Therefore, an investigation of the vertical pressure variation is necessary as it will lead to differential forcing on the bottom surfaces of the structures. Figure 17 shows the vertical pressure coefficient on the surfaces of free ends of FC, FCHP, TCHP, and OC4 at Re = 2900 and 200,000, respectively. Across all the structures, the negative coefficient shows that the pressure is in the downward direction. The highest pressure is experienced at Re = 200,000, and pressure varies across the surface in the streamwise direction, with the highest force acting near the leading edge and gradually decreasing with the lowest pressure near the rear edge. In the cases of TCHP and OC4, the bottom surfaces of upstream members experience lower negative pressure than the members at the rear, with some parts of the cross-braces of OC4 experiencing very high pressure. The gradient increases across the surfaces as the Re increases. This pressure variation on surfaces depends on the area covered by the recirculation zone over the bottom surfaces. Similarly, vertical pressure variation is also observed at Re = 43,000 for all structures and is included in Appendix A.3. For FOWT platforms, quantifying this differential forcing on the members is essential for accounting for the stress acting on the mooring system, designing individual members, and choosing construction materials.

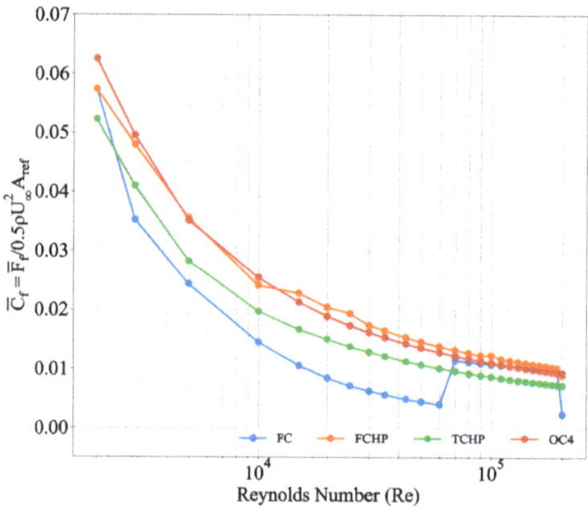

(a)

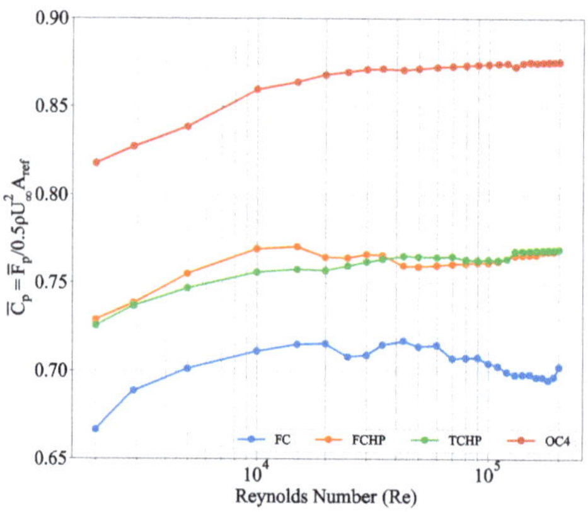

(b)

Figure 9. (**a**) Plot of time-averaged viscous drag coefficient ($\overline{C}_f$) for all structures at Re = 2000 to 200,000. $\overline{C}_f$ decreases with increasing Re. The addition of heave plates significantly increases friction drag compared to a finite cylinder. (**b**) Plot of time-averaged pressure drag coefficient ($\overline{C}_p$) for all structures at Re = 2000 to 200,000. Clearly, pressure drag dominates over friction drag as Re increases. Also, $\overline{C}_p$ increases with increasing structural members.

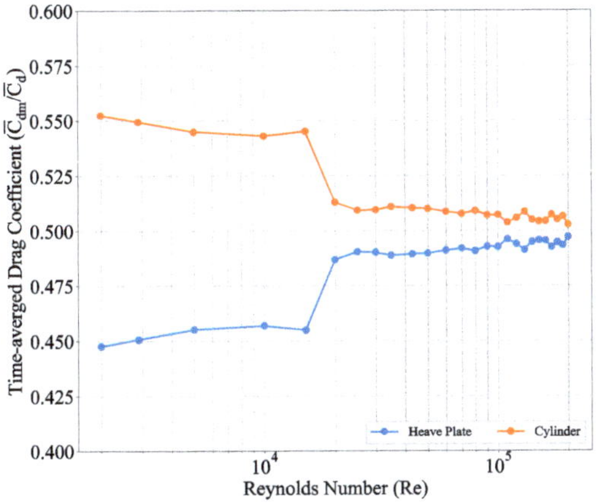

(a)

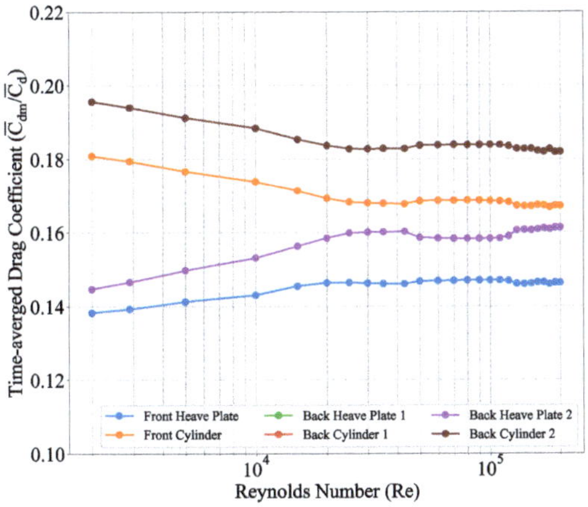

(b)

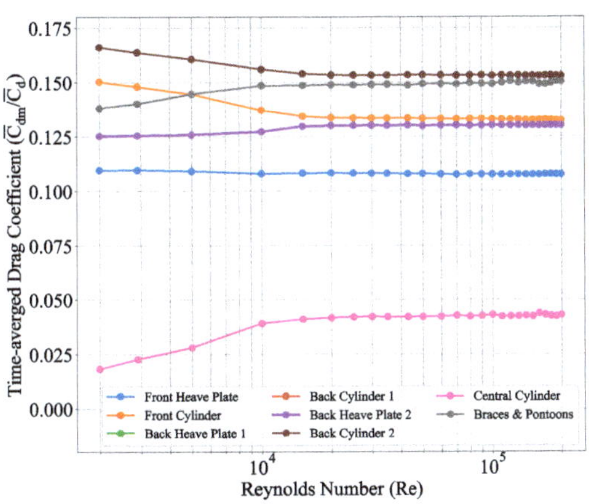

(c)

Figure 10. Time-averaged drag coefficient ($\overline{C}_{dm}/\overline{C}_{d}$) contributions of individual structural members normalised by the $\overline{C}_{d}$ of the entire structure for Re = 2000 to 200,000: FCHP (**a**), TCHP (**b**), and OC4 (**c**).

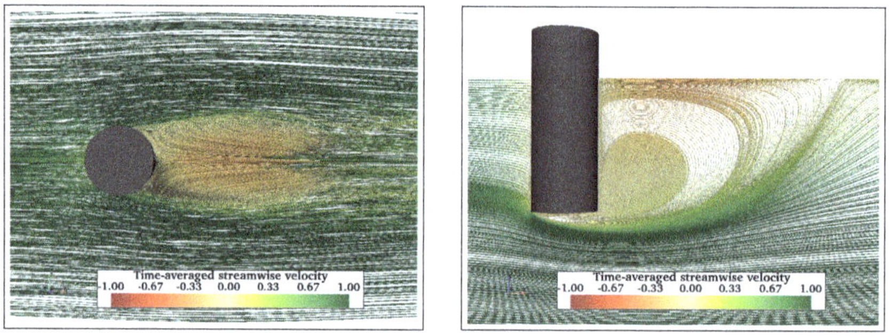

(a) (b)

Figure 11. Streamlines of time-averaged streamwise velocity: plan view of the FC from top (**a**) and profile of streamlines on a vertical plane through the centre of FC (**b**) at $Re = 2900$.

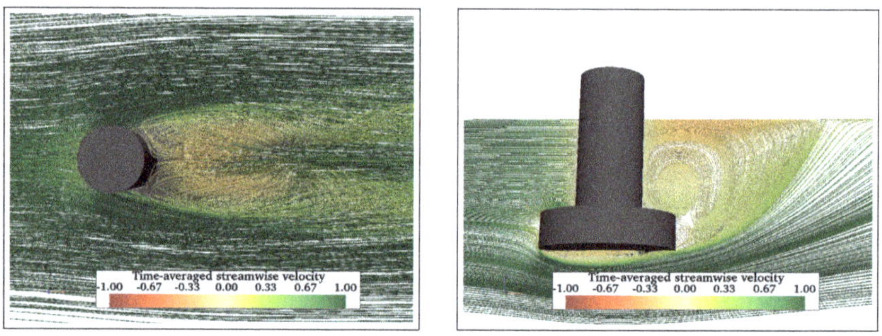

(a) (b)

Figure 12. Time-averaged streamwise velocity streamlines around the FCHP at $Re = 2900$: plan view of the FCHP from top (**a**) and profile view on a vertical plane (**b**).

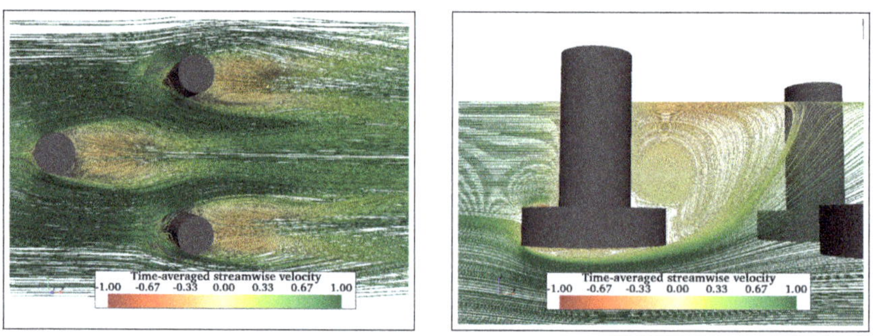

(a) (b)

Figure 13. Similar to FC, symmetric vortices are formed behind each of the cylinders of TCHP. However, the effect of the upstream cylinder can be seen on the vortices behind back cylinders (**a**). Recirculation regions are formed behind the cylinder and under the heave plate of the front cylinder (**b**).

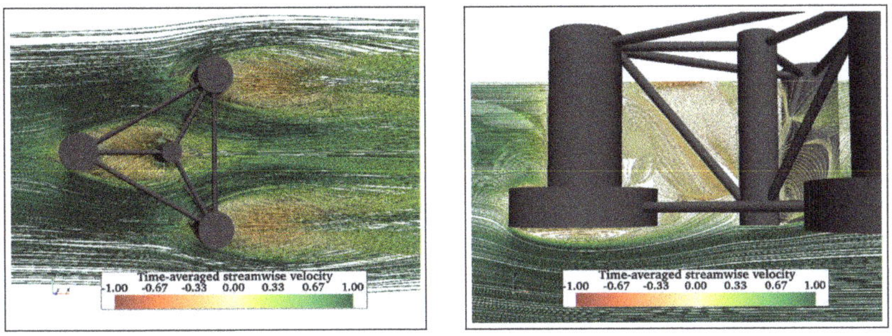

(**a**) (**b**)

Figure 14. Mean streamwise velocity streamlines in the wake of OC4 at $Re = 2900$. Two vortex cores are formed behind each of the back cylinders of OC4 (**a**). The recirculation region behind the front cylinder is restricted between the cylinder and cross-brace (**b**).

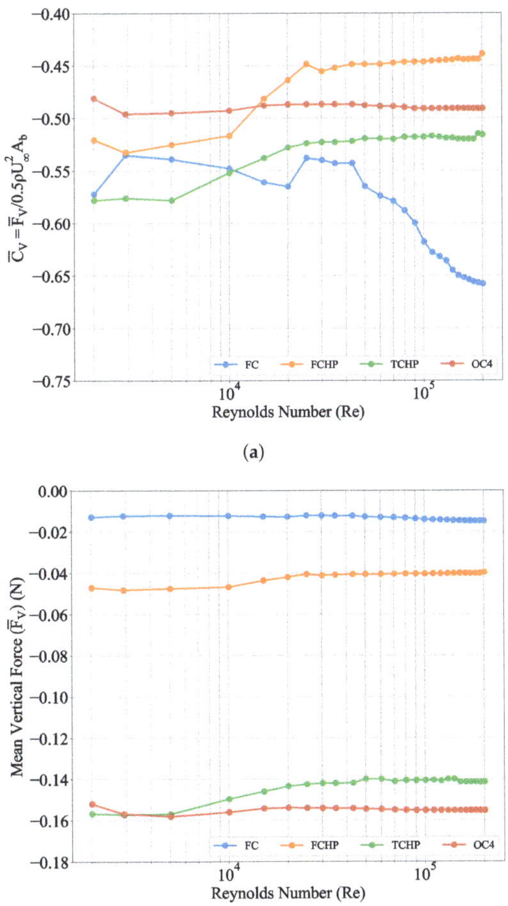

(**a**)

(**b**)

Figure 15. Plots of time-averaged vertical force coefficient ($\overline{C}_V$) normalised by the density, free stream velocity, diameter, and projected area of structures in the vertical direction (**a**) and vertical force ($\overline{F}_V$) (**b**) for all Re cases ranging from 2000 to 200,000. The addition of more structural members substantially increases the vertical force.

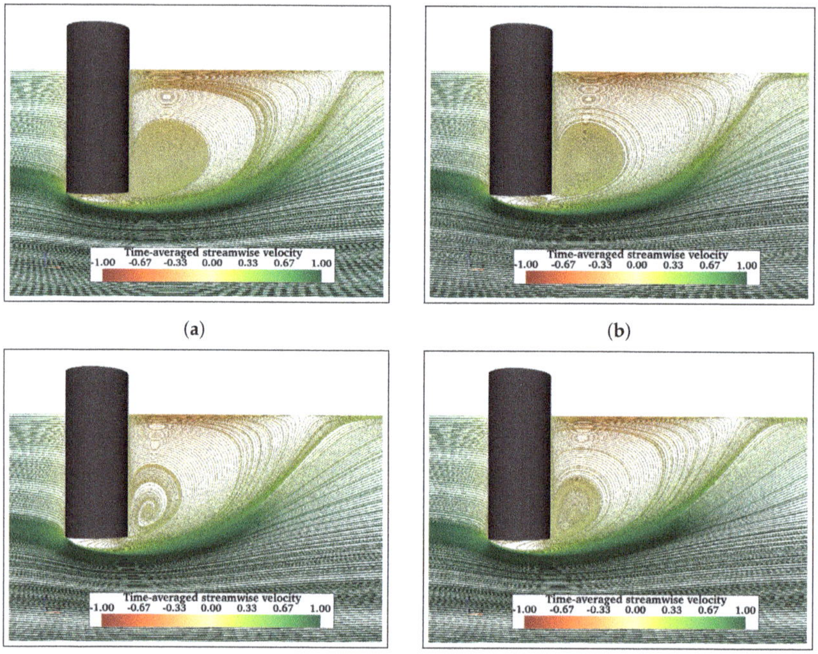

(**a**) (**b**)

(**c**) (**d**)

Figure 16. Time-averaged velocity streamlines in the wake of FC at *Re* = 2900 (**a**), *Re* = 10,000 (**b**), *Re* = 100,000 (**c**), and *Re* = 200,000 (**d**).

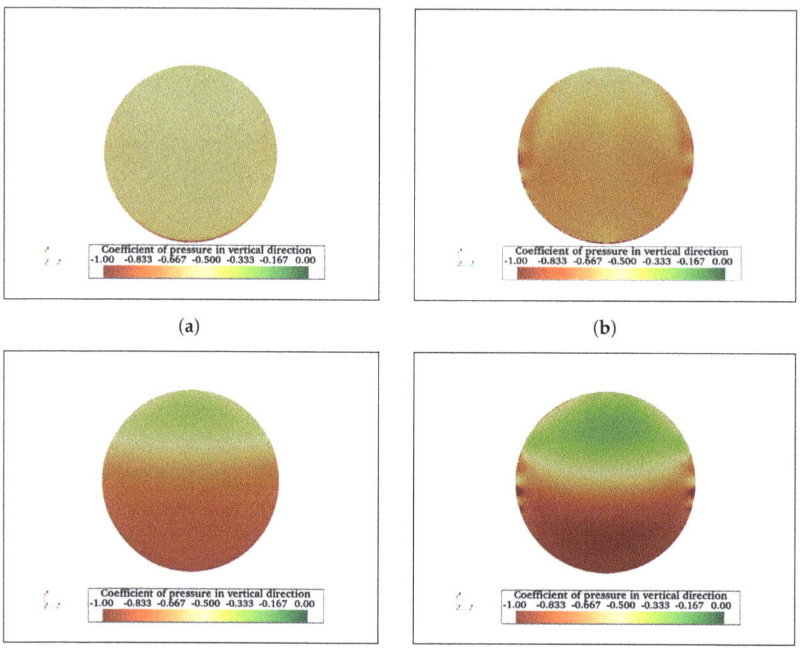

(**a**) (**b**)

(**c**) (**d**)

Figure 17. *Cont.*

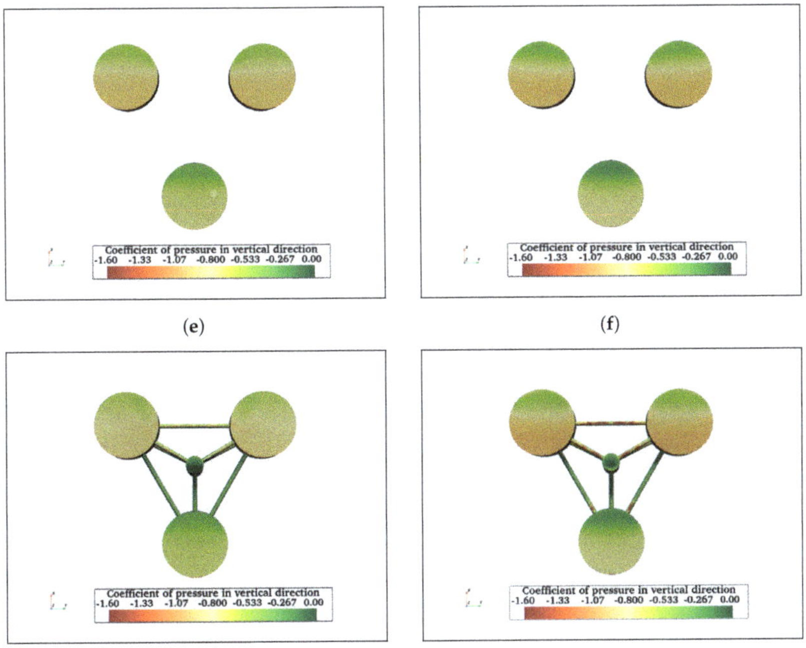

Figure 17. Vertical pressure coefficient acting on the bottom surface of FC at $Re = 2900$ (**a**) and at $Re = 200{,}000$ (**b**). Similarly, (**c**,**d**) show the vertical pressure coefficient on the bottom surface of the heave plate of FCHP at $Re = 2900$ and 200,000, respectively. In both cases, the pressure varies on the surface in the streamwise direction, with lower pressure near the leading edge and higher near the rear edge. Vertical pressure coefficients at $Re = 2900$ and 200,000 for TCHP are shown in (**e**) and (**f**), respectively, and for OC4 in (**g**) and (**h**), respectively. The vertical pressure varies across the bottom surfaces of each heave plate, with higher negative pressure on the rear heave plates than on the leading one for both Re. In the case of OC4, the leading braces experience lower pressure than the rear one.

In addition to pressure in the vertical direction, FOWT foundations are also subjected to streamwise pressure and friction, and both are expected to vary on the surfaces of the structures. Figure 18 shows the streamwise pressure coefficient acting on FC, FCHP, TCHP, and OC4 at $Re = 200{,}000$. The pressure coefficient acting on the surface of cylinders can be divided into three zones: the green zone where the pressure is highest, the red zone on the side where pressure is lowest, and the yellow zone in between where the pressure transitions. At the front surface, where the streamwise flow impinges orthogonally, the velocity is minimum; therefore, pressure is maximum. As the fluid flows on either side of the cylinders, it accelerates, decreasing the pressure (red zone). The thin vertical zone between these two zones is where the pressure transitions. Similar to cylinders, the front surfaces of heave plates of FCHP, TCHP, and OC4 experience considerable pressure. However, the width of the high-pressure zone on the front surfaces of heave plates is not uniform. This is attributed to the incoming flow separating with a portion moving downward towards the lower edge and the rest towards the upper edge of the plate. In the case of OC4, the pressure on the central cylinder is negative, indicating that the cylinder is within the wake of the front cylinder. The pressure on braces varies; for example, the pressure on the upper half of the inclined brace connecting the central cylinder to the outer cylinder is higher than the lower half. The sections of horizontal braces which are within the wakes of the front cylinder or central cylinder experience negative pressure compared with the rest of the members.

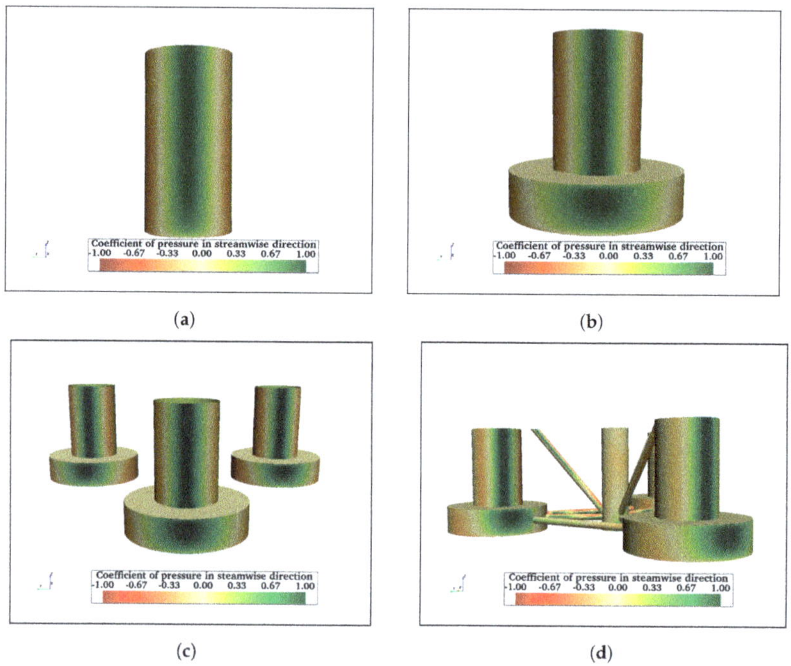

Figure 18. Streamwise pressure coefficient acting on FC (**a**), FCHP (**b**), TCHP (**c**), and OC4 (**d**) at Re = 200,000. The front surface of cylinders and heave plates can be divided into three zones: the green zone in the middle where the pressure is maximum; the red zones on either side where the pressure is minimum; the transition zone in between. The variation of pressure on the surface of the heave plate is due to separation of incoming flow.

An investigation of friction acting on the surfaces of structures is also important. It has implications on the type of corrosion and bio-fouling preventive coating to be selected, as well as general maintenance of the floating structures. Figure 19 shows the friction coefficient acting on the front and bottom surfaces of FC, FCHP, TCHP, and OC4 at Re = 200,000. The streamwise friction on the cylinder surfaces can be distinguished into two sections: the section in the middle where the friction is least due to velocity being lowest and the two zones on either side where friction is maximum. These are the two sections where the flow accelerates along the surface, increasing friction. Similarly, the sides of heave plates of FCHP, TCHP, and OC4 experience higher friction than the front surfaces. In the case of FCHP, TCHP, and OC4, the highest friction is experienced on the top and sides of heave plates. The negative friction on top of heave plates is due to a small upstream flow from recirculating flows formed at the junction of the heave plate and cylinder (Figure 13b). Only the lower half of the central cylinder of OC4 experiences significant friction. This is due to the rest of the cylinder being within the wake of the front cylinder. Clearly, the friction on the surfaces of the cross-braces varies; for example, the rear half of the horizontal braces connecting the front and back heave plates is where friction is highest. Meanwhile, the friction on the leading half is very low due to the wake influence of the front cylinder. This is also the case for the horizontal brace between the front heave plate and the central cylinder. The friction on the brace connecting the central cylinder to the back heave plates and the brace between the two back heave plates is substantially high. This is because these members are outside the wake effect of upstream members. Also, on their surfaces, small sections of negative friction can be observed; this is due to small pockets of recirculating flow being formed. The negative friction on the front and back surfaces of the inclined brace connecting the front and central cylinder shows the influence of recirculating flow in the wake of the front cylinder. Also, similar to the central cylinder, high friction is restricted to the sides of its lower half. For the inclined braces at the rear, the highest friction is on the sides compared with the front surface facing the flow orthogonally.

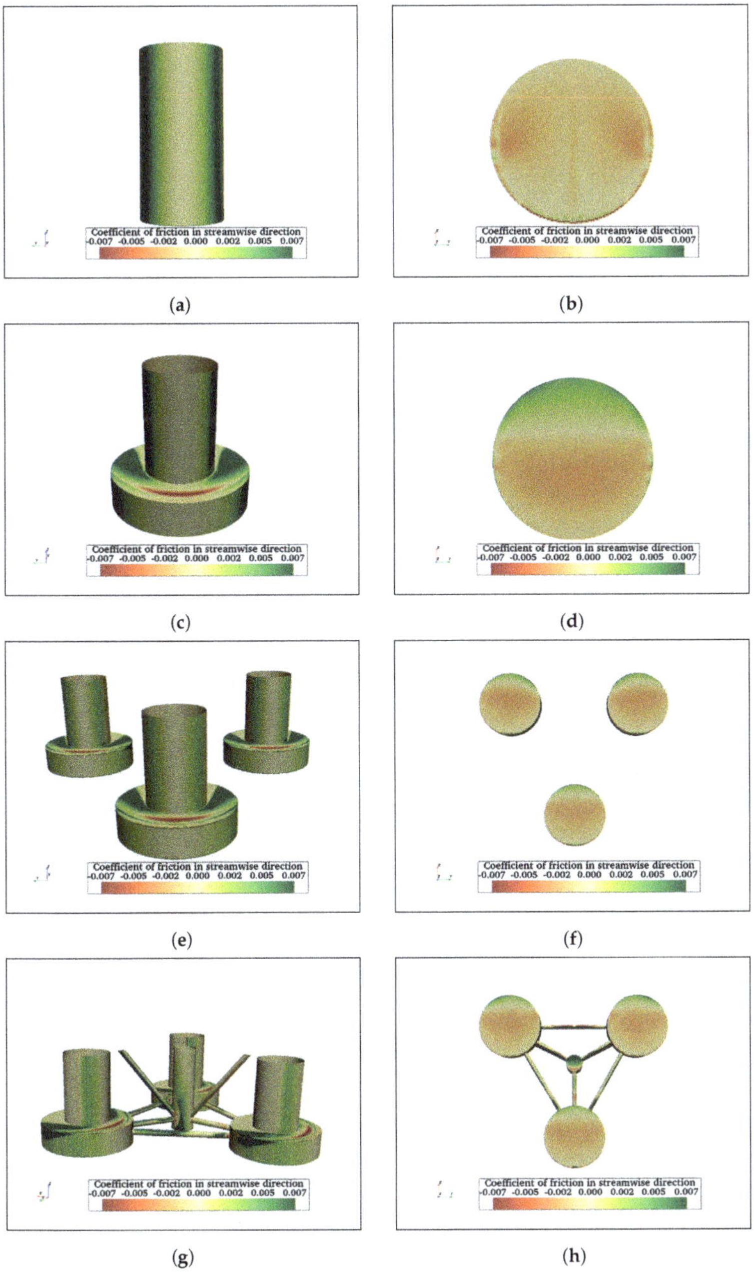

Figure 19. Streamwise friction coefficient acting on front and bottom surfaces of FC (**a**,**b**), FCHP (**c**,**d**), TCHP (**e**,**f**), and OC4 (**g**,**h**) at $Re = 200{,}000$. Friction is maximum on the two regions adjacent to the front section of the cylinder and heave plates where the flow is orthogonal. In cases of TCHP and OC4, friction is highest on the top surfaces of heave plates.

The friction on the free end is due to flow along the surface moving upstream in the recirculation zone (Figure 19b), hence negative friction. In the case of FC, the friction on the free surface can be divided into two symmetric sections. This is due to fluid in the vortex cores behind the cylinder flowing down along the two sides and then under the free surface. The central ridge is where fluid from either side meets, increasing the friction. For TCHP and OC4, friction on the bottom surfaces of heave plates can be divided into red and green zones. The recirculating flow under the heave plates is responsible for high friction (red zone), with the flow being upstream along the surface. The friction is positive near the rear edge because the flow is downstream and attaches behind the recirculation region, as seen in Figure 13b. Also, the friction varies from the middle section moving towards the leading edge. This is because the flow velocity of the recirculation flow is higher near the middle section, and the flow decelerates moving upstream. The pressure and friction coefficient variation of all three structures at Re = 2900 and 43,000 are presented in Appendix A.3.

4. Discussion

In this study, the OC4 semi-submersible platform developed by NREL for the DeepCWind project is chosen as a baseline FOWT foundation to numerically study the influence of structural members on hydrodynamic loading due to uni-directional current flow on a static structure. The investigation has been divided into four stages: in the first stage, a simple truncated/finite cylinder (FC) and in the second stage, a finite cylinder with a heave plate (FCHP) are considered. In the third stage, three cylinders with heave plates (TCHP) arranged in an equilateral triangular arrangement are considered. During the final stage, a central cylinder and braces are added to the TCHP to obtain the OC4 semi-submersible platform. Force analyses are carried out for each structure for three different Reynolds numbers ($Re = uD/\nu$, where u is the velocity, D is the diameter, and ν is the kinematic viscosity of the fluid), Re = 2900, 43,000, and 200,000.

The addition of members at each stage increases the drag coefficient ($\overline{C}_d$) acting on the structure for each Re. Although $\overline{C}_d$ has remained relatively constant with increasing Re, analyses of viscous drag coefficient ($\overline{C}_f$) and pressure drag coefficient ($\overline{C}_p$) have shown that $\overline{C}_f$ decreases with increasing Re, whereas $\overline{C}_p$ increases with Re. This indicates the dominance of pressure drag over friction drag at higher Re. The FCHP experiences significantly high $\overline{C}_f$ compared to FC, comparable with that of OC4. TCHP experiences lower $\overline{C}_f$ than FCHP because of low viscous drag on rear members due to the wake effect. An investigation of individual contributions of members of FCHP, TCHP, and OC4 show that the contributions vary with Re. The cylinders' contribution is higher than that of the heave plates, showing that the contribution is directly dependent on AR. In the case of TCHP and OC4, the contribution of the rear members is higher than the leading members due to wake suppression of rear members. The contribution of heave plates increases until $Re \leq 30{,}000$ due to three-dimensionality in the wake of the structures. In the case of OC4, the contribution of the central cylinder is the least among all members due to the wake effect of the front cylinder, whereas the contribution of braces and pontoons is similar to that of the back cylinders. Further, as observed in the works of Kawamura et al. [55], Okamoto et al. [56], and Benitz et al. [44], in this work, no vortex shedding is observed for finite structures with an $AR = 2$. Benitz et al. [44] observed vortex shedding increases with increasing AR of the finite cylinder, with vortex shedding completely suppressed for cylinders with $AR \leq 3$. Flow visualisation in the vicinity of the structures showed two symmetric vortex cores and recirculating flow in the wake of FC. Similar vortices are formed behind each of the cylinders of FCHP, TCHP, and OC4, with recirculation zones formed under the heave plates. The recirculating flow under the structures exerts vertical force in a downward direction, and the addition of heave plates substantially increases this vertical force. Further, this vertical force is observed to vary across the bottom surfaces of structures. On the bottom surfaces of the free end of FC and heave plates, the highest vertical force is near the leading edge of the surface and gradually decreases in a downstream direction such that the lowest force is observed close to the rear end. Also, in the case of TCHP and OC4, leading members are subject to higher vertical force than rear members.

The streamwise pressure coefficient on the surface of the FC can be distinguished into three different sections: a green zone on the front where the force is highest, a red zone on the sides of the FC where the pressure is negative, and a transition zone in between. Three similar zones are also observed on the cylinder surfaces of FCHP, TCHP, and OC4. However, the pressure on most of the central cylinders is negative because of the wake effect of the front cylinder. The pressure coefficient on the braces varies from negative to positive depending on whether the member is within the wake of an upstream member or in the free stream. Similar to the streamwise pressure coefficient, analyses of the streamwise friction coefficient are also presented. In the case of FC, the friction is highest on the sides of the front surface where the streamwise flow accelerates. The friction on the surface of the free end is negative due to the flow along the surface being upstream. Similarly, friction on the sides of the cylinders of FCHP, TCHP, and OC4 is substantial; however, the highest friction is on the top surfaces and sides of heave plates. The friction on the bottom surfaces of heave plates varies in the streamwise direction, with friction being negative on the leading and middle sections where friction is positive on the rear edge. The negative friction is due to recirculating flow along the bottom surfaces of heave plates upstream. Near the rear edge, streamwise flow is attached to the surfaces behind the recirculation zone. Only the sides of the lower half of the central cylinder experience high friction compared with the rest of the cylinder. The friction on the braces varies depending on their location in the OC4, with parts of the braces experiencing higher friction than the rest.

5. Conclusions

In this work, the flow past floating offshore wind turbine platforms is detailed. Numerical studies are used to quantify the change in hydrodynamic loading with increasing structural complexity. It is shown that subject to uni-directional flow, increasing geometric complexity leads to an increase in drag coefficient ($\overline{C}_d$). Compared to FC, the addition of cylinders, heave plates, braces, and pontoons increased $\overline{C}_d$ by ∼23.6%, averaging Re = 2000 to 200,000. However, there is no significant change in the drag coefficient with increasing Re for any of the structures. However, analyses of viscous ($\overline{C}_f$) and pressure drag ($\overline{C}_p$) coefficients show that where the former decreases with increasing Re, where the latter shows a reverse trend. Also, $\overline{C}_p$ increases with increasing complexity. Although $\overline{C}_f$ increases significantly with the addition of a heave plate to the finite cylinder, there is no further increase with increasing structural complexity, and $\overline{C}_f$ of FCHP is comparable with that of OC4. An investigation into individual contributions of members towards total drag shows that contributions vary with Re. The contributions of cylinders are higher than those of heave plates, showing that contributions are directly related to AR. The contributions of heave plates vary until $Re \approx$ 30,000 due to three-dimensionality in the wake. In the case of TCHP and OC4, the contribution of rear members toward $\overline{C}_d$ is higher than that of leading members. The combined contribution of one rear cylinder and heave plate of TCHP towards $\overline{C}_d$ is ∼8.5% higher than the combined contribution of the leading cylinder and heave plate, and the same for OC4 is ∼16.84%. This indicates that the wake effect of rear members is stronger than that of leading members. The low contribution of the central cylinder is attributed to the wake effect of the front cylinder, whereas the contribution of braces and pontoons is ∼14.86% of total $\overline{C}_d$. Further, for a typical minimum shelf sea tidal velocity of 0.02 m/s [57], OC4 cylinder diameter of 12 m, and $\overline{C}_d = 0.884$ at Re = 200,000, the drag force acting on the OC4 is ∼202 N. Flow visualisation in the vicinity of the structures showed unique flow behaviour, such as upwash, recirculating flow, and vortices in the wake of the structures. The additional structural members significantly increase vertical force acting on the structures, with the vertical force coefficient varying strongly with Re due to changes in recirculating flow in the wake of the structure. Variation of vertical pressure on the free end along the streamwise direction is observed across all structures, with downward pressure on the leading edge and upward near the rear.

Streamwise pressure is highest on the front surfaces of members facing orthogonally to the flow. The highest friction is on the sides of the cylinder and the top surfaces of heave plates. Also, friction varies on the bottom surfaces of heave plates due to recirculation flow. The friction is negative in the leading and central regions due to the flow being upstream, whereas near the rear end, where the flow is reattached and is in the downstream direction, experiences positive friction. These findings are important for designing improvements not only of individual components and whole FOWT platforms but also of their mooring and anchoring systems. Furthermore, this can feed into the choice of structural material and designing of corrosion and biofouling-resistant paints for offshore structures. This, in turn, can help in decision-making for the construction and maintenance of FOWT platforms, leading towards the development of cost-effective FOWT platforms.

Author Contributions: N.D. wrote and coordinated contributions to the paper. C.J.L. contributed to the Abstract, Introduction, and Methodology. R.M.D. and J.W. provided suitable corrections, comments, and suggestions. All authors have read and agreed to the published version of the manuscript.

Funding: R.M.D. acknowledges the support of NERC, grant number NE/S014535/1. C.J.L. was supported by an Early Career Fellowship funded by the Leverhulme Trust. N.D. acknowledges the support of Offshore Renewable Energy Catapult and EPSRC, grant number EP/S023763/1 and project reference 2852987. We acknowledge the Viper High Performance Computing facility of the University of Hull and its support team.

Institutional Review Board Statement: The study is conducted as part of PhD research at the Energy and Environment Institute, University of Hull. This research does not involve humans and/or animals.

Data Availability Statement: The data used to support this study can found at the following: https://zenodo.org/records/14161879?preview=1&token=eyJhbGciOiJIUzUxMiJ9.eyJpZCI6ImNmO\TQzYTQ5LWZlYTEtNGE1ZS1iMDFlLTFjYmI3ZWVlNmFjYiIsImRhdGEiOnt9LCJyYW5kb20iOiI\2YThkMDk1NjliYzljNGVlMmY4NTFkZmE2ZDA5MTY1YyJ9.D56x2CXJxkB5aGj8GNUOKJq0hOo\Uw6dfS6q2D05LhzScBBGQ63IVTh6_W0xk3VGpzqM-mah8f_RpTGeA8WwXZg, accessed on 14 November 2024.

Conflicts of Interest: The authors declare no conflicts of interest.

Abbreviations

The following abbreviations are used in this manuscript:

MDPI	Multidisciplinary Digital Publishing Institute
DOAJ	Directory of Open Access Journals
FOWT	Floating Offshore Wind Turbine
FC	Finite Cylinder
FCHP	Finite Cylinder with Heave Plate
TCHP	Three Cylinders with Heave Plates
OC4	OC4 Semi-submersible Platform
Re	Reynolds Number
NREL	National Renewable Energy Laboratory
TLP	Tension-Leg Platform
$\overline{C}_d$	Time-averaged Drag Coefficient
$\overline{C}_f$	Time-averaged Viscous Drag Coefficient
$\overline{C}_p$	Time-averaged Pressure Drag Coefficient
$\overline{C}_v$	Time-averaged Vertical Force Coefficient

Appendix A

Appendix A.1. Numerical Validation of the Eddy Viscosity Model

The flow past an infinite cylinder at $Re = 3900$ has been extensively studied experimentally and numerically, providing a wide range of datasets. Therefore, the accuracy of the $k - \omega$ SST model is numerically verified for unidirectional flow past an infinite cylinder at $Re = 3900$ against the works of [58,59]. The size of the computational domain is

$-10 \leq x/D \leq 40$ in streamwise, $-12 \leq x/D \leq 12$ in cross-stream, and $0 \leq x/D \leq 3$ in vertical directions, where D is the diameter of the cylinder. The simulations are carried out with five hexahedral meshes of increasing resolution from $\sim$ 500k to $\sim$ 3012k cells. The turbulent quantities k and ω are initialised based on turbulence intensity (I) of 0.2% and a ratio of turbulent to molecular kinematic viscosity (ν_t/ν) of 10^{-3}. A uniform streamwise velocity is specified at the inlet and Dirichlet boundary conditions at the cylinder wall. For pressure, a Neumann boundary condition is adopted at the inlet and outlet, whereas it is set to zero at the cylinder wall. The initial conditions for turbulent kinetic energy k and ω are calculated based on Equation (A1).

$$k = \frac{3}{2}(IU_\infty)^2, \qquad \omega = \frac{\rho k}{\mu}\left(\frac{\mu_t}{\mu}\right)^{-1}, \tag{A1}$$

where I is the turbulent intensity, U_∞ is the free-stream velocity, ρ is the density of the fluid, and μ_t/μ is the ratio between the turbulent and molecular velocity. Near the wall, the k is expected to be very low, whereas ω increases; therefore, a low value for k and a high value for ω is specified at the wall. At the top, bottom, and side boundaries, a symmetry boundary condition is specified. The simulations were run on OpenFOAM, which is a leading open-source computational fluid dynamics software developed by OpenFOAM Foundation [40]. A second-order, implicit scheme is used for time discretisation. The gradient terms are discretised using the Gauss linear scheme, whereas a total variation diminishing (TVD) scheme, which is first/second-order accurate, is specified for the convective terms. A second-order Gauss linear scheme is specified for Laplacian terms.

The time-averaged drag coefficient ($\overline{C}_d$) acting on the cylinder is calculated as follows:

$$\overline{C}_d = \frac{\overline{F}_d}{\frac{1}{2}\rho U_\infty^2 A_{ref}}, \tag{A2}$$

where $\overline{F}_d$ is the ensemble-averaged drag force, ρ is the dynamic density of the fluid, U_∞ is the free stream velocity of the fluid, and A_{ref} is the projected area of the cylinder. The $\overline{C}_d$ acting on the cylinder ranged between $\sim$ 1.00 and 0.998 across all the simulations, where the $\overline{C}_d$ in [58,59] are 1.245 and 0.98, respectively. Karman vortex street, a feature observed in the wake of infinite cylinders, is vortices shed alternatively from either side of the cylinder. The non-dimensionalised frequency Strouhal number (S_t) associated with this periodic vortex shedding is given as follows:

$$S_t = \frac{fD}{U_\infty}, \tag{A3}$$

where f is the vortex shedding frequency, D is cylinder diameter, and U_∞ is free stream velocity. S_t for the simulations is $\sim$0.19, whereas S_t reported by [58,59] are 0.211 and 0.208 ± 0.002, respectively. Further, velocity deficit profiles in the wake of the cylinder at $x/D = 1.06$ and 1.54 are compared against those reported in [59]. Figure A1 shows the velocity profiles for all five simulation runs in [59]. Clearly, at $x/D = 1.06$, the simulation results agree very well with the literature; however, at $x/D = 1.54$, although the simulation results converge, the velocity profiles are under-predicted compared with [59]. This is also observed in the numerical work of [58]. Nevertheless, URANS with the $k - \omega$ SST model can resolve past cylinders and, therefore, have been adopted in further numerical study.

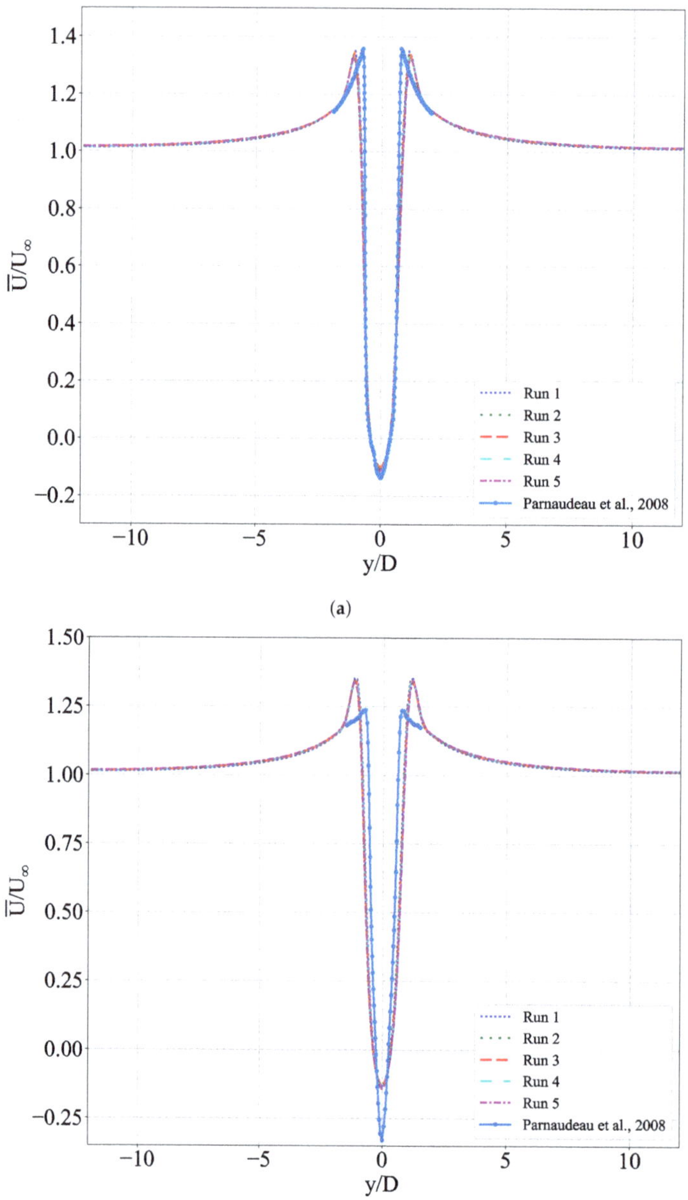

(**a**)

(**b**)

Figure A1. Time-averaged streamwise velocity profiles in the cross-stream direction normalised by free-stream velocity ($\overline{U}/U_\infty$) at $x/D = 1.06$ (**a**) and 1.54 (**b**) for each simulation run and the results reported in [59]. The velocity profiles at $x/D = 1.06$ agree very well with the results in literature. However, at $x/D = 1.54$, although simulation results converge, simulations underestimate velocity deficit compared with [59]. This difference is also reported in [58].

Appendix A.2. Flow Visualisation at Re = 43,000

This section presents flow visualisation in the wake of FC, FCHP, TCHP, and OC4 at *Re* = 43,000 and 200,000. Similar to the *Re* = 2900 case, vortices and recirculation zones are formed in the near wake of the structures. The influence of upstream members can be seen on the vortices behind the back cylinders of TCHP and OC4. In the case of OC4, only one vortex core is formed behind each back cylinder.

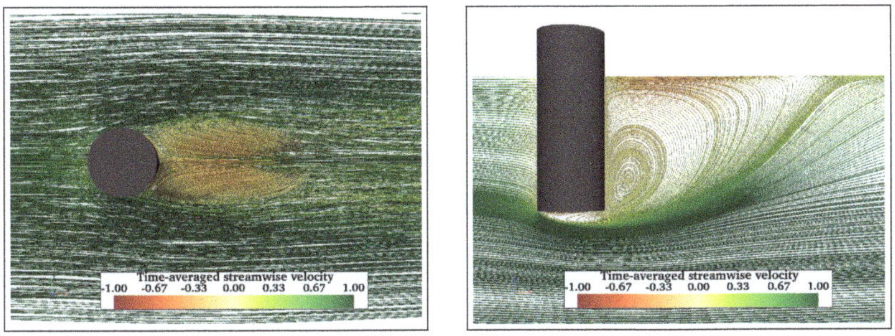

Figure A2. Flow visualisation in the wake of FC at *Re* = 43,000 showing the formation of two vortex cores (**left**) and recirculating flow (**right**).

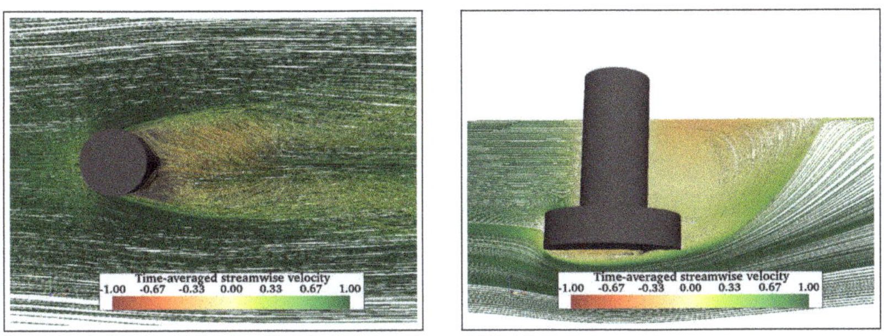

Figure A3. Flow visualisation in the wake of FCHP at *Re* = 43,000 showing symmetric vortices in the near wake (**left**) and recirculation zones behind the cylinder and under the heave plate (**right**).

Figure A4. Flow visualisation wake of TCHP at *Re* = 43,000 showing symmetric vortex cores behind each cylinder (**left**), though the influence of upstream member on vortices behind back cylinders can be seen. The effect of the heave plate reduces the size of the recirculation zone behind the cylinder (**right**).

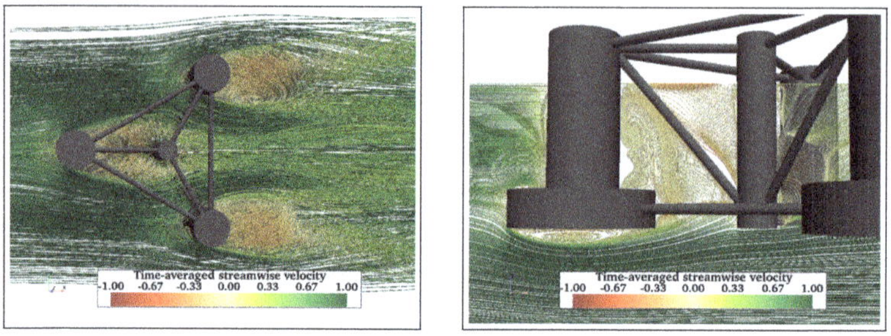

Figure A5. Flow visualisation in wake of OC4 at Re = 43,000. Only one vortex core is formed behind each back cylinder (**left**). Recirculation flow behind the front cylinder is restricted by the inclined cross-brace (**right**).

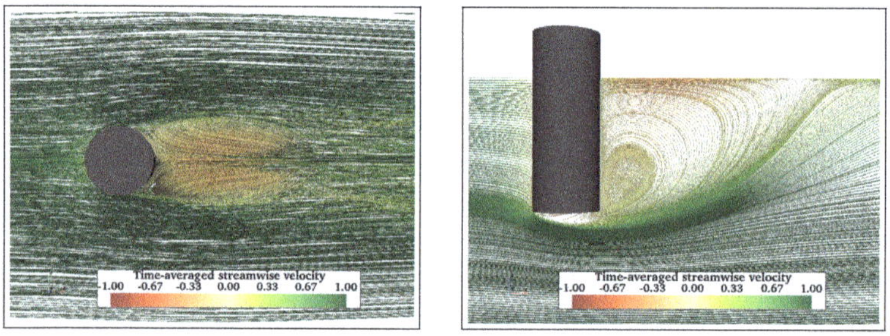

Figure A6. Time-averaged streamwise velocity streamlines around the FC at Re = 200,000: plan view of the FC from top (**left**) and profile view on a vertical plane (**right**).

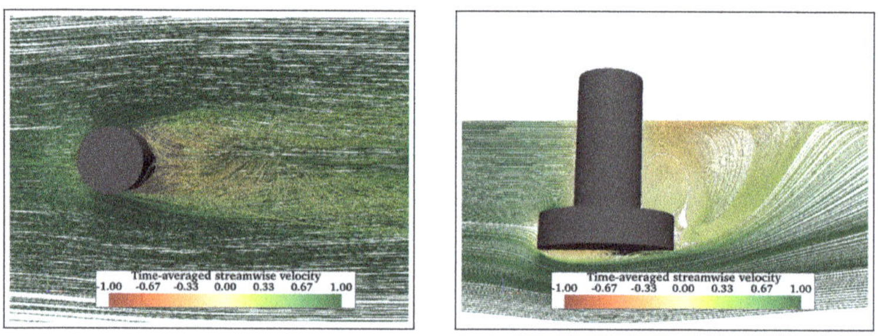

Figure A7. Flow visualisation in the wake of FCHP at Re = 200,000. Unlike Re = 2900 and 43,000, vortices in the wake are less distinct (**left**). However, similar to the other two cases, recirculation regions are formed under the heave plate and behind the cylinder (**right**).

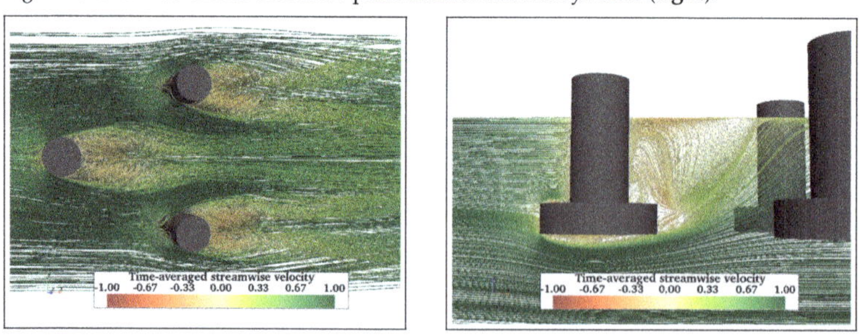

Figure A8. Flow visualisation in the wake of TCHP at Re = 200,000 showing the symmetric vortices behind cylinders (**left**) and recirculating flows (**right**).

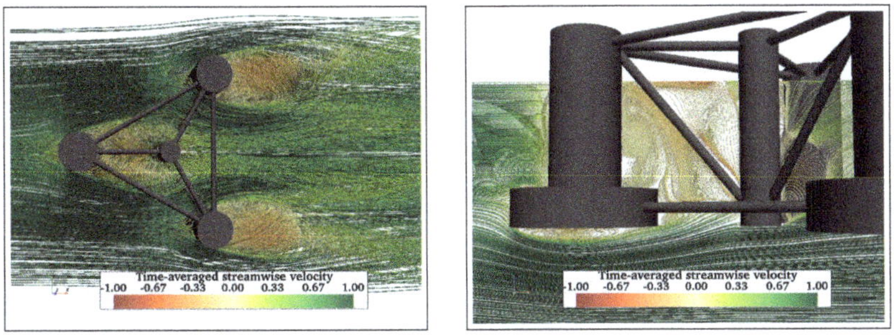

Figure A9. Mean streamwise velocity streamlines in the wake of OC4 at *Re* = 200,000. A single vortex core is formed behind each of the back cylinders of OC4 (**left**). Similar to TCHP, recirculating flow can be observed behind the front cylinder and the heave plate (**right**).

Appendix A.3. Pressure and Friction Coefficient Results

In this section, vertical pressure coefficients for all structures are presented at *Re* = 43,000. Similar to the other two *Re* cases, vertical pressure varies in a streamwise direction on the free end of FC and bottom surfaces of TCHP and OC4.

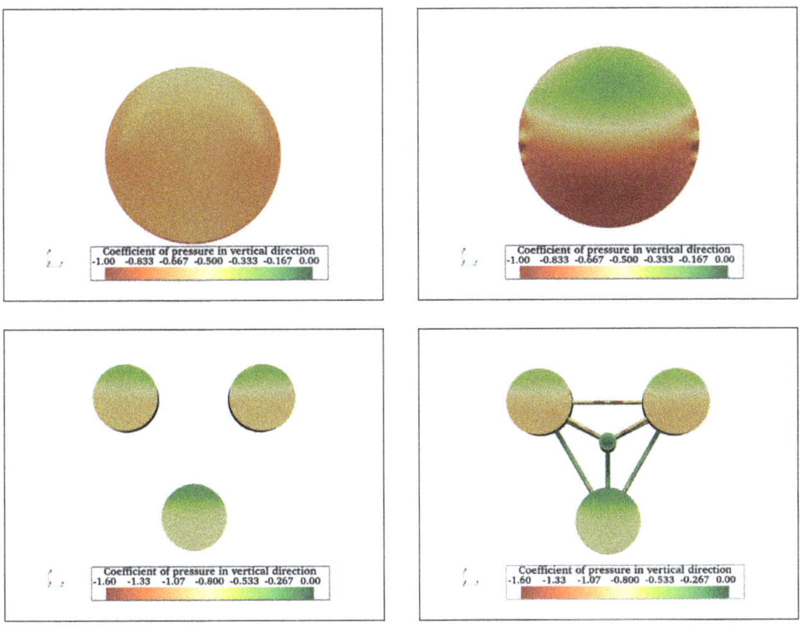

Figure A10. Vertical pressure coefficient acting on FC (**top left**), (**top right**), TCHP (**bottom left**), and OC4 (**bottom right**) at $Re = 43,000$. The pressure varies across the surfaces of the free end of FC and heave plates of FCHP, TCHP, and OC4. This is attributed to recirculating flows under the heave plates.

Also, the streamwise pressure coefficient for all structures is presented in Figure A11 at $Re = 2900$ and in Figure A12 at *Re* = 43,000. Similar to the *Re* = 200,000 cases, the highest pressure is on the front surfaces of FC and cylinders of TCHP and OC4. Also, the two red zones on either side are where the pressure is minimal as the flow accelerates.

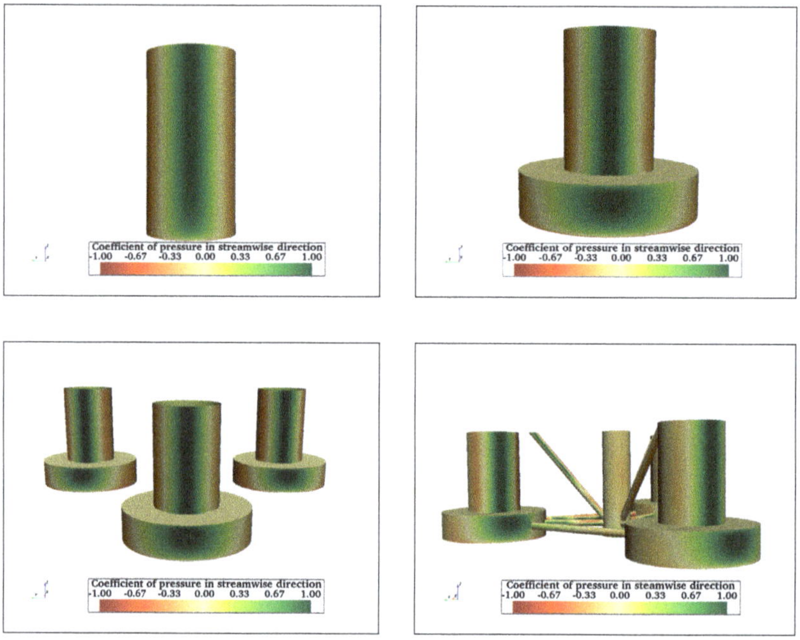

Figure A11. Streamwise pressure coefficient at $Re = 2900$ on FC (**top left**), FCHP (**top right**), TCHP (**bottom left**), and OC4 (**bottom right**). The front surfaces of FC and cylinders of FCHP, TCHP, and OC4 are where pressure is at its maximum.

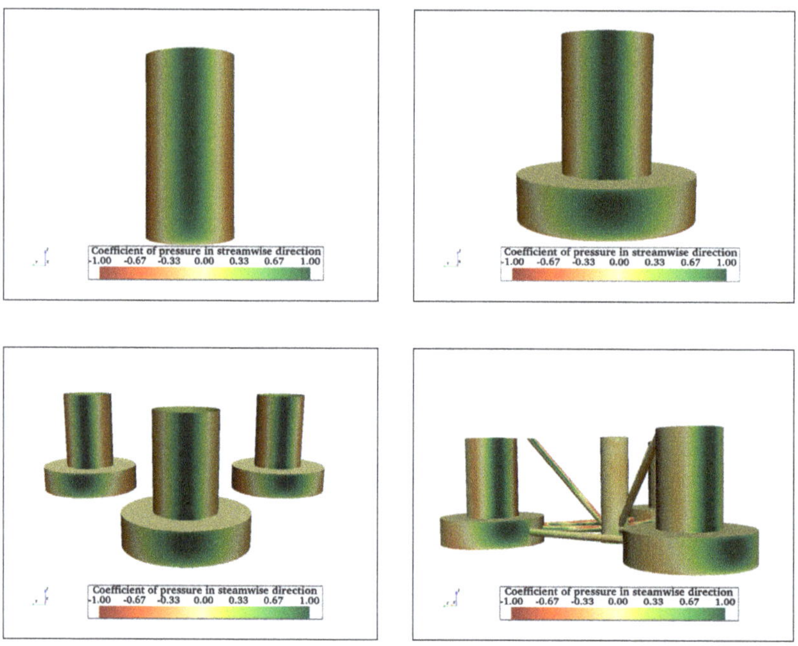

Figure A12. Streamwise pressure coefficient at Re = 43,000 on FC (**top left**), FCHP (**top right**), TCHP (**bottom left**), and OC4 (**bottom right**). Similar to $Re = 2900$ cases, the front surfaces of FC and cylinders of FCHP, TCHP, and OC4 are where pressure is at its maximum.

Figures A13 and A14 show the streamwise friction coefficients on FC at Re = 2900 and 43,000, respectively. Negative friction on free surfaces of free ends is due to flow being upstream. At Re = 2900, the highest friction is at the leading edge; however, at Re = 43,000, friction is high on the rear half of the free end. The friction is least on the front surface where pressure is maximum, and friction is maximum on either side due to flow acceleration. The streamwise friction coefficient for TCHP for Re = 2900 and 43,000 are presented in Figures A17 and A18, respectively. Similar to the Re = 200,000 case, the friction varies on the free ends on heave plates due to recirculation zones. There is considerable friction on the top surfaces of heave plates and sides of cylinders. A similar trend can be seen in the case of OC4 at Re = 2900 (Figure A19) and at Re = 43,000 (Figure A20). Similar to Re = 200,000, the friction on braces varies depending on whether sections of braces are within the wake of the upstream member.

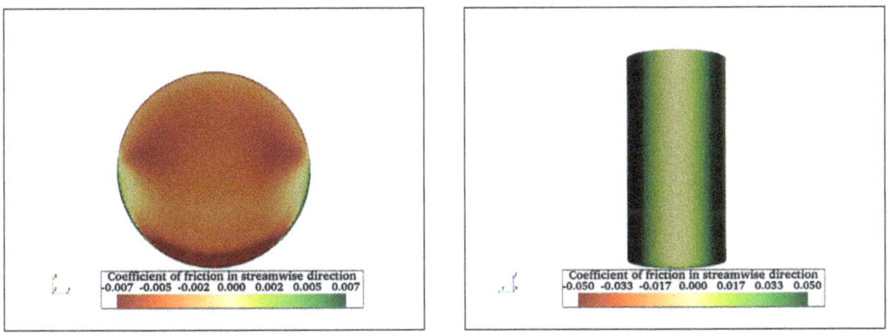

Figure A13. Streamwise friction coefficient on the free end (**left**) and front surface of FC (**right**) at Re = 2900. The highest friction is at the leading edge of the free end whereas on front surface friction is highest where the flow accelerates.

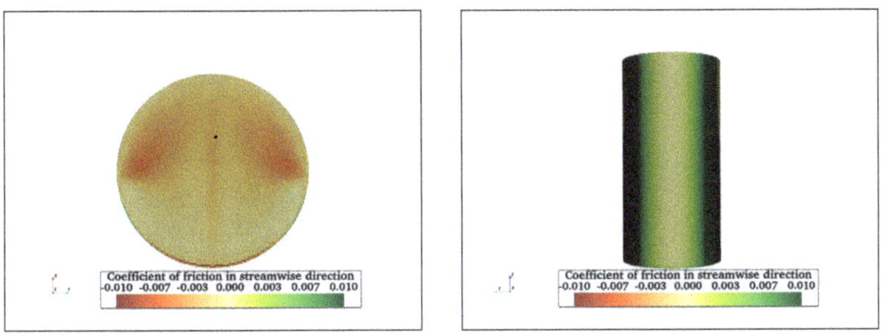

Figure A14. Streamwise friction coefficient on the free end (**left**) and front surface of FC (**right**) at Re = 43,000. Friction increases on the surface of the free end in a streamwise direction.

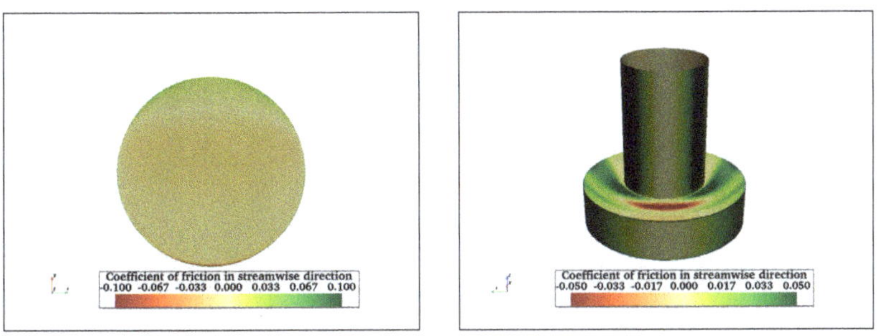

Figure A15. Streamwise friction coefficient on the bottom surface of the heave plate (**left**) and cylinder and the top surface of the heave plate (**right**) at Re = 2900.

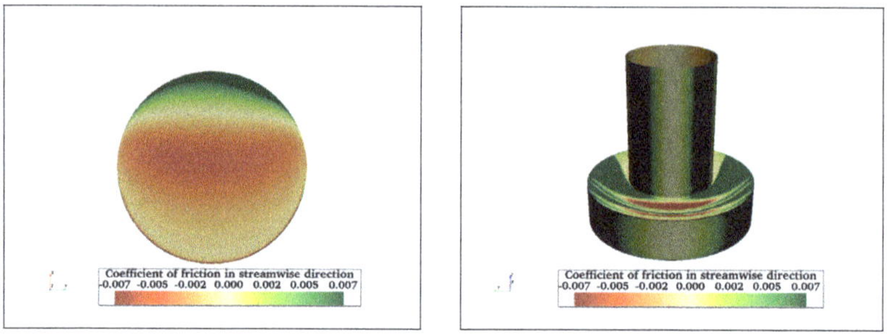

Figure A16. Streamwise friction coefficient on the bottom surface of the heave plate (**left**) and cylinder and the top surface of the heave plate (**right**) at Re = 43,000.

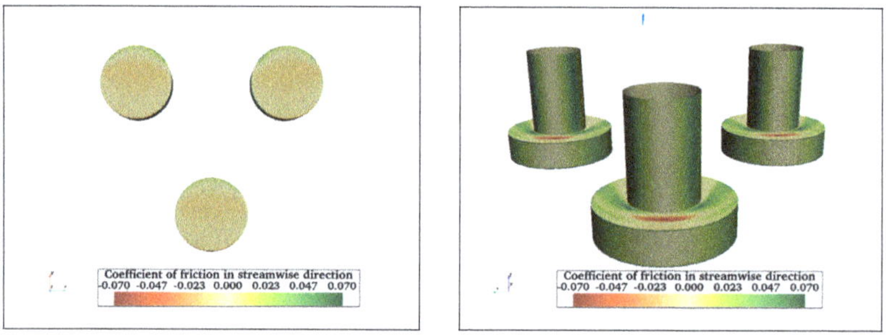

Figure A17. Streamwise friction coefficient at $Re = 2900$ on bottom surfaces of TCHP (**left**) and the rest of TCHP (**right**).

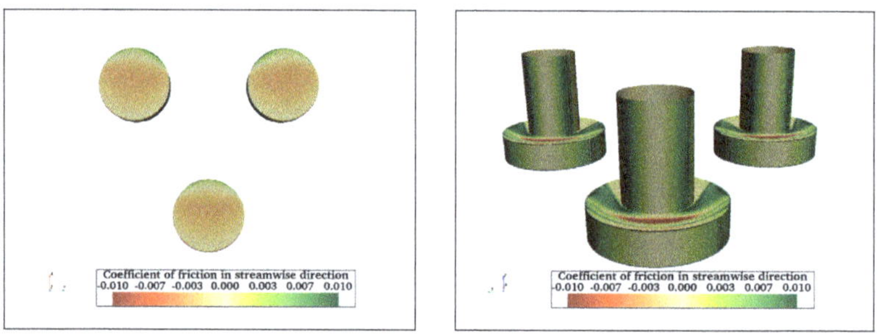

Figure A18. Streamwise friction coefficient at Re = 43,000 on bottom surfaces of TCHP (**left**) and the rest of TCHP (**right**).

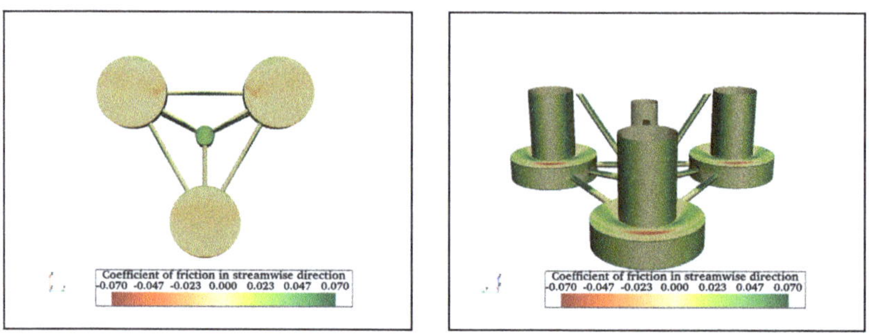

Figure A19. Streamwise friction coefficient at $Re = 2900$ on bottom surfaces of OC4 (**left**) and the rest of OC4 (**right**).

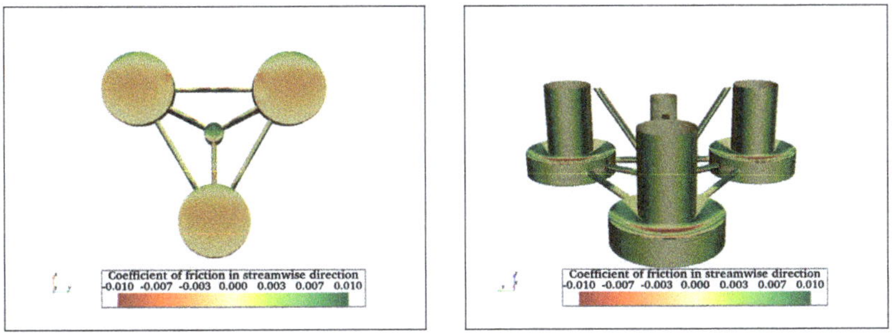

Figure A20. Streamwise friction coefficient at Re = 43,000 on bottom surfaces of OC4 (**left**) and the rest of OC4 (**right**).

References

1. Global Wind Energy Council. Global Offshore Wind Report 2024. 2024. Available online: https://gwec.net/wp-content/uploads/2024/04/GWR-2024_digital-version_final-1.pdf (accessed on 29 July 2024).
2. Sun, X.; Huang, D.; Wu, G. The current state of offshore wind energy technology development. *Energy* **2012**, *41*, 298–312. [CrossRef]
3. Jacobsen, H.K.; Hevia-Koch, P.; Wolter, C. Nearshore and offshore wind development: Costs and competitive advantage exemplified by nearshore wind in Denmark. *Energy Sustain. Dev.* **2019**, *50*, 91–100. [CrossRef]
4. Edwards, E.C.; Holcombe, A.; Brown, S.; Ransley, E.; Hann, M.; Greaves, D. Evolution of floating offshore wind platforms: A review of at-sea devices. *Renew. Sustain. Energy Rev.* **2023**, *183*, 113416. [CrossRef]
5. Schneider, J.A.; Senders, M. Foundation design: A comparison of oil and gas platforms with offshore wind turbines. *Mar. Technol. Soc. J.* **2010**, *44*, 32–51. [CrossRef]
6. Jacobsen, A.; Godvik, M. Influence of wakes and atmospheric stability on the floater responses of the Hywind Scotland wind turbines. *Wind Energy* **2021**, *24*, 149–161. [CrossRef]
7. Castro-Santos, L.; Silva, D.; Bento, A.R.; Salvação, N.; Soares, C.G. Economic feasibility of floating offshore wind farms in Portugal. *Ocean Eng.* **2020**, *207*, 107393. [CrossRef]
8. Uzunoglu, E.; Soares, C.G. Hydrodynamic design of a free-float capable tension leg platform for a 10 MW wind turbine. *Ocean Eng.* **2020**, *197*, 106888. [CrossRef]
9. Leimeister, M.; Kolios, A.; Collu, M. Critical review of floating support structures for offshore wind farm deployment. *J. Phys. Conf. Ser.* **2018**, *1104*, 012007. [CrossRef]
10. Risch, D.; Favill, G.; Marmo, B.; van Geel, N.; Benjamins, S.; Thompson, P.; Wittich, A.; Wilson, B. *Characterisation of Underwater Operational Noise of Two Types of Floating Offshore Wind Turbines*; Scottish Association for Marine Science (SAMS): Oban, UK, 2008.
11. Jonkman, J.M. *Dynamics Modeling and Loads Analysis of an Offshore Floating Wind Turbine*; University of Colorado at Boulder: Denver, CO, USA, 2007.
12. Wei, K.; Arwade, S.R.; Myers, A.T. Incremental wind-wave analysis of the structural capacity of offshore wind turbine support structures under extreme loading. *Eng. Struct.* **2014**, *79*, 58–69. [CrossRef]
13. Cui, W.; Fu, S.; Hu, Z. *Encyclopedia of Ocean Engineering*; Springer: Berlin/Heidelberg, Germany, 2022.
14. Bauer, J. NREL Floats New Offshore Wind Cost Optimization Vision. 2020. Availabe online: https://www.nrel.gov/news/program/2020/nrel-floats-new-offshore-wind-cost-optimization-tool.html (accessed on 1 August 2024).
15. Tran, T.T.; Kim, D.H. The coupled dynamic response computation for a semi-submersible platform of floating offshore wind turbine. *J. Wind Eng. Ind. Aerodyn.* **2015**, *147*, 104–119. [CrossRef]
16. Tran, T.T.; Kim, D.H. Fully coupled aero-hydrodynamic analysis of a semi-submersible FOWT using a dynamic fluid body interaction approach. *Renew. Energy* **2016**, *92*, 244–261. [CrossRef]
17. Robertson, A.N.; Wendt, F.; Jonkman, J.M.; Popko, W.; Dagher, H.; Gueydon, S.; Qvist, J.; Vittori, F.; Azcona, J.; Uzunoglu, E.; et al. OC5 project phase II: Validation of global loads of the DeepCwind floating semisubmersible wind turbine. *Energy Procedia* **2017**, *137*, 38–57. [CrossRef]
18. Coulling, A.J.; Goupee, A.J.; Robertson, A.N.; Jonkman, J.M.; Dagher, H.J. Validation of a FAST semi-submersible floating wind turbine numerical model with DeepCwind test data. *J. Renew. Sustain. Energy* **2013**, *5*, 023116. [CrossRef]
19. Wang, Y.; Chen, H.C.; Vaz, G.; Burmester, S. CFD simulation of semi-submersible floating offshore wind turbine under pitch decay motion. In Proceedings of the International Conference on Offshore Mechanics and Arctic Engineering, Scotland, UK, 9–14 June 2019; American Society of Mechanical Engineers: New York, NY, USA, 2019; Volume 59353, p. V001T01A002.
20. Wang, L.; Robertson, A.; Jonkman, J.; Yu, Y.H. OC6 phase I: Improvements to the OpenFAST predictions of nonlinear, low-frequency responses of a floating offshore wind turbine platform. *Renew. Energy* **2022**, *187*, 282–301. [CrossRef]
21. Wang, Y.; Chen, H.C.; Koop, A.; Vaz, G. Verification and validation of CFD simulations for semi-submersible floating offshore wind turbine under pitch free-decay motion. *Ocean Eng.* **2021**, *242*, 109993. [CrossRef]

22. Li, H.; Bachynski-Polić, E.E. Experimental and numerically obtained low-frequency radiation characteristics of the OC5-DeepCwind semisubmersible. *Ocean. Eng.* **2021**, *232*, 109130. [CrossRef]
23. Robertson, A.N.; Gueydon, S.; Bachynski, E.; Wang, L.; Jonkman, J.; Alarcon, D.; Amet, E.; Beardsell, A.; Bonnet, P.; Boudet, B.; et al. OC6 Phase I: Investigating the underprediction of low-frequency hydrodynamic loads and responses of a floating wind turbine. *J. Phys. Conf. Ser.* **2020**, *1618*, 032033. [CrossRef]
24. Robertson, A.; Bachynski, E.E.; Gueydon, S.; Wendt, F.; Schünemann, P. Total experimental uncertainty in hydrodynamic testing of a semisubmersible wind turbine, considering numerical propagation of systematic uncertainty. *Ocean Eng.* **2020**, *195*, 106605. [CrossRef]
25. Burmester, S.; Vaz, G.; Gueydon, S.; el Moctar, O. Investigation of a semi-submersible floating wind turbine in surge decay using CFD. *Ship Technol. Res.* **2020**, *67*, 2–14. [CrossRef]
26. Burmester, S.; Vaz, G.; el Moctar, O.; Gueydon, S.; Koop, A.; Wang, Y.; Chen, H. High-fidelity modelling of floating offshore wind turbine platforms. In Proceedings of the International Conference on Offshore Mechanics and Arctic Engineering, Virtual, 3–7 August 2020; American Society of Mechanical Engineers: New York, NY, USA, 2020; Volume 84416, p. V009T09A068.
27. Wang, Y.; Chen, H.C.; Vaz, G.; Burmester, S. CFD simulation of semi-submersible floating offshore wind turbine under regular waves. In Proceedings of the ISOPE International Ocean and Polar Engineering Conference (ISOPE), Shanghai, China, 11–16 October 2020; p. ISOPE–I.
28. Liu, Y.; Ge, D.; Bai, X.; Li, L. A CFD study of vortex-induced motions of a semi-submersible floating offshore wind turbine. *Energies* **2023**, *16*, 698. [CrossRef]
29. Liu, Y.; Liu, F.; Wang, E.; Xiao, Q.; Li, L. The effect of base column on vortex-induced vibration of a circular cylinder with low aspect ratio. *Ocean Eng.* **2020**, *196*, 106822. [CrossRef]
30. Benitz, M.A.; Schmidt, D.P.; Lackner, M.A.; Stewart, G.M.; Jonkman, J.; Robertson, A. Validation of hydrodynamic load models using CFD for the OC4-DeepCwind semisubmersible. In Proceedings of the International Conference on Offshore Mechanics and Arctic Engineering, St John's, NF, Canada, 31 May–5 June 2015; American Society of Mechanical Engineers: New York, NY, USA, 2015; Volume 56574, p. V009T09A037.
31. Ma, S.; Xu, D.k.; Duan, W.y.; Chen, J.k.; Liao, K.p.; Wang, H. The numerical study of viscous drag force influence on low-frequency surge motion of a semi-submersible in storm sea states. *Ocean Eng.* **2020**, *213*, 107511. [CrossRef]
32. Berthelsen, P.A.; Baarholm, R.; Pa´ kozdi, C.; Stansberg, C.T.; Hassan, A.; Downie, M.; Incecik, A. Viscous drift forces and responses on a semisubmersible platform in high waves. In Proceedings of the International Conference on Offshore Mechanics and Arctic Engineering, Honolulu, HI, USA, 31 May–6 June 2009; Volume 43413, pp. 469–478.
33. Apolinario, M.; Coutinho, R. Understanding the biofouling of offshore and deep-sea structures. In *Advances in Marine Antifouling Coatings and Technologies*; Elsevier: Amsterdam, The Netherlands, 2009; pp. 132–147.
34. Maduka, M.; Schoefs, F.; Thiagarajan, K.; Bates, A. Hydrodynamic effects of biofouling-induced surface roughness–Review and research gaps for shallow water offshore wind energy structures. *Ocean Eng.* **2023**, *272*, 113798. [CrossRef]
35. Edyvean, R. Biodeterioration problems of North Sea Oil and gas production—A review. *Int. Biodeterior.* **1987**, *23*, 199–231. [CrossRef]
36. Polman, H.; Verhaart, F.; Bruijs, M. Impact of biofouling in intake pipes on the hydraulics and efficiency of pumping capacity. *Desalin. Water Treat.* **2013**, *51*, 997–1003. [CrossRef]
37. Robertson, A.; Jonkman, J.; Masciola, M.; Song, H.; Goupee, A.; Coulling, A.; Luan, C. *Definition of the Semisubmersible Floating System for Phase II of OC4*; Technical Report; National Renewable Energy Lab. (NREL): Golden, CO, USA, 2014.
38. Menter, F.R.; Kuntz, M.; Langtry, R. Ten years of industrial experience with the SST turbulence model. *Turbul. Heat Mass Transf.* **2003**, *4*, 625–632.
39. Goupee, A.J.; Koo, B.; Lambrakos, K.; Kimball, R. Model tests for three floating wind turbine concepts. In Proceedings of the Offshore Technology Conference (OTC), Houston, TX, USA, 30 April–3 May 2012; p. OTC–23470.
40. Greenshields, C. *OpenFOAM v10 User Guide*; The OpenFOAM Foundation: London, UK, 2022.
41. Alkishriwi, N.; Meinke, M.; Schröder, W. A large-eddy simulation method for low Mach number flows using preconditioning and multigrid. *Comput. Fluids* **2006**, *35*, 1126–1136. [CrossRef]
42. Greenshields, C.; Weller, H. *Notes on Computational Fluid Dynamics: General Principles*; CFD Direct Ltd.: Reading, UK, 2022.
43. Issa, R.I. Solution of the implicitly discretised fluid flow equations by operator-splitting. *J. Comput. Phys.* **1986**, *62*, 40–65.
44. Benitz, M.A.; Carlson, D.; Seyed-Aghazadeh, B.; Modarres-Sadeghi, Y.; Lackner, M.; Schmidt, D. CFD simulations and experimental measurements of flow past free-surface piercing, finite length cylinders with varying aspect ratios. *Comput. Fluids* **2016**, *136*, 247–259. [CrossRef]
45. Rosetti, G.F.; Vaz, G.; Hoekstra, M.; Gonçalves, R.T.; Fujarra, A.L. CFD calculations for free-surface-piercing low aspect ratio circular cylinder with solution verification and comparison with experiments. In Proceedings of the International Conference on Offshore Mechanics and Arctic Engineering, Nantes, France, 9–14 June 2013; American Society of Mechanical Engineers: New York, NY, USA, 2013; Volume 55416, p. V007T08A056.
46. He, J.w.; Zhao, W.w.; Wan, D.c.; Wang, Y.q. Numerical study of free end effect of cylinder with low aspect ratios on vortex induced motion. *J. Hydrodyn.* **2022**, *34*, 106–115. [CrossRef]
47. Frederich, O.; Wassen, E.; Thiele, F. Prediction of the flow around a short wall-mounted finite cylinder using LES and DES1. *JNAIAM* **2008**, *3*, 231–247.

48. Frederich, O.; Wassen, E.; Thiele, F.; Jensch, M.; Brede, M.; Hüttmann, F.; Leder, A. Numerical simulation of the flow around a finite cylinder with ground plate in comparison to experimental measurements. In *New Results in Numerical and Experimental Fluid Mechanics VI: Contributions to the 15th STAB/DGLR Symposium, Darmstadt, Germany, 2006*; Springer: Berlin/Heidelberg, Germany, 2006; pp. 348–355.
49. Gonçalves, R.T.; Franzini, G.R.; Rosetti, G.F.; Meneghini, J.R.; Fujarra, A.L.C. Flow around circular cylinders with very low aspect ratio. *J. Fluids Struct.* **2015**, *54*, 122–141. [CrossRef]
50. Schlichting, H.; Gersten, K. *Boundary-Layer Theory*; Springer: Berlin/Heidelberg, Germany, 2016.
51. Gu, Z.; Sun, T. Classifications of flow pattern on three circular cylinders in equilateral-triangular arrangements. *J. Wind Eng. Ind. Aerodyn.* **2001**, *89*, 553–568. [CrossRef]
52. Bao, Y.; Zhou, D.; Huang, C. Numerical simulation of flow over three circular cylinders in equilateral arrangements at low Reynolds number by a second-order characteristic-based split finite element method. *Comput. Fluids* **2010**, *39*, 882–899. [CrossRef]
53. Zheng, S.; Zhang, W.; Lv, X. Numerical simulation of cross-flow around three equal diameter cylinders in an equilateral-triangular configuration at low Reynolds numbers. *Comput. Fluids* **2016**, *130*, 94–108. [CrossRef]
54. Wang, R.; He, Y.; Chen, L.; Zhu, Y.; Wei, Y. Numerical simulations of flow around three cylinders using momentum exchange-based immersed boundary-lattice Boltzmann method. *Ocean Eng.* **2022**, *247*, 110706. [CrossRef]
55. Kawamura, T.; Hiwada, M.; Hibino, T.; Mabuchi, I.; Kumada, M. Flow around a finite circular cylinder on a flat plate: Cylinder height greater than turbulent boundary layer thickness. *Bull. JSME* **1984**, *27*, 2142–2151. [CrossRef]
56. Okamoto, S.; Sunabashiri, Y. Vortex shedding from a circular cylinder of finite length placed on a ground plane. *J. Fluids Eng.* **1992**, *11*, 512–521. [CrossRef]
57. Vindenes, H.; Orvik, K.A.; Søiland, H.; Wehde, H. Analysis of tidal currents in the North Sea from shipboard acoustic Doppler current profiler data. *Cont. Shelf Res.* **2018**, *162*, 1–12. [CrossRef]
58. Pereira, F.S.; Vaz, G.; Eça, L. Evaluation of RANS and SRS methods for simulation of the flow around a circular cylinder in the sub-critical regime. *Ocean Eng.* **2019**, *186*, 106067. [CrossRef]
59. Parnaudeau, P.; Carlier, J.; Heitz, D.; Lamballais, E. Experimental and numerical studies of the flow over a circular cylinder at Reynolds number 3900. *Phys. Fluids* **2008**, *20*, 085101. [CrossRef]

Journal of
Marine Science and Engineering

MDPI

Article

Experimental Study of the Hydrodynamic Forces of Pontoon Raft Aquaculture Facilities Around a Wind Farm Monopile Under Wave Conditions

Deming Chen [1], Mingchen Lin [1], Jinxin Zhou [2], Yanli Tang [1], Fenfang Zhao [1], Xinxin Wang [1,*], Mengjie Yu [1], Qiao Li [3] and Daisuke Kitazawa [2]

1 Department of Marine Fisheries, Fisheries College, Ocean University of China, Qingdao 266003, China; chendeming@stu.ouc.edu.cn (D.C.); linmingchen1130@stu.ouc.edu.cn (M.L.); tangyanli@ouc.edu.cn (Y.T.); zhaoff@ouc.edu.cn (F.Z.); yumengjie@ouc.edu.cn (M.Y.)

2 Institute of Industrial Science, The University of Tokyo, 5-1-5 Kashiwanoha, Chiba 277-8574, Japan; jxzhou@iis.u-tokyo.ac.jp (J.Z.); dkita@iis.u-tokyo.ac.jp (D.K.)

3 Faculty of Engineering, Yokohama National University, Yokohama 240-8501, Japan; li-qiao-yt@ynu.ac.jp

* Correspondence: wxinxin@ouc.edu.cn; Tel.: +86-159-6420-1390

Abstract: The integrated development of offshore wind power and marine aquaculture represents a promising approach to the sustainable utilization of ocean resources. The present study investigated the hydrodynamic response of an innovative combination of a wind farm monopile and pontoon raft aquaculture facilities (PRAFs). Physical water tank experiments were conducted on PRAFs deployed around a wind farm monopile using the following configurations: single- and three-row arrangements of PRAFs with and without a monopile. The interaction between the aquaculture structure and the wind farm monopile was examined, with a particular focus on the mooring line tensions and bridle line tensions under different wave conditions. Utilizing the wind farm monopile foundation as an anchor, the mooring line tension was reduced significantly by 16–66% in the single-row PRAF. The multi-row PRAF arrangement experienced lower mooring line tension in comparison with the single-row PRAF arrangement, with the highest reduction of 73%. However, for the bridle line tension, the upstream component was enhanced, while the downstream one was weakened with a monopile, and they both decreased in the multi-row arrangement. Finally, we developed numerical models based on flume tank tests that examined the interactions between the monopile and PRAFs, including configurations of a single monopile, along with single- and three-row arrangements of PRAFs. The numerical simulation results confirmed that the monopile had a dampening effect on the wave propagation of 5% to 20%, and the impact of the pontoons on the monopile was negligible, implying that the integration of aquaculture facilities around wind farm infrastructure may not significantly alter the hydrodynamic loads experienced by the monopile.

Keywords: multi-row arrangement; numerical simulation; raft aquaculture; tank test; wave propagation; wind farm monopile

1. Introduction

Mussels represent an affordable and environmentally sustainable source of protein, with minimal greenhouse gas emissions [1]. It is estimated that approximately 94% of global mussel production is currently derived from aquaculture, with raft culture representing one of the most efficient techniques, enabling consistent year-round production [2]. However, the expansion of raft culture is constrained by spatial limitations [3]. Concurrently, the accelerated growth of renewable energy sources has established offshore wind farms as a crucial element in the global drive toward a clean energy future. However, as offshore wind installations expand, they increasingly occupy limited coastal waters, which may result in potential conflicts with existing aquaculture activities. To address these spatial challenges, integrating wind energy infrastructure with raft aquaculture facilities represents a promising solution [4]. By deploying aquaculture structures within the confines of a wind farm, the utilization of ocean space can be optimized. Meanwhile, turbine foundations can serve as dual-purpose anchoring points for aquaculture, potentially functioning as artificial reefs that support marine biodiversity [5]. This integrated approach represents a compelling opportunity for economic synergy. Despite the existence of numerous conceptual designs for integrated systems of this nature, only a small number have progressed to the stage of commercial operation. The majority of countries remain at the pilot development stage [6–8].

The combination of offshore wind farms with aquaculture systems has the potential to offer several benefits, including the optimization of marine space usage and advancement of sustainable development. Specifically, integrating mussel raft culture with wind farms could facilitate efficient, year-round production by leveraging shared infrastructure. However, a notable gap exists in the literature regarding the effective layouts and structural interactions between mussel raft systems and offshore wind turbines. Despite the proposal by Buck et al. (2004) that combines bivalve raft culture with offshore wind turbines [7], detailed research on the structural interactions between wind turbine foundations and aquaculture facilities remains scarce. The potential for wind turbine monopiles to be effectively utilized as anchor points for mussel raft structures has yet to be investigated, and the optimal structural arrangements for aquaculture within offshore wind farms have not been thoroughly explored.

The majority of studies have concentrated on the integration of offshore wind farms with aquaculture cages rather than raft systems. Since cage aquaculture requires extensive marine areas, it is typically combined with floating wind turbines. Floating platforms for floating wind turbines can be categorized into the following three types [9]: (1) spar buoy platform, (2) tension leg platform, and (3) semi-submersible platform. A semi-submersible platform is the most suitable type of floating platforms to be integrated with aquaculture structures due to its high stability and ample space. For example, Zheng et al. (2018) evaluated the dynamic response of an innovative floating offshore wind turbine integrated with a steel fish-farming cage (FOWT-SFFC) under diverse environmental conditions, confirming its exceptionally dynamic performance [10]. Similarly, Zhao et al. (2023) employed OpenFOAM software to simulate the structural load interactions between the combined wind turbine and fish cage systems, with a particular focus on the effects of the wave height and water depth on the load amplitude [11]. Further investigation of the dynamic behavior of wind–cage structures was conducted by Zhang et al. (2022) and Cao et al. (2022), who observed the damping effects of cage nets on platform dynamics [12,13]. Yi et al. (2024) demonstrated that the aquaculture cage serves to reduce the oscillation amplitude and acts as a damper in an integrated offshore wind–wave–aquaculture (WWA) system [14]. Overall, the hydrodynamic interactions and coupling effects between wind turbines and aquaculture cage systems under wind and wave conditions have been adequately studied [15–17]. These findings underscore the focus on fish cage systems and highlight the lack of equivalent research on raft culture within offshore wind farms.

In the past, studies tended to focus on the hydrodynamic characteristics of wind turbines and raft aquaculture structures as separate entities. In recent years, an increasing number of studies underscore the value of numerical simulations based on computational fluid dynamics (CFD) for evaluating the performance of floating wind turbines and aquaculture structures. CFD-based numerical models have been widely applied in the study of wave–structure interactions. For example, by solving the Navier–Stokes equations (i.e., viscous flow solver), it is possible to effectively investigate the gap resonance problems between two boxes [18–20], a stationary box and a vertical wall [21], and two side-by-side barges [22]. By solving the Reynolds-averaged Navier–Stokes (RANS) equations coupled with advanced turbulence models, CFD can account for highly nonlinear effects, such as wave breaking, vortex shedding, and transient loads, which become particularly significant under extreme environmental conditions (e.g., large wave heights and strong currents). For example, Zhong et al. (2023) analyzed the dynamics of a semi-submersible platform designed for a 15 MW wind turbine in the context of freak wave conditions, with a particular focus on the platform's motion responses and mooring line tension [23]. Tang et al. (2021) employed CFD to examine the nonlinear wave load impacts on a 10 MW scale monopile offshore wind turbine by investigating the variations in wave pressure and horizontal forces under diverse sea conditions [24]. Investigations into raft aquaculture structures have concentrated on the hydrodynamic behavior of isolated structures. In a recent study, Wang et al. (2023) employed numerical simulations to investigate the hydrodynamic characteristics of an attenuator-type wave energy converter with varying raft shapes, with a particular focus on parameters such as the wave period and structural design [25]. Wang et al. (2022) proposed that the scale and spatial layout of the floating raft aquaculture structures should be considered when assessing the impact of aquaculture activities on the hydrodynamic environment [26]. The aforementioned studies provide valuable insights into the hydrodynamic characteristics of monopile foundations or raft facilities, addressing various factors such as the wave loads, structural integration, and dynamic responses under different environmental conditions. The wind monopiles in offshore wind farms significantly affect the forces on raft aquaculture structures by altering the flow field, wave field, and their coupling with environmental factors. This impact is multifaceted, including the drag force, wave force, dynamic loads, resonance effects, and foundation stability. To accurately assess the force conditions of raft aquaculture facilities, it is necessary to conduct an in-depth study on the coupling effects between monopiles, waves, and currents. However, current research on the hydrodynamic characteristics of raft aquaculture facilities within offshore wind farms is notably insufficient, particularly in revealing the interaction mechanisms of the current–wave structure in complex wind farm environments. This highlights the necessity for further systematic research in this field.

Mooring line tension directly affects the stability of offshore structures in the marine environment. The potential consequences of high mooring line tension are multifaceted. Increased tension may lead to a higher risk of mooring line failure and shorten the service life of the mooring system. Additionally, the structural integrity of the mooring system itself may be compromised, impacting the overall stability and safety of the structure. Tomasicchio et al. (2018) conducted experimental research on a spar-type floating offshore wind turbine under different wave–wind conditions, and the results indicate that waves have a more significant impact on structural dynamics compared with wind [27]. Lauria et al. (2024) proposed that the peak frequency of mooring line tension is primarily influenced by the wave frequency [28]. Therefore, fully understanding and accurately assessing the impact of wave action on the mooring line tension of aquaculture facilities within offshore wind farms is crucial for ensuring their long-term structural reliability and safety.

At present, the applied research on the combination of a wind monopile and aquaculture structures is still in the preliminary stage, although some studies have been conducted on the integration of floating wind turbines with fish cages. There is a lack of practical technical guidance on how to arrange raft facilities around the wind monopile. Therefore, this study investigated the hydrodynamic forces of pontoon raft aquaculture facilities (PRAFs) that used the wind farm monopile foundation as an anchor. Our previous paper (Ai et al., 2024) considered the mooring line tension [29]. In this study, we extend our previous work by incorporating both the mooring line tension and bridle line tension to achieve a more comprehensive understanding of the hydrodynamic loads acting on the PRAFs. This dual focus represents a novel perspective, as prior research has rarely examined the internal connection forces (bridle lines) alongside mooring forces. Moreover, the hydrodynamic performance of the single-row PRAF arrangement and multi-row PRAF arrangement were compared. The effects of the wind monopile on wave propagation were assessed through numerical simulations. By integrating water tank experiments and numerical simulations, this work clarified wave–structure interaction mechanisms in the presence of a monopile—an area that has not been extensively addressed in the literature. Ultimately, this study advanced the understanding of co-locating wind and aquaculture structures, laying the groundwork for safer and more efficient offshore operations. The findings can provide valuable insights into the design and optimization of integrated wind–aquaculture areas.

2. Physical Experiment Methods

2.1. Study Area

The wind farm studied in this paper was located in the southern area of Laizhou Bay of Changyi City, Shandong Province, China (Figure 1), which was 15.5 km from the coastline. Local water depths of 6.7 m to 9.2 m are suitable for the installation of wind power. Waves in this area have significant wave heights varying from 0.5 m to 2.0 m and average periods varying from 2.3 s to 7.1 s [29]. Moreover, the wind farm was oriented toward the integration of offshore aquaculture, with the aspiration of establishing a model for the development of multifunctional areas. The foresight of the installation of mariculture facilities within an offshore wind farm is depicted in Figure 2.

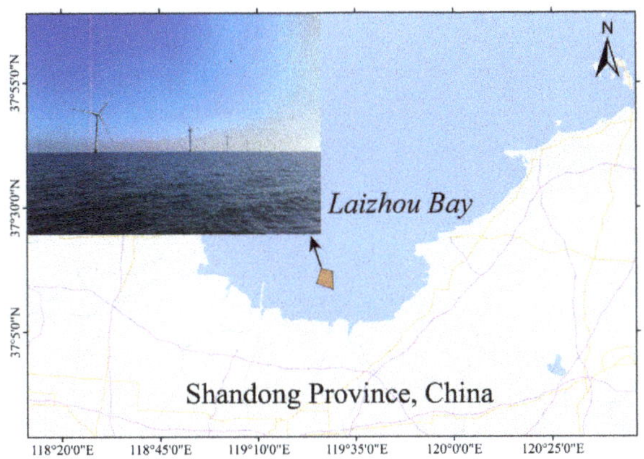

Figure 1. Location of the offshore wind farm in Laizhou Bay of Changyi City, Shandong Province, China.

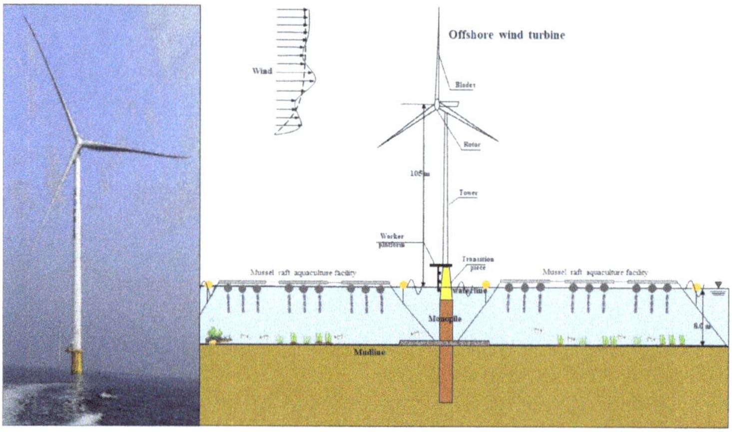

Figure 2. Schematic diagram of 'Offshore Wind-Raft Aquaculture Integration'.

2.2. Wave Tank

Experimental studies were carried out at the Ocean Engineering Basin, Institute of Industrial Science, University of Tokyo. The wave tank was 50 m in length, 10 m in width and 5 m in depth, as shown in Figure 3. The experimental water depth can be adjusted using a platform. The facility is equipped with a piston-type wavemaker, which can generate regular incident waves with periods of 0.5–5.0 s and wave heights of 0.02–0.30 m. An absorbing metallic ramp opposite to the wavemaker can dissipate the incident waves and avoid the reflected waves.

Figure 3. Experimental wave tank.

2.3. Monopile and PRAF Model

A Froude principle scale of 1:8 was adopted in this experiment according to the dimensions of the water tank and PRAFs. As shown in Figure 4a, the monopile was 2 m in height and 0.63 m in diameter, and was composed of Polyvinyl Chloride (PVC) to ensure stability in the waves. The PRAF is referenced from our previous studies in Wang et al. (2015) and Ai et al. (2024) [29,30]. A PRAF is a structure composed of three rafts (Figure 4f) combined together, with adjacent rafts connected by two bridle lines. Each raft system primarily consists of three PVC pontoons and three steel angles (Figure 4c) fixedly connected. Mussel rope models (Figure 4b) are suspended beneath the pontoons. The depth of the PRAF immersed in the water was 3.8 cm (1/3 times the height of the PRAF), as shown in Figure 4f,g. The 3D-printed mussel ropes were weighted to equal the mass of the actual mussel ropes in the water. More detailed parameters of the PRAF model are provided in Table 1.

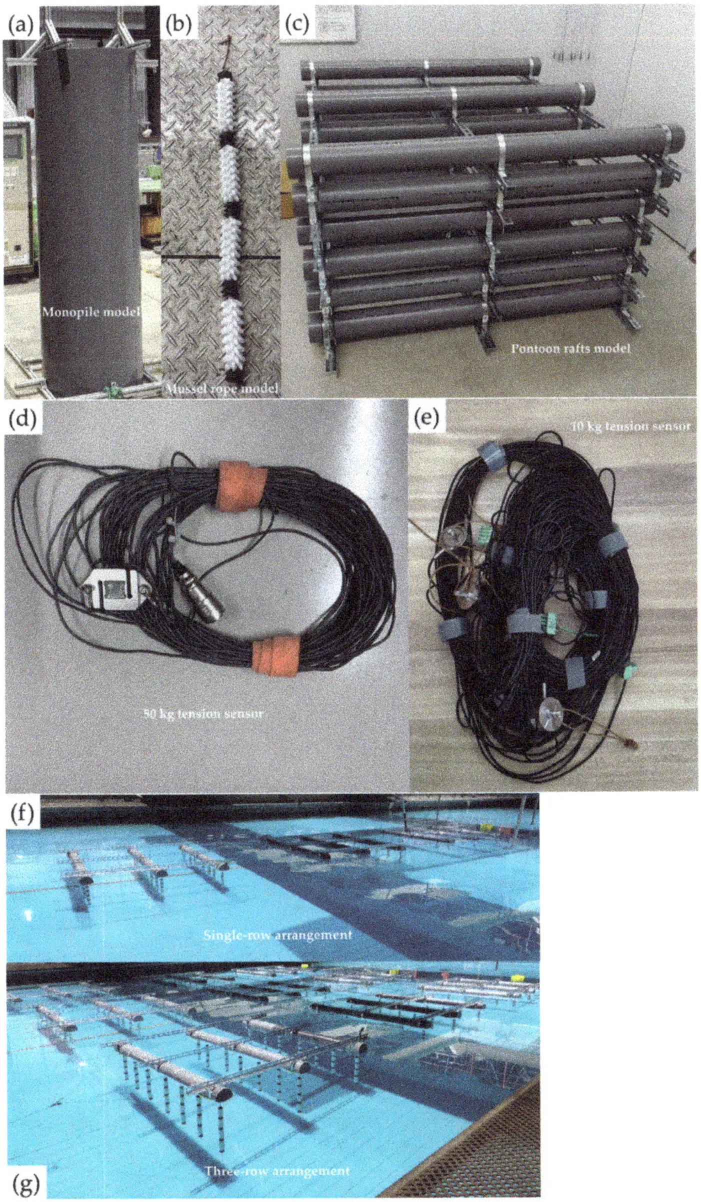

Figure 4. Experimental models and arrangement: (**a**) wind turbine monopile model; (**b**) mussel rope model; (**c**) pontoon rafts model; (**d**) 50 kg tension sensor; (**e**) 10 kg tension sensor; (**f**) picture of single-row PRAF model arrangement; (**g**) picture of three-row PRAF model arrangement.

Table 1. Model structural parameters of PRAF [29].

Components (Number)	Description
Pontoons (9)	1.5 m long, 11.4 cm diameter, buoyancy 12.9 kgf
Steel angle (9)	1.8 m long, 4.1 cm wide, 2.1 cm high, 1.0 mm thick, 2.38 kg weight
Pipe strap (27)	12.4 cm diameter, 2.0 mm thick, 0.15 kg weight
Screw and nut (54)	24.0 g total weight
Mussel ropes (63)	0.52 m long, 4.05 cm diameter, 0.50 kg weight (in water)
Buoy (18)	2.0–9.5 cm diameter (half of an ellipsoid), buoyancy 325 gf
Buoy tether line (1)	0.18 m long, 0.30 cm diameter
Bridle lines (4)	0.92 m long, 0.30 cm diameter
Mooring lines (2)	2.46 m long, 0.30 cm diameter

2.4. Experimental Setup

As illustrated in Figure 5, this is a configuration of the single-row PRAF with a monopile in the experiments [29]. The monopile was secured to the middle line of the bottom of the water tank at 16.63 m from the upstream wavemaker to act as an anchoring element for the upstream mooring line of the single-row PRAF. The PRAF was installed directly behind the monopile with a mooring line length-to-water-depth ratio of 3 (water depth was 1 m).

Figure 5. The installed monopile model and single-row PRAF model.

In this study, three configurations of PRAFs were designed and analyzed [29]:

(1) Single-row PRAF with a monopile: The water depth was 1 m. The PRAF was located 2.83 m directly behind the monopile and was aligned along the middle line of the wave tank, with the monopile serving as an anchor for the upstream mooring line of the PRAF (Figure 6a,b). The purpose of this arrangement was to study the effect of the monopile on the mooring line tension and bridle line tension of the PRAF under different wave conditions.

(2) Single-row PRAF without a monopile: Similar to configuration (1), but an iron block replaced the monopile as the anchor. The tests of this arrangement were designed as control experiments.

(3) Three-row PRAFs without a monopile: Three PRAFs were installed in parallel with each other with a spacing of 2.5 m (the inner facility was arranged along the middle line of the wave tank), and their anchors were all iron blocks (Figure 6c,d). This

experimental setup aimed to compare the hydrodynamic forces of three-row and single-row PRAF configurations.

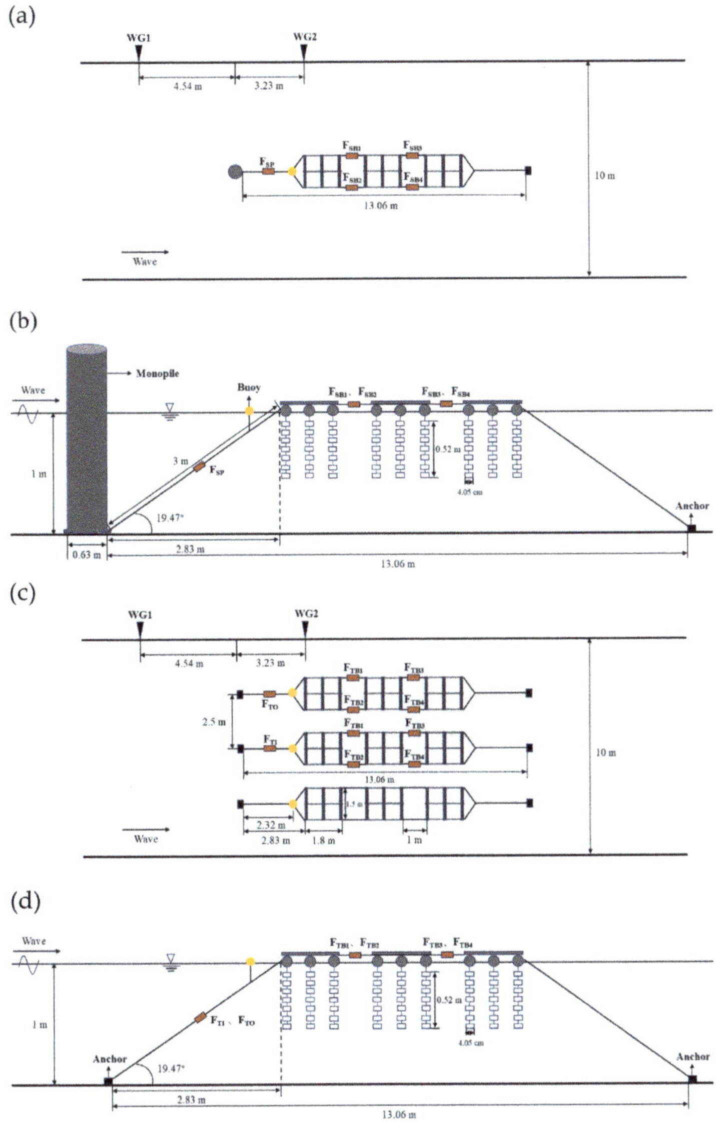

Figure 6. The layouts of PRAFs in the wave tank: (**a**) the single-row arrangement with a monopile in bird's-eye view; (**b**) the single-row arrangement with a monopile in side view; (**c**) the three-row arrangement in bird's-eye view; (**d**) the three-row arrangement in side view.

Representative layouts in the experiments are shown in Figure 6. A 50 kg tension sensor (TLP-10KS, full capacity of ±50 kgf, Tokyo Measuring Instruments Co., Ltd., Tokyo, Japan) (Figure 4d) was affixed to the middle of the upstream mooring line to measure the mooring line tension under wave conditions. Four 10 kg tension sensors (DYMH-107, full capacity of ±10 kgf, DAYSENSOR Engineering Co., Ltd., Bengbu, China) (Figure 4e) were affixed to the middle of the bridle lines to obtain the internal forces between the rafts. Two wave gauges (CH-601, Linearity error ±0.5% of full scale, KENEK Co., Ltd.,

Tokyo, Japan) were placed. WG1 was positioned 4.54 m upstream of the monopile to collect incident wave data and WG2 was positioned 3.23 m downstream of the monopile to monitor transmitted wave data [29]. In summary, all the instruments were equipped with corresponding signal amplifiers to stabilize and enhance the voltage signals, and the data were collected by the same data acquisition unit. The sampling time was set to 70 s, with a sampling frequency of 1000 Hz.

Based on the Froude similarity, the wave conditions for the experiments were designed accordingly. Tests were conducted at a water depth of 1 m, where the wave heights ranged from 0.063 to 0.188 m and the periods from 1.0 to 1.4 s. The experimental wave conditions are listed in Table 2. When scaled to actual sea conditions at a depth of 8 m, these ranges corresponded to wave heights of 0.504 to 1.504 m and wave periods of 2.8 to 4.0 s, which closely matched the marine environment at the Changyi City wind farm [29].

Table 2. Experimental wave conditions set up.

Case	1	2	3	4	5	6	7	8	9	10
Wave Periods T (s)	1.0	1.1	1.2	1.3	1.4	1.237	1.237	1.237	1.237	1.237
Wave Heights H (m)	0.1125	0.1125	0.1125	0.1125	0.1125	0.063	0.100	0.125	0.156	0.188
Wave Length L (m)	1.56	1.88	2.23	2.59	2.97	2.36	2.36	2.36	2.36	2.36

3. Numerical Simulation Methods

This study employed numerical simulations to investigate how monopile foundations influence wave transmission, which offered valuable insights for optimizing the layout of raft aquaculture facilities. STAR-CCM+ (19.02.012-R8), which is commercial computational fluid dynamics software, is well-suited for multidisciplinary problems involving fluid and solid continuum mechanics. It serves as a powerful tool for analyzing complex fluid dynamics. In this study, the Reynolds-averaged Navier–Stokes (RANS) equations were solved using the finite volume method based on the SST k-ω turbulence model in STAR-CCM+ to determine the wave attenuation after interacting with a wind turbine monopile foundation [31]. The free surface was modeled using the Volume of Fluid (VOF) method.

3.1. Governing Equation

Fluid motion must adhere to three fundamental laws: the law of conservation of mass, the law of conservation of momentum, and the law of conservation of energy. For incompressible fluid flow, the energy conservation equation can be neglected due to negligible heat transfer. In this numerical simulation, wave motion is governed by the continuity equation and the momentum equation, Equation (1) is expressed through the mass balance of the control volume, and Equation (2) shows that the time rate of change of linear momentum is equal to the resultant force of the continuous forces, as follows:

$$\frac{\partial \rho}{\partial t} + \nabla \cdot (\rho \overline{v}) = 0 \tag{1}$$

$$\frac{\partial (\rho \overline{v})}{\partial t} + \nabla \cdot (\rho \overline{v} \otimes \overline{v}) = -\nabla \cdot \overline{p}_{mod} I + \nabla \cdot \left(\overline{T} + T_{RANS} \right) + f_b \tag{2}$$

where ρ is the fluid density; $\overline{v}$ is the mean velocity; $\overline{p}_{mod} = \overline{p} + \frac{2}{3}\rho k$ is the corrected mean pressure, where $\overline{p}$ is the mean pressure and k is the turbulent kinetic energy; I is the unit tensor; $\overline{T}$ is the mean viscous stress tensor; f_b is the combined volume force (e.g., gravity and centrifugal force); and T_{RANS} is the stress tensor extra term, which can be obtained from Equation (3):

$$T_{RANS} = \mu_t\left(\nabla\overline{v} + \nabla\overline{v}^T\right) - \frac{2}{3}(\mu_t\nabla\cdot\overline{v})I \tag{3}$$

where μ_t is the turbulent eddy viscosity, obtained through the solution of the turbulence model. In numerical simulations, turbulence is a complex phenomenon typically modeled using various turbulence models. To accurately simulate this phenomenon, this study employed the SST k-ω turbulence model [32]. Compared with the standard k-ω model, the SST k-ω model effectively combines the k-ε model in the far field with the k-ω model near the wall by introducing a blending function. This allows the SST k-ω model to retain the high accuracy of the standard k-ω model near the wall while addressing its sensitivity to inlet boundary condition settings.

3.2. Free Surface Tracking

This study employed the VOF method to determine the interface between the air and water. A second-order upwind scheme was utilized for the discretization of the volume-fraction transport equation and momentum equations, which helped to minimize the numerical diffusion across the interface. To maintain a sharp interface, the High-Resolution Interface Capturing (HRIC) method of STAR-CCM+ was employed to ensure that the free surface profile remained well-defined, even under nonlinear wave conditions. The VOF modeling approach assumes that the grid resolution is sufficient to meet the solution requirements, and tracks the motion of the fluid interface by defining a function of volume fraction to characterize the volume share of a phase of fluid in the grid cell and solving the transport equation for the volume fraction [33]. This phase volume fraction α_i is defined as shown in Equation (4):

$$\alpha_i = \frac{V_i}{V} \tag{4}$$

where V_i represents the volume of phase i in the grid cell; V is the grid cell volume.

The sum of the volume fractions of all Euler phases within a grid is 1. Based on the value of the Euler phase volume fraction α_i, the distribution of Euler phases in each grid can be determined:

(1) $\alpha_i = 0$, indicating that there is no Euler phase i in the grid cell;
(2) $\alpha_i = 1$, indicating that the grid cell is filled with Euler phase i;
(3) $0 < \alpha_i < 1$, indicating that there is an interface within the grid cell.

In grid cells containing interfaces, the material properties depend on the properties of each component and their respective proportions. The fluid within the cell is treated as a mixture. The density and dynamic viscosity of the mixture are as follows:

$$\rho = \sum\nolimits_i \rho_i\alpha_i \tag{5}$$

$$\mu = \sum\nolimits_i \mu_i\alpha_i \tag{6}$$

where ρ_i is the density; μ_i is the dynamic viscosity. The distribution of phase i is obtained from the phase mass conservation equation, as shown in Equation (7):

$$\frac{\partial}{\partial t}\int_V \alpha_i \mathrm{d}V + \oint_A \alpha_i v\cdot \mathrm{d}a = \int_V\left(S_{\alpha_i} - \frac{\alpha_i}{\rho_i}\frac{D_{\rho_i}}{D_t}\right)\mathrm{d}V - \int_V \frac{1}{\rho_i}\nabla\cdot(\alpha_i\rho_i v_{d,i})\mathrm{d}V \tag{7}$$

where v is the average mixing velocity; a is the surface area vector; S_{α_i} is the customized source term; $\frac{D_{\rho_i}}{D_t}$ is the Lagrange derivative of the phase density ρ_i; and $v_{d,i}$ is the diffusion velocity. By solving Equation (7), the volume fraction of each phase is obtained, thereby identifying the grid cells containing interfaces. These interfaces are connected to form the free surface. Additionally, the HRIC scheme is employed in the simulation to enhance the accuracy of the free surface prediction. When two phases are present, the volume fraction transport is solved for the first phase only, and the volume fraction of the second phase is then adjusted in each grid cell so that the sum of the volume fractions of the two phases is 1.

3.3. Boundary Conditions

The wave flume utilized in the physical experiment was similarly employed in the numerical simulations, with appropriate simplifications to reduce the computational demand. Figure 7 presents the established numerical wave flume, which has dimensions of 22.6 m in length, 10 m in width, and 1.7 m in height. A wind turbine monopile foundation, with a diameter of 0.63 m and a height of 1.4 m, was 4.6 m from the wave inlet. The water depth in the flume was set at 1 m. Fifth-order waves were generated at the flume inlet using Stokes' fifth-order wave theory [34], and the inlet boundary conditions were specified as a velocity inlet.

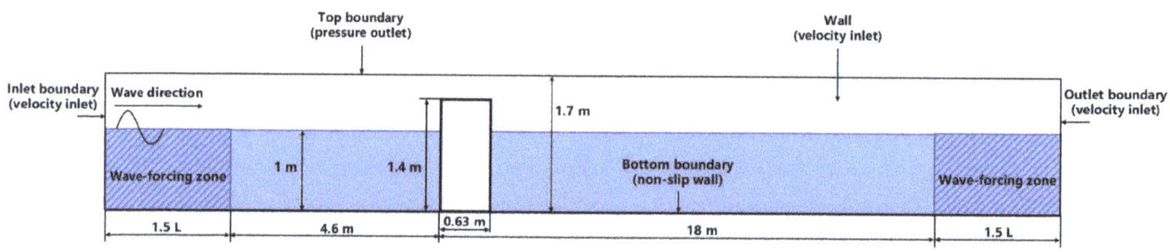

Figure 7. Boundary conditions of the numerical flume.

To further optimize the computational efficiency, a symmetry plane was defined along the centerline of the flume. While the primary wave inlet was at one end of the flume, the outlet and opposite side were also set as velocity inlets with wave-attenuation zones to reduce any potential wave reflection returning into the domain, as confirmed by Zhang et al. [35]. Practically, this approach enabled the outgoing waves to exit the numerical flume without spuriously reflecting and ensured the stability of the air–water interface during the simulations. As a result, the air phase was not assigned a separate inlet velocity since it mainly followed the free-surface motion generated by the imposed wave field. We used a wave-forcing method based on the Euler Overlay Method (EOM) proposed by Kim et al. [36] to reduce the wave reflection from the boundary. According to Kim et al.'s method, the discrete Navier–Stokes equations can be forced to converge to a theoretical solution in the setting region, thus preventing wave reflection from the far-field by using a smaller computational domain. The wave-forcing zones were activated at both the inlet and outlet, extending the wave-absorbing region by 1.5 times the wavelength (Figure 7). Moreover, a source term, as expressed in Equation (8), was introduced into the momentum equation within the wave-absorbing region to simulate wave forces [36]:

$$S_{\phi}^{*} = -\gamma\rho(\phi - \phi_E) \tag{8}$$

where γ is the force coefficient, ϕ is the solution to the transport equation, and ϕ_E is the value approximated by the tension solution. The solution of the Navier–Stokes equations was made to converge to the set numerical solution within the specified wave dissipation region.

3.4. Grid Settings

Mesh sensitivity analysis is an indispensable part of numerical simulation, which not only verifies the accuracy of the simulation results but also optimizes the computational resources and reduces error [37,38]. The mesh sensitivity analysis is provided in Appendix B. During the grid-setting process, the cut volume mesh was selected as the mesh model. The refined region was created within one wave height above and below the free surface. To ensure the mesh was sufficiently fine to capture wave variations, based on the mesh sensitivity analysis results (Appendix B), 24 grids were divided along the wave height direction, and 100 grids were divided along the wavelength direction. The base size of the background mesh was set to 0.8 times the wave height, and the mesh growth rate was set to be gradual to ensure smooth transitions between the grids. The time step was set to 0.002 s, and the Courant number was maintained below 0.3 to avoid phase differences. Additionally, an overlapping mesh zone was established around the monopile model, with the background mesh size in this zone set to 0.01 m to further refine the computational domain surrounding the monopile model. A prism mesh was applied to the model walls to accurately resolve the near-wall fluid behavior, which is crucial for capturing fluid separation characteristics. The actual grid division is illustrated in Figure 8.

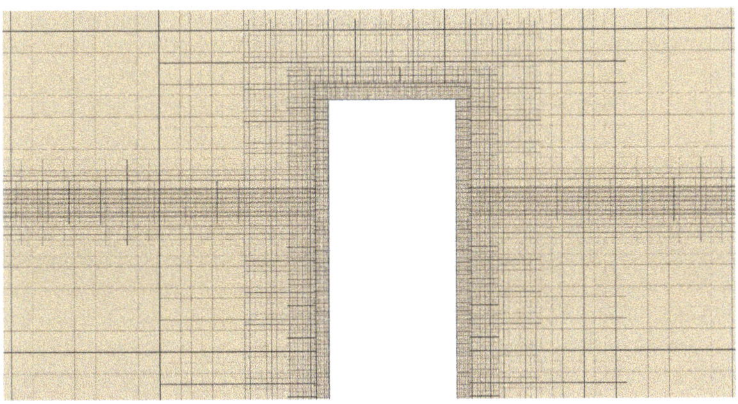

Figure 8. The scheme of grid settings around the wind monopile model.

3.5. Wave Height Monitoring-Points Setup

The validation of the numerical flume is provided in Appendix A. This section investigated the wave propagation characteristics in the wake region behind the monopile foundation subjected to regular waves. Figure 9 shows four monitoring points established in the numerical flume to calculate the wave heights at specific locations. R1 was located 4.6 m in front of the wind turbine monopile foundation to verify the incident wave conditions, while the remaining points (R2, R3, and R4) were distributed 2.8 m behind the monopile, located slightly ahead of the PRAFs in the experiments. The wave height values at the representative points (R2, R3, and R4) were used to evaluate the effect of the wind monopile foundation on the surrounding wave field. The wave height at R1 was defined as H_i, and the average of the wave heights at points (R2, R3, and R4) was defined as H_t.

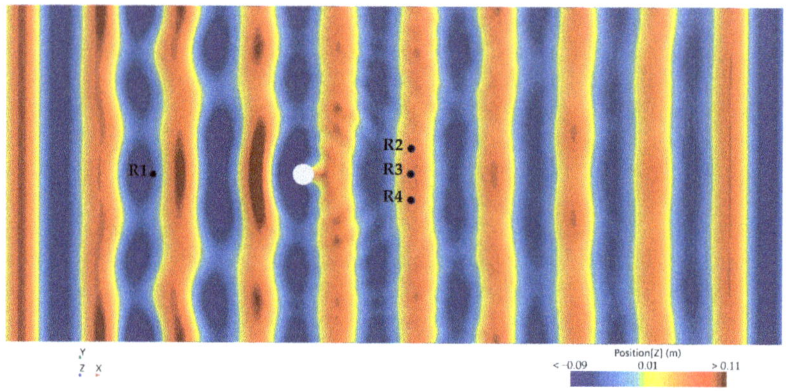

Figure 9. The locations of four set points to monitor the wave heights in a numerical flume with the monopile.

Detailed numerical simulations were conducted for all the experimental scenarios to examine its influence on the wave propagation. The numerical results provide insights into why the presence of a wind turbine monopile foundation reduced the mooring line tension of the raft aquaculture facility. This reduction in mooring line tension can be attributed to the attenuation and redistribution of the wave energy as the waves interacted with the monopile, thereby altering the hydrodynamic loads acting on the aquaculture system.

To further investigate the impact of the PRAF on the monopile, as well as the changes in the surrounding wave fields, we established three wave height monitoring points (P1, P2, and P3) positioned 10 cm from the monopile foundations. Numerical simulations were conducted under three wave conditions (Case 6, Case 8, and Case 10) to assess both wave heights at the monitoring points and the wave forces acting on the monopile foundation. Following this, we arranged single-row and three-row PRAFs in the lee zone of the monopile foundation to ensure that the mooring lines were not connected to the monopile itself (Figure 10). The primary objective of this work was to indicate how different PRAF configurations influenced the wave forces exerted on the monopile foundation and to analyze the alterations in wave propagation patterns around these structures.

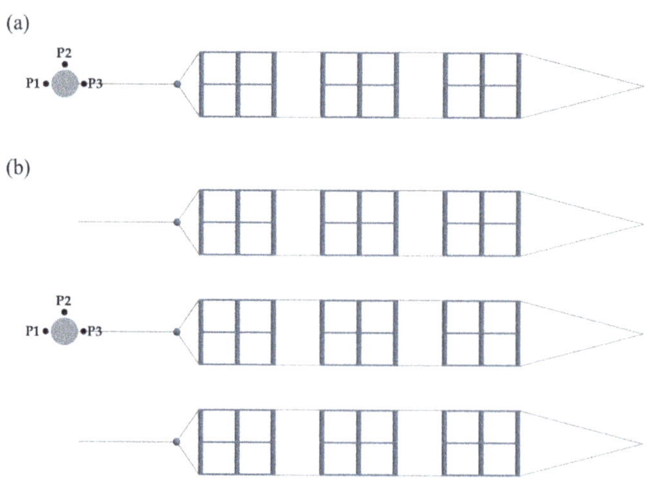

Figure 10. The locations of three set points to monitor the wave heights around the monopile: (**a**) single-row PRAF; (**b**) three-row PRAFs.

4. Results and Discussion

In this study, multiple wave parameters (e.g., wave height, wave period, and raft layouts) were considered to evaluate their respective impacts on the hydrodynamic forces (mooring line tension and bridle line tension) on the PRAFs. Moreover, to better visualize the actual hydrodynamic performance of the aquaculture facilities at sea, the model test results were converted to full-scale results according to the model criterion. The following subsections present the results of this study and summarize our expanded discussion regarding the physical mechanisms at play.

4.1. Mooring Line Tensions of PRAF

The structural integrity of aquaculture facilities relies heavily on the mooring system. Mooring line failure is an important factor in exposing the structure to risks. This study analyzed the effect of different arrangements on the mooring system of the PRAF, which can provide an important reference for the design and construction of raft aquaculture facilities, as well as the decision-making of wind turbine foundations as anchorage points.

4.1.1. The Maximum Mooring Line Tensions of the Single-Row PRAF Under Wave Conditions

Figure 11 shows the maximum tensions of the upstream mooring line for a single-row PRAF under different wave periods and heights. Here, F_{SN} and F_{SP} represent the maximum mooring line tension without and with a monopile, respectively. The results indicate that regardless of the presence of the wind turbine monopile in the wave tank, the maximum mooring line tension decreased in the wave periods from 1.0 s to 1.4 s when the wave height was 0.1125 m (Figure 11a). Under longer wave periods, the incident wave energy was distributed over a slower timescale, meaning that each crest exerted a force over a longer duration but generally had a lower peak value. As a result, both the mooring cables and the raft had more time to adjust to the motion of the waves, which could reduce the maximum tension in the cables. Meanwhile, shorter wave periods increased the frequency of the crest impacts within the same timeframe, which led to higher acceleration and velocity variations for the structure. This could result in amplified instantaneous forces, which caused increases in the mooring line tensions and more pronounced high-frequency oscillation. When the wave period was 1.0 s and the wave height was 0.1125 m (Case 1), the wave steepness was at its maximum ($H/L = 0.084$), which corresponded to the highest recorded mooring line tension F_{SN} of 80.73 kN.

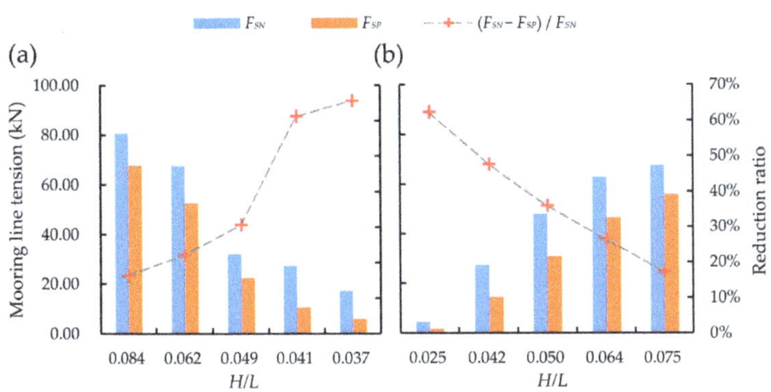

Figure 11. The maximum tensions of the upstream mooring line of the single-row PRAF under different wave periods and wave heights in the presence (F_{SP}) and absence (F_{SN}) of the monopile: (**a**) $H = 0.1125$ m, and T varies from 1.0 s to 1.4 s; (**b**) $T = 1.237$ s, and H varies from 0.063 m to 0.188 m.

The wave height determines the overall magnitude of the wave energy, while the wave steepness reflects its nonlinearity. Higher waves carry more energy; consequently, the inertia and viscous drag forces on the PRAF system and mooring lines are elevated. When the wave period was 1.237 s, the maximum mooring line tension increased as the wave height rose from 0.063 m to 0.188 m (Figure 11b). This trend is consistent with previous research findings [30,39]. Our results indicate that as the wave height increased, the maximum tension in the mooring lines also grew. This enhanced motion contributed further to the increased dynamic loads on the mooring lines. These observations follow a well-established pattern in marine engineering: as the wave steepness increases, the structural stability decreases and the wave-induced force on the structure grows, thereby leading to higher tension levels in the mooring system [40,41].

By comparing and analyzing the effects of the wind monopile on the PRAF, $(F_{SN} - F_{SP})/F_{SN}$ represents the reduction ratio of the maximum mooring line tension in the presence of the wind monopile relative to the condition without it. The results show that when the monopile was present, the maximum mooring line tension was decreased by 16% to 66%, and the reduction ratio increased as the wave steepness decreased. This effect was the most pronounced when the wave steepness was 0.037. As a type of marine structure, a wind turbine monopile disrupts wave propagation, with wave diffraction and reflection being the main mechanisms of wave energy loss when waves encounter an obstacle. The likelihood and intensity of diffraction effects are closely related to the length scale ratio between the monopile diameter (D) and the wavelength (L): at small D/L ratios, the flow separation is important while diffraction is minimal, whereas at large D/L ratios, diffraction becomes dominant and the flow separation is less significant. de Leon et al. (2011) [42] noted that the monopile creates a shadow zone downstream with a lower significant wave height (H_s). Meanwhile, the interaction between waves and the monopile leads to a weakened and more confined oscillating flow in the vicinity of the monopile [43]. These wave–structure interaction effects contribute to a reduction in mooring line tension compared with conditions without a monopile.

Figure 12 compares the F_{SN} at full scale for a single-row PRAF with three rafts (each raft supported 31.05 t of mussel weight) with a raft facility described in [39]. As the wave steepness increased from 0.025 to 0.084, the maximum mooring line tension rose from 4.60 kN to 80.73 kN. This trend aligns with the numerical simulation results of a raft facility reported by Wang et al. (2017), where the maximum mooring line tension increased from 5.96 kN to 21.13 kN at wave steepnesses that ranged from 0.013 to 0.077 [39]. The consistency suggests that the relationship between the wave steepness and the mooring line tension is robust for different configurations, supporting both the validity of the present experimental setup and the importance of wave steepness as a key parameter in pontoon raft aquaculture systems [41]. The comparison also indicates that the mooring line tension increases with the addition of each pontoon raft, yet not in a manner strictly proportional to the number of rafts. The increased tension suggests complex hydrodynamic interactions between the additional pontoons, which could amplify the structural loads beyond a simple summation effect. Moreover, Masoudi et al. (2024) showed that the number of pontoons significantly affects diffraction wave amplitudes, with multi-pontoon structures improving the breakwater efficiency by up to 45% within the same cross-sectional area [44]. These findings highlight the importance of considering such interactions when designing multi-pontoon raft aquaculture systems.

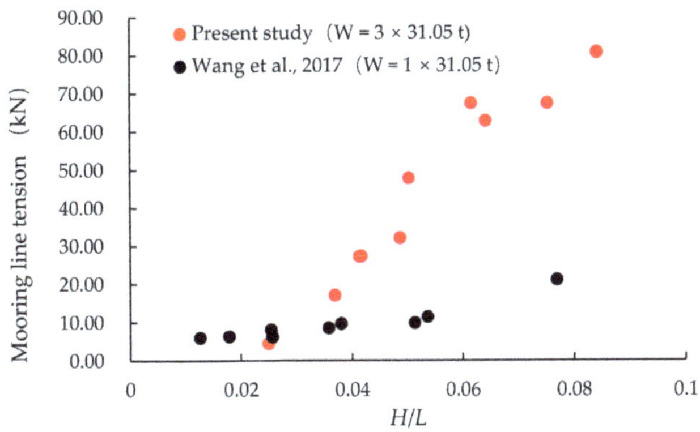

Figure 12. Comparison of the F_{SN} at prototype scale between a single-row PRAF with three rafts in H/L from 0.025 to 0.084 and a raft facility in H/L from 0.013 to 0.077 [39].

4.1.2. The Maximum Mooring Line Tensions of the Three-Row PRAFs Under Wave Conditions

Figure 13 compares the maximum tensions of the upstream mooring line for the outer (F_{TO}) and the inner (F_{TI}) raft facility in the three-row PRAFs. Similar to the single-row PRAF, when the wave height was 0.1125 m, the maximum mooring line tension decreased as the wave period increased from 1.0 s to 1.4 s. However, at the wave period of 1.237 s, the maximum mooring line tension increased in response to wave heights that ranged from 0.063 m to 0.188 m. The comparison showed no significant difference between F_{TO} and F_{TI}, indicating that under these test conditions, whether the raft facility was located in the outer or inner position did not notably affect the mooring line tension. This finding suggests that the PRAFs in the multi-row configuration may be equally exposed to wave impact under different wave conditions.

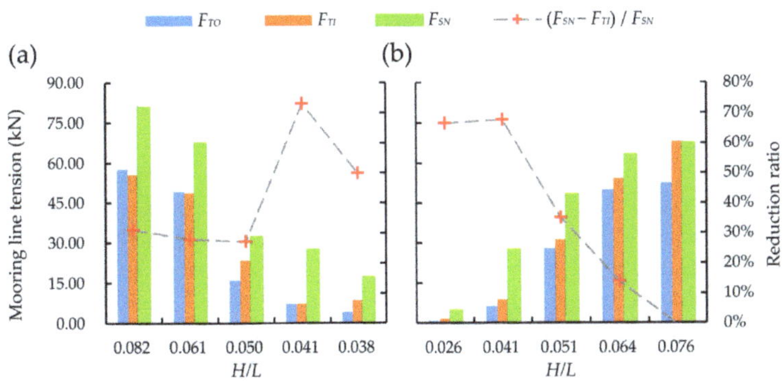

Figure 13. The maximum tensions of the upstream mooring line of single-row (F_{SN}) and three-row (F_{TO} and F_{TI}) PRAFs without a monopile under different wave periods and wave heights: (**a**) H = 0.1125 m, and T varies from 1.0 s to 1.4 s; (**b**) T = 1.237 s, and H varies from 0.063 m to 0.188 m.

Moreover, it can be observed in Figure 13 that the maximum tensions of the upstream mooring line in the three-row configuration were generally smaller than those in the single-row configuration under the same wave conditions. When the wave steepness was 0.041, the maximum mooring line tension had a maximum reduction ratio of 73%. This finding suggests that the multi-row PRAFs could effectively reduce the impact of the wave forces by altering the wave propagation and promoting energy dissipation around these structures. The reduction in the maximum mooring line tension for the three-row configuration can be attributed to the intensified energy dissipation that resulted from the interference between multiple rows of pontoons, which collectively scattered and dampened the wave energy more effectively than a single-row arrangement. These results align with the work of Masoudi et al. (2024) and Li et al. (2022), who showed that the multi-row arrangement significantly reduces both the wave energy and mooring loads compared with the single-row system [44,45]. Furthermore, the presence of multiple rows of pontoons may create constructive and destructive interference patterns within the wave field, thereby reducing the overall force acting on individual mooring lines.

Compared with the single-row setup, the multi-row configuration distributed the hydrodynamic loads more evenly, which reduced the concentration of forces on the individual mooring lines and enhanced the overall system stability. By lowering the mooring line tension, the multi-row arrangement underscored the importance of optimizing pontoon raft layouts to improve the wave attenuation performance. Moreover, the multi-row configuration provided greater stability across a range of wave conditions and mitigated the wave-induced loads effectively, which contributed to the long-term sustainability and safety of offshore aquaculture facilities. In future work, we will continue to conduct both experimental and numerical studies to explore how different spacing arrangements affect multi-row setups. These findings will ultimately provide practical guidance for the spatial layout of multi-row raft aquaculture in offshore wind farms.

4.2. Bridle Line Tensions of PRAFs

The combination of raft aquaculture facilities offers several advantages, including enhanced comprehensive utilization of aquaculture waters, reduced equipment investment costs, and improved feasibility for large-scale management. In addition to the mooring lines, the connecting lines between individual rafts within the PRAFs are also subjected to hydrodynamic loads. This section presents the variation in bridle line tensions for both the single-row PRAF (e.g., F_{SB1}, F_{SB2}, F_{SB3}, F_{SB4}) and three-row PRAFs (e.g., F_{TB1}, F_{TB2}, F_{TB3}, F_{TB4}).

4.2.1. The Maximum Bridle Line Tensions of the Single-Row PRAF Under Wave Conditions

Figure 14 shows the relationship between the bridle line tensions F_{SB1}–F_{SB4} and the wave steepness under different wave conditions, with and without the presence of a wind monopile. Similar to the mooring line tension, the bridle line tension increased as the wave steepness increased. The bridle line tensions had maximums when the wave steepness approached values near 0.08. The maximum values of F_{SB1}–F_{SB4} were 25.82 kN, 24.96 kN, 32.80 kN, and 29.94 kN, respectively. It should be noted that the maximum values of F_{SB1} and F_{SB2} occurred with the monopile, while the maximum values of F_{SB3} and F_{SB4} occur without the monopile.

Figure 14a,b show that the presence of the monopile increased the bridle line tensions F_{SB1} and F_{SB2} with rising wave steepness, which exhibited similar sensitivities to both the wave height and wave period. In contrast, Figure 14c,d show the presence of the monopile weakened the tensions F_{SB3} and F_{SB4}, and that these tensions were more responsive to changes in the wave height than to the wave period. This outcome aligns with the understanding that the wave height primarily affects the maximum force amplitude, whereas the wave period plays a larger role in determining the force distribution. Consequently, the increase in the wave height led to a noticeable rise in the maximum bridle line tension. Furthermore, Figure 15a,c also demonstrate that in the presence of the monopile, F_{SB1} and F_{SB2} exceeded F_{SB3} and F_{SB4} across Cases 1–5. This was precisely the opposite of the situation without the monopile. The difference between F_{SB1}–F_{SB2} and F_{SB3}–F_{SB4} can be attributed to the distinct positions of the connecting lines within the raft system. The monopile altered the surrounding wave field, particularly the upstream region of the raft facilities, which intensified the hydrodynamic loads on the upstream bridle lines closest to it. Miles et al. (2017) found that oscillatory velocities decreased immediately behind the pile but recovered and even exceeded the background levels rapidly within a distance of approximately 1.65 to 3.5 times the pile diameter [46]. Consequently, the first raft in the wave propagation direction (located 3.15 m from the monopile center, i.e., beyond 3.5 times the diameter of the pile) experienced greater wave forces than in the no-monopile scenario while consuming more wave energy, and thus, exposing subsequent rafts to lower significant wave heights (H_s). This phenomenon underlay the marked differences in F_{SB1}–F_{SB2} and F_{SB3}–F_{SB4} between the monopile and no-monopile configurations. In the future, further research is required to investigate the effects of the geometric parameters of the monopile, wave characteristics, and other factors on the hydrodynamic performance. This will facilitate a more comprehensive understanding and enable improved prediction of the hydrodynamic responses of raft aquaculture facilities in offshore wind farms.

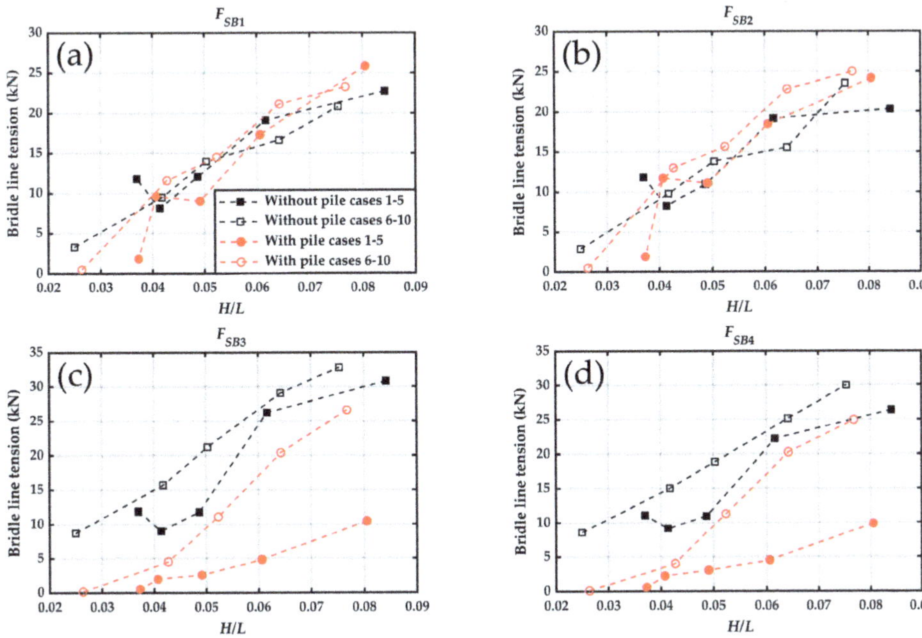

Figure 14. The maximum tensions of the bridle line F_{SB1}–F_{SB4} for different wave periods and wave heights in the presence and absence of a monopile: (**a**) the maximum tension of bridle line 1; (**b**) the maximum tension of bridle line 2; (**c**) the maximum tension of bridle line 3; (**d**) the maximum tension of bridle line 4.

Figure 15 also shows the mooring and bridle line tensions for a single-row PRAF under different wave conditions with and without the monopile. The tensions F_{SB1}–F_{SB4} remained lower than F_{SN} or F_{SP}, and the mooring line tensions rose more sharply than the bridle line tensions as the wave period decreased or the wave height increased. These observations differed from the findings of Martinelli et al. (2008) on the floating breakwaters [47]. This was because in this study, both the mooring line and bridle line were in a slack state initially, and the mooring line was exposed to the incident wave earlier. Their peak values did not coincide due to phase differences between the rafts. In contrast to the mooring lines, which largely responded to the incident wave, the bridle line tensions arose from interactions between the adjacent rafts. F_{SB1}–F_{SB2} and F_{SB3}–F_{SB4} measured the connecting forces between pairs of rafts, where each raft experienced distinct motion states. Furthermore, the wave energy was dissipated through impacts and reflections on the mussel ropes and pontoons, which led to varying hydrodynamic loads on each raft. According to Liao et al. (2017), combined raft structures under regular wave conditions can substantially reduce mooring forces, suggesting that combined raft structures may outperform single raft structures in terms of wave-induced loads if the ropes of the system are strong enough [48].

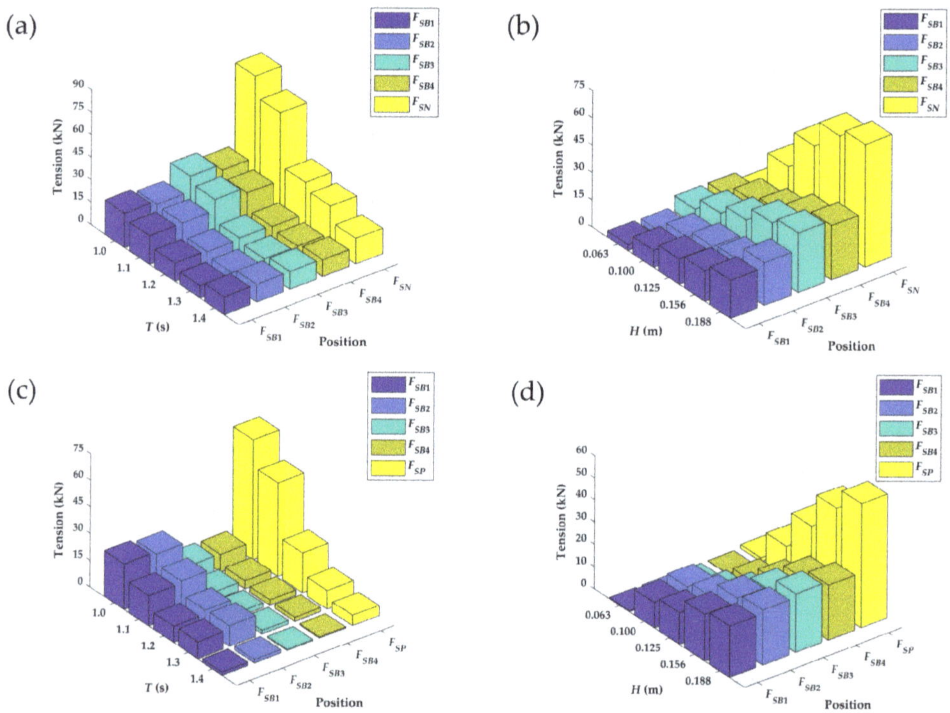

Figure 15. Comparison of the mooring line tensions and bridle line tensions for the single-row PRAF under different wave conditions with and without the monopile: (**a**) without monopile, $H = 0.1125$ m; (**b**) without monopile, $T = 1.237$ s; (**c**) with monopile, $H = 0.1125$ m; (**d**) with monopile, $T = 1.237$ s.

4.2.2. The Maximum Bridle Line Tensions of the Three-Row PRAFs Under Wave Conditions

Figure 16 shows the effects of different wave periods and wave heights on the bridle line tensions F_{TB1}–F_{TB4}, including the outer structure and inner structure of the three-row PRAFs. F_{TB1}–F_{TB4} tended to increase with increasing wave steepness. As the wave steepness approached 0.08, F_{TB1}–F_{TB4} had maximum values of 31.79 kN, 23.30 kN, 21.47 kN, and 18.95 kN, respectively. Figure 16a,b indicate that F_{TB1} and F_{TB2} of the outer structure were slightly higher than those of the inner structure with rising wave steepness. The sensitivities of F_{TB1} and F_{TB2} to the wave height and wave period were not obvious. Figure 16c,d indicate that F_{TB3} and F_{TB4} of the outer structure were notably greater than F_{TB3} and F_{TB4} of the inner structure with increasing wave steepness, and the sensitivities of F_{TB3}–F_{TB4} to changes in the wave height were stronger than in the wave period.

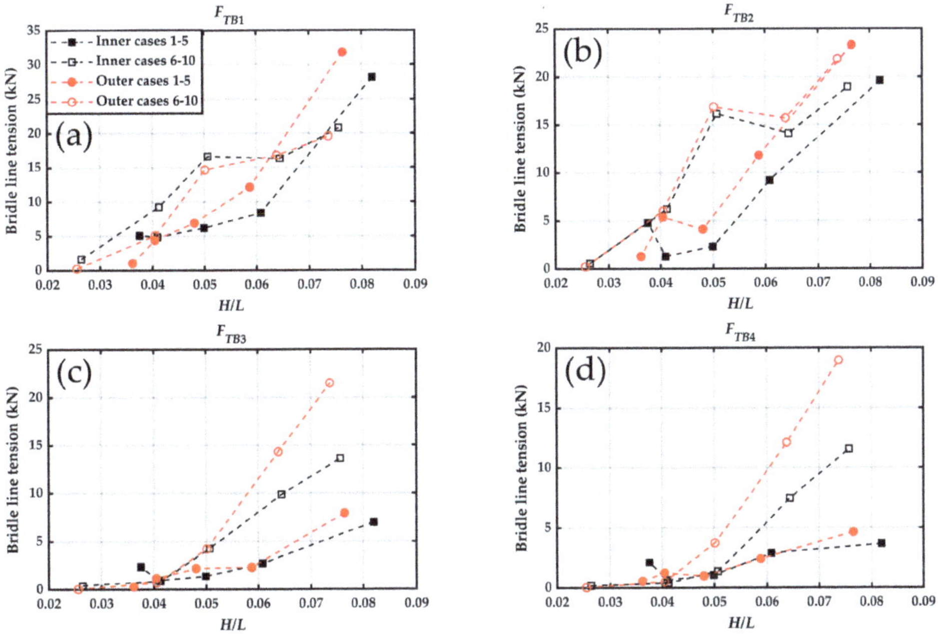

Figure 16. The maximum tensions of the bridle line F_{TB1}–F_{TB4} for different wave periods and wave heights: (**a**) the maximum tension of bridle line 1; (**b**) the maximum tension of bridle line 2; (**c**) the maximum tension of bridle line 3; (**d**) the maximum tension of bridle line 4.

In the three-row PRAFs, significant interactions between the structures were observed. The bridle line tension results suggest that the inner structure exhibited greater stability under varying wave conditions. This can likely be attributed to the redistribution of the wave field after the waves encountered the raft system. In the three-row configuration, the outer structures diffracted the incoming waves around both sides, which caused these wave components to converge upon the inner structure. Because the converging waves often arrived with varying phases, partial destructive interference occurred, which effectively reduced the local wave energy around the inner structure and led to lower hydrodynamic loads, especially under stronger wave conditions (e.g., high wave height or greater wave steepness). These findings highlight the importance of wave interference and energy reduction for multi-row configurations, which inspires the spatial deployment of offshore aquaculture.

Figure 17 compares the mooring line tensions (F_{TO}, F_{TI}) and bridle line tensions (F_{TB1}–F_{TB4}) for the three-row PRAFs under different wave conditions. Overall, these tensions (F_{TB1}–F_{TB4}) tended to remain lower than the upstream mooring line tensions F_{TO} or F_{TI}. However, under more energetic wave conditions (e.g., larger wave heights or steeper waves), especially at shorter wave periods, the tensions in F_{TB1} and F_{TB2} became significantly higher than F_{TB3} and F_{TB4}. This phenomenon indicates that shorter-period, high-steepness waves imparted greater localized loads on the forward bridle lines than on those positioned further inward.

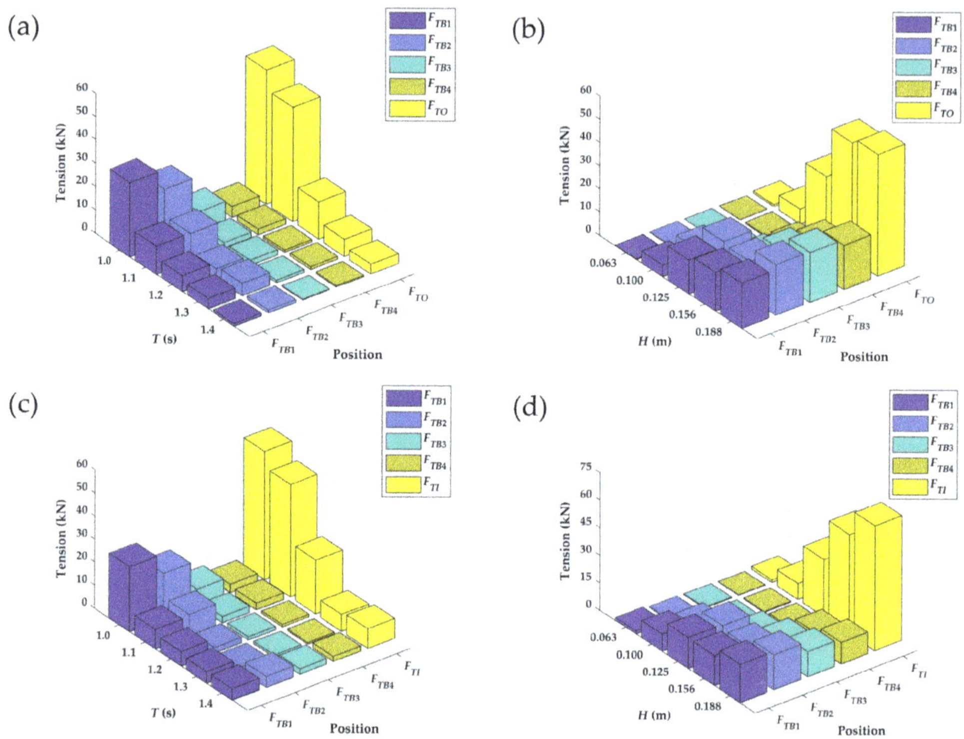

Figure 17. Comparison of the mooring line tensions and bridle line tensions for the three-row PRAFs under different wave conditions: (**a**) outer structure, H = 0.1125 m; (**b**) outer structure, T = 1.237 s; (**c**) inner structure, H = 0.1125 m; (**d**) inner structure, T = 1.237 s.

The primary reason for the observed tension difference was the attenuation and the redistribution of wave energy caused by the multi-row configuration. In the single-row configuration, the rear rafts were directly exposed to waves that had undergone minimal dissipation, which led to cumulative wave impacts on the trailing rafts. By contrast, in the three-row configuration, the upstream raft directly encountered the undisturbed incident waves and experienced higher hydrodynamic loads on the front bridle lines. Acting much like a small breakwater, these upstream rafts dissipated a portion of the wave energy through diffraction, viscous effects, and local turbulence, and thus, reduced the wave forces transmitted to the subsequent rafts.

Furthermore, the constructive and destructive interference patterns that arose from the wave diffraction at the edges of the upstream rafts redistributed the hydrodynamic loads between the inner and rear rows. In particular, the destructive interference could diminish the wave amplitudes in the lee of the front row, which lowered the bridle line tensions downstream. Additionally, the shadowing effect of the front-row rafts provided further protection for the interior and rear rows from the full incident wave energy.

In summary, the difference in tension distribution between the single- and three-row raft systems can be attributed to the combined mechanisms of the wave energy dissipation, interference, and shadowing effects. As illustrated in Figure 13, the three-row configuration effectively decreased the wave forces on the downstream rafts, which resulted in lower overall mooring line tensions. In contrast, a single-row arrangement exposed the raft directly to the undistributed wave field, which led to higher cumulative forces on its mooring line.

Figure 18 shows the comparison of the bridle line tensions of the single-row PRAF (F_{SB1} and F_{SB3}) and those of the three-row raft facilities (F_{TB1} and F_{TB3}). The results indicate that the three-row configuration generally reduced the bridle line tensions compared with the single-row configuration, except under the condition of T = 1.0 s and H = 0.1125 m, where F_{TB1} exceeded F_{SB1}. As shown in Figure 17a, F_{TB1} under this specific condition appeared to be an abnormal value. One possible explanation was that certain components of the three-row PRAFs may have a natural vibration period similar to the incident wave period (T = 1.0 s), which led to localized resonance and generating the higher-than-expected bridle line tensions. This phenomenon suggests that resonance effects in a multi-row configuration can significantly impact structure loads under specific wave conditions. Martinelli et al. (2008) also observed that under perpendicular wave conditions, mooring and intermodal connector forces can vary significantly with increasing layout complexity [47]. Therefore, the findings highlight the importance of optimizing the spatial layout of aquaculture facilities to enhance their hydrodynamic performance and reduce structural loads, particularly in multi-row configurations.

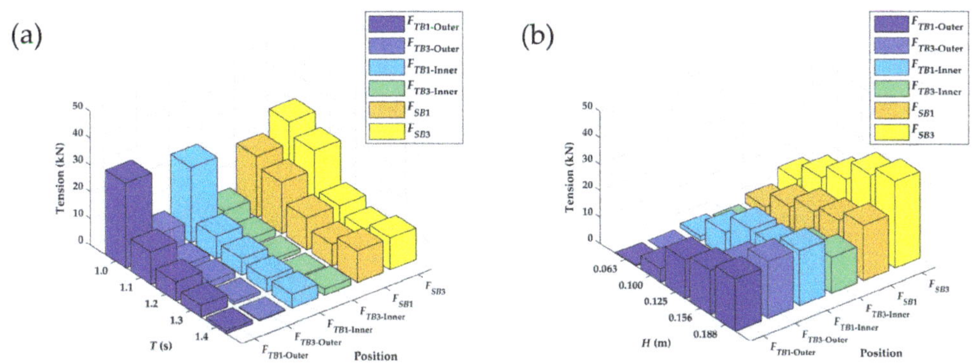

Figure 18. Comparison of the bridle line tensions for the single-row and three-row PRAFs under different wave conditions: (**a**) H = 0.1125 m; (**b**) T = 1.237 s.

4.3. The Impact of the Monopile Foundation on the Wave Propagation

In this section, we examined how the presence of the wind turbine monopile foundation influenced the wave propagation and, consequently, the mooring line tensions of the raft aquaculture system. Detailed validation of the numerical flume model against the experiment by Mo et al. [49] is provided in Appendix A, which confirmed that our simulation setup reliably captured key wave–structure interactions.

The numerical results are shown in Table 3. The transmission coefficient ($K_t = H_t/H_i$) is an important coefficient used to reflect the wave attenuation performance of the monopile. In Cases 1–5, K_t decreased with the increase in D/L and K_t was greater than 0.799. When $D/L > 0.2$, the diffraction effects of waves start to become significant [50]. In the design of a breakwater, $D/L > 0.25$ is considered as a reference to be able to effectively resist waves [44]. This rule coincides with our results. The presence of the monopile helped to reduce the wave height in the transmitted region. This study concentrates on the K_t. As shown in Figure 19, it can be seen that K_t increased with increasing wave period and decreased with increasing wave height.

Table 3. Numerical simulation results of H_i and H_t.

Case	H_i (m)	H_t (m)	D/L	K_t	K_a
1	0.1313	0.1049	0.404	0.799	0.201
2	0.1161	0.955	0.335	0.823	0.177
3	0.1086	0.953	0.283	0.878	0.122
4	0.1073	0.1012	0.243	0.943	0.057
5	0.1097	0.104	0.212	0.948	0.052
6	0.59	0.532	0.267	0.902	0.098
7	0.987	0.882	0.267	0.894	0.106
8	0.1189	0.1053	0.267	0.886	0.114
9	0.1518	0.1331	0.267	0.877	0.123
10	0.1781	0.1543	0.267	0.866	0.134

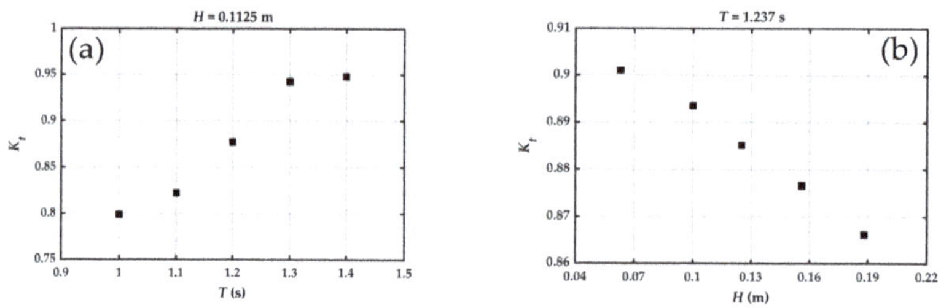

Figure 19. K_t of the monopile under different wave conditions: (**a**) K_t varied with the wave period increasing from 1.0 s to 1.4 s (H = 0.1125 m); (**b**) K_t varied with the wave height increasing from 0.063 m to 0.188 m (T = 1.237 s).

Figure 20 shows the relationship between the reduction in the mooring line tension ΔF ($\Delta F = F_{SN} - F_{SP}$) and the attenuation coefficient K_a ($K_a = 1 - K_t$). For Cases 1–5, when $K_a < 0.07$ ($D/L < 0.25$), ΔF could not be used as a reference for the change in the mooring line tension because the monopile was not a means of wave attenuation in this condition. ΔF increased initially from 0.12 to 0.18 of K_a and decreased from 0.18 to 0.2 of K_a. For Cases 6–10, ΔF increased initially from 0.1 to 0.11 of K_a and decreased from 0.11 to 0.13 of K_a. Consequently, the response of ΔF to K_a was more sensitive to the wave height compared with the wave period, which suggests that when the wave period was constant, the wave height condition was more restrictive to maintain a high ΔF. As shown in Figure 20, when K_a ranged from 0.11 to 0.20, the monopile could effectively reduce the mooring line tension of the PRAF.

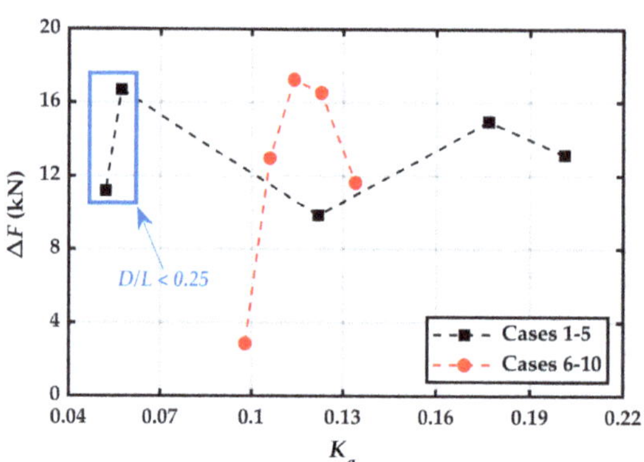

Figure 20. The relationship between ΔF and K_a.

Therefore, the numerical results corroborate the experimental finding that the mooring line tensions were reduced in the presence of the wind turbine monopile. For instance, under moderate wave steepness (H/L = 0.03), we observed an up to 40–50% tension reduction on the mooring lines behind the monopile compared with no monopile. This reduction can be explained by Figure 21: (i) wave diffraction around the monopile edges, where we can see a downstream "shadow zone" with lower wave amplitude, and (ii) partial reflection and turbulence-induced dissipation near the monopile surface, where high waves were formed in front of the monopile due to the superposition of reflected and incident waves. As the wave steepness increased, the local flow separation and vortex shedding enhanced these dissipation effects, which further decreased the transmitted wave energy. This wave energy reduction was particularly beneficial for the multi-row raft arrangements. The rafts located behind the monopile were subjected to smaller wave heights, which experienced reduced hydrodynamic loading and lower peak tensions in the mooring system. Combined with the physical experiments on the single-row and three-row raft systems, the numerical analysis confirmed that strategically placing a monopile upstream could serve as a mini-breakwater, which shielded the aquaculture rafts from fully developed waves. As the wave energy was partially blocked or scattered, the inner and rear rows experienced reduced the hydrodynamic loads and consequently lowered the mooring line tensions. Thus, if multiple rows of PRAFs are placed behind a monopile, the wave field will first undergo diffraction at the pile, then interact with the leading PRAF row, further attenuating the wave energy for the subsequent rows. The second or third rafts will experience a significantly reduced disturbance, lowering both the motion amplitude of the rafts and the peak loads on the mooring line. These mechanisms are crucial for optimizing large-scale aquaculture system configurations in wind farm areas, contributing to both the economic efficiency and structural safety considerations.

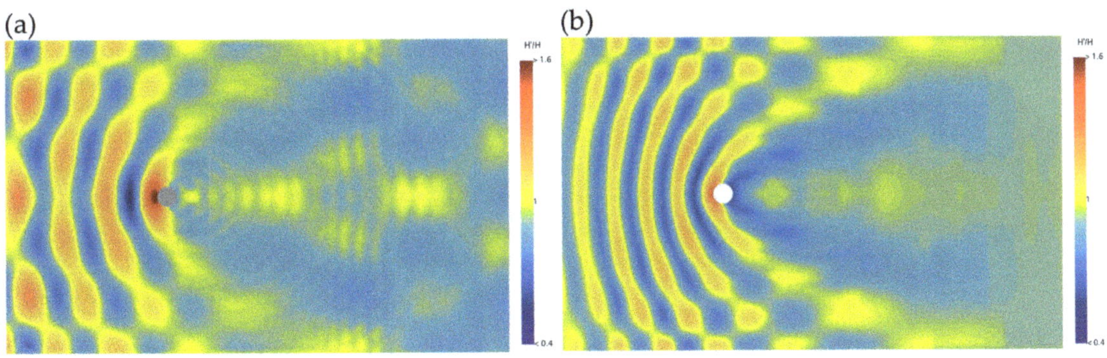

Figure 21. Distribution of the ratio of the measured wave height H′ to the incident wave height H around the monopile: (**a**) when T = 1.4 s and H = 0.1125 m; (**b**) when T = 1.237 s and H = 0.188 m.

4.4. The Impact of the PRAF on the Wave Force Acting on the Monopile Foundation and the Surrounding Wave Field

In actual offshore wind farms, a monopile is subjected to multiple simultaneous loads, including those from wind, waves, and currents. However, this study specifically focused on the interaction between the monopile and the PRAF within the integrated structure. When the monopile foundation was used as the anchorage point for the PRAF, the influence of the PRAF on the monopile force was mainly reflected in two aspects: one was the mooring line tension, as the PRAF exerted a direct tension on the monopile foundation, and the magnitude of this part of the force had already been measured by the model flume test; the second was the additional hydrodynamic loading on the monopile foundation from reflected waves generated by the PRAF. To study the mechanism underlying these interactions, we conducted a comparative analysis of the forces that acted on the monopile foundation in two scenarios: "with PRAF" and "without PRAF". This approach enabled us to quantify the extent to which the PRAF influenced the wave loads on the monopile foundation. Through this comparative study, we aimed to provide a comprehensive assessment of the additional effects associated with the deployment of the PRAF around the monopile foundation.

As shown in Table 4, the deployment of the single-row PRAF behind the monopile foundation resulted in a reduction in the wave forces that acted on the monopile. Specifically, the forces decreased by 8.75%, 1.59%, and 1.35% for the incident wave period of 1.237 s and wave heights of 0.063 m, 0.125 m, and 0.188 m, respectively. However, when the three-row PRAFs were deployed, an increase in the wave forces on the monopile foundation was observed, with increases of 3.77%, 1.00%, and 2.46% corresponding to the same wave conditions. These findings indicate that the influence of the PRAF on the wave force experienced by the monopile foundation was relatively minimal.

Table 4. Effects of single-row and three-row PRAFs on monopile forces and nearby wave heights.

Layout	Case	H_{P1} (m)	H_{P2} (m)	H_{P3} (m)	F_P (N)
Monopile foundation	$T = 1.237$ s, $H = 0.063$ m	0.098	0.059	0.054	65.57
	$T = 1.237$ s, $H = 0.125$ m	0.209	0.121	0.111	117.14
	$T = 1.237$ s, $H = 0.188$ m	0.320	0.197	0.160	183.06
Monopile foundation + single-row PRAF	$T = 1.237$ s, $H = 0.063$ m	0.095	0.065	0.057	59.83
	$T = 1.237$ s, $H = 0.125$ m	0.189	0.134	0.101	115.28
	$T = 1.237$ s, $H = 0.188$ m	0.295	0.204	0.168	180.59
Monopile foundation + three-row PRAFs	$T = 1.237$ s, $H = 0.063$ m	0.100	0.064	0.057	68.14
	$T = 1.237$ s, $H = 0.125$ m	0.201	0.132	0.109	118.32
	$T = 1.237$ s, $H = 0.188$ m	0.292	0.211	0.176	187.68

Figure 22 shows the effects of the monopile foundation and single-row PRAF on the wave dynamics. As the waves propagated, there were obvious diffraction and reflection phenomena that occurred around the monopile foundation and the PRAF, which resulted in corresponding changes in the distribution of wave heights in both the upwind and downwind areas that surrounded these structures. It can be observed that the wave height behind the monopile foundation was weakened, which formed a localized wave shadow area, which showed a typical diffraction phenomenon. The PRAF that was positioned downstream of the monopile foundation exhibited a significant distortion in the wave color bands in its vicinity. This indicates that the reflected waves from the PRAF were superimposed with the waves diffracted by the monopile foundation, which led to a wave interference phenomenon. Additionally, as the wave height increased, the expansion of the wave color band areas that surrounded both the monopile foundation and the PRAF became increasingly pronounced, which signified an enhancement in the disturbance intensity of the wave field caused by these structures. As presented in Table 4, the wave height H_{P1} in front of the pile was significantly larger than the incident wave height, and the wave run-up increased as the incident wave height rose. Due to wave diffraction, the wave height H_{P2} along the side of the monopile experienced only slight variations, whereas the wave height H_{P3} behind the pile decreased. These numerical results support the observations made in Figure 21. When comparing the scenario in which only the monopile was present to the case that included the PRAF, it is evident that the presence of the PRAF slightly reduced H_{P1} while it increased both H_{P2} and H_{P3}. And the PRAF further changed the wave propagation characteristics behind the pile, which made the distribution of the wave field around the pile more complicated. Therefore, the PRAF had a certain weakening or redistribution effect on the local wave field, although this effect caused only a slight change in the wave height.

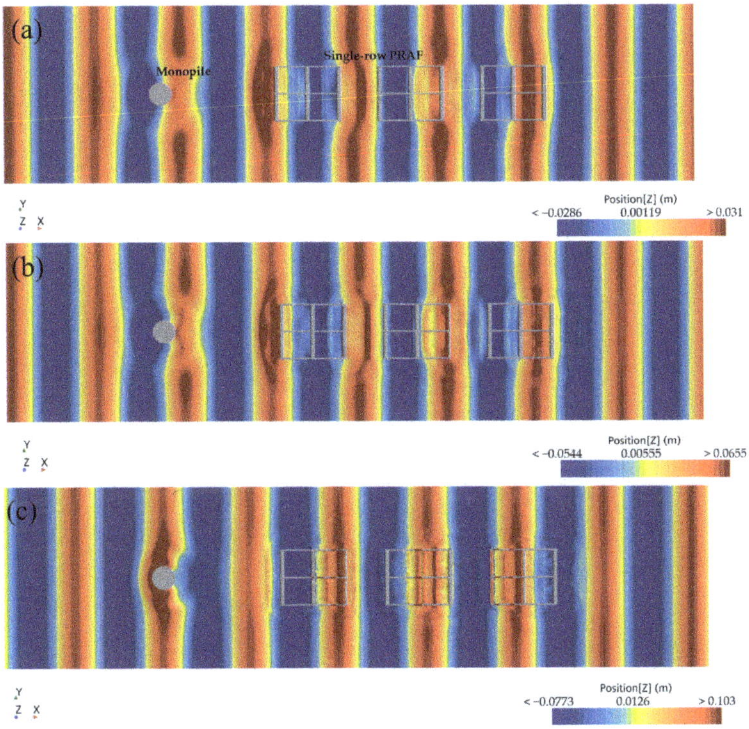

Figure 22. The wave fields around the monopile foundation and single-row PRAF: (**a**) $T = 1.237$ s, $H = 0.063$ m; (**b**) $T = 1.237$ s, $H = 0.125$ m; (**c**) $T = 1.237$ s, $H = 0.188$ m.

Figure 23 illustrates the impacts of the monopile foundation and the three-row pontoon raft aquaculture facilities (PRAFs) on the wave dynamics. The variations in wave heights depicted in this figure reveal that both the monopile foundation and the three-row PRAFs created distinct sheltered areas within the overall wave field. Compared with the single-row PRAF, the three-row configuration exhibited more pronounced reflection and diffraction phenomena, which affected a larger area of the wave field. Notably, the wave heights in the regions between the rafts showed a regular ripple-like distribution. As the wave height increased, the fluid–structure coupling effects became more pronounced, which led to an increase in local wave heights due to enhanced reflection and diffraction. However, the presence of the three-row PRAFs exerted only a minimal effect on the wave heights around the monopile foundation itself.

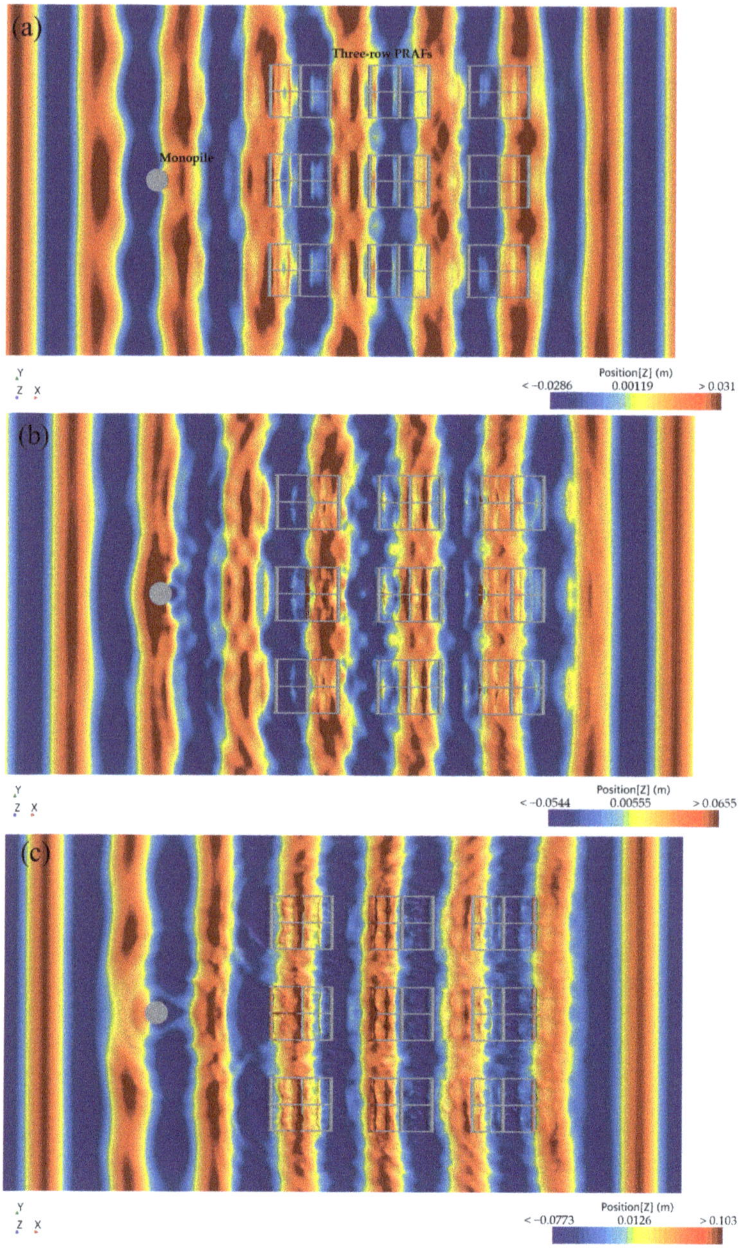

Figure 23. The wave fields around the monopile foundation and three-row PRAFs: (**a**) T = 1.237 s, H = 0.063 m; (**b**) T = 1.237 s, H = 0.125 m; (**c**) T = 1.237 s, H = 0.188 m.

5. Conclusions

In this study, we investigated the hydrodynamic performance of a PRAF placed around a wind farm monopile under different wave conditions. Our experimental and numerical analyses focused on the mooring line and bridle line tensions, with an additional emphasis on how wave height attenuation impacts these forces. The results yield several key insights that have broader significance for designing and deploying integrated offshore wind–aquaculture systems:

(1) Influence of wave parameters

The mooring line and bridle line tension of the PRAF consistently decreased with the increasing wave period and increased with increasing wave height. This finding underscores the critical role of wave steepness (H/L) in determining the magnitude of the hydrodynamic forces. By highlighting specific wave conditions, particularly within the range of 0.02 to 0.09 of H/L, this study helps designers anticipate and manage peak loads more effectively.

(2) Effects of the wind farm monopile

Both the experimental and computational results demonstrate a significant sheltering effect behind the monopile. Part of the incoming wave energy was reflected or dissipated at the front side of the monopile, which decreased the wave energy available in the downstream region. The presence of a wind monopile significantly impacted both the mooring line tension and the bridle line tension for the PRAF. The mooring line tension was reduced, and the reduction ratio decreased when the wave steepness increased, where it ranged from 16% to 66%. Furthermore, there was an increase in the bridle line tension on the upstream component, with a decrease in the bridle line tension on the downstream component.

(3) Significance for multi-row PRAF configurations

The results from the single-row and multi-row PRAF arrangements highlight a clear attenuation effect and a more even distribution of hydrodynamic loading in multi-row configurations. This is particularly relevant for large-scale offshore aquaculture operations, where a multi-row layout can enhance the resilience to extreme waves and reduce the peak stresses on individual lines to improve both the safety and long-term durability. Thus, in the future, we have an opportunity to achieve robust structural performance by optimizing the raft layouts, a crucial step for large-scale applications in offshore wind farms.

(4) Wave height loss of waves passing around the monopile

Wave height attenuation occurred as the waves propagated around the monopile, where the wave height loss (K_a) ranged from 5% to 20%. It was observed that the monopile's influence on the wave height diminishes with an increasing wave period. Additionally, a greater wave height corresponded to an increased K_a. When K_a ranged from 0.11 to 0.20, the presence of the monopile could effectively reduce the mooring line tension of the PRAF. Meanwhile, our results show that using the monopile itself as a mooring point may induce a high-level mooring load, potentially affecting the monopile's stability and safety. We therefore advise caution and do not recommend directly attaching the PRAF mooring lines to the monopile in practical applications.

(5) Influence of the PRAF on the force on the monopile foundation

The numerical simulation results show that a single-row PRAF deployed behind the monopile could reduce the wave forces on the foundation by up to 8.75% for certain conditions, while the three-row PRAF setups produced minor increases in forces (up to 4%). Overall, the impact on the monopile wave forces remained minimal, indicating the viability of rationally deploying PRAFs in offshore wind farms.

In conclusion, these research findings provide valuable insights into the co-location of offshore wind and aquaculture systems. By placing PRAFs near a wind monopile, existing infrastructure can be effectively utilized, thus conserving marine space and mitigating hydrodynamic loads. Nevertheless, due to the inherent complexity of marine environments, this study primarily examined wave-induced responses. A more comprehensive investigation of wind, combined wave–current, and combined wind–wave effects remains an important avenue for future work. Moreover, the geometric dimensions of the monopile and the configuration of aquaculture facilities have a significant influence on the force distribution and motion behavior. Consequently, subsequent investigations should extend to multiple environmental loading scenarios and conduct deeper analyses of monopile geometry and aquaculture layout. From a modeling perspective, developing refined hydrodynamic and structural coupling models for PRAFs in offshore wind farms will optimize the facility design under a wider range of sea states, ensuring both structural stability and operational efficiency. Advancing these research directions will further support the sustainable development of offshore wind–aquaculture integration.

Author Contributions: Conceptualization, X.W. and Y.T.; methodology, D.C. and X.W.; software, M.L. and D.C.; validation, X.W. and D.C.; formal analysis, D.C.; investigation, D.C., J.Z., F.Z., X.W. and Q.L.; resources, Y.T.; data curation, D.C. and X.W.; writing—original draft preparation, D.C.; writing—review and editing, X.W.; visualization, F.Z. and X.W.; supervision, X.W. and M.Y.; project administration, Y.T.; funding acquisition, Y.T. and D.K. All authors have read and agreed to the published version of the manuscript.

Funding: This research was mainly funded by the China Three Gorges Renewable (Group) Co., Ltd., No. 37022078, and partially supported by the JSPS Grants-in-Aid for Scientific Research No. 23H01619.

Data Availability Statement: The code for the data analysis is available upon request from the authors.

Acknowledgments: The authors would like to express gratitude to J. Han from the Yokohama National University for the water tank experiment assistance.

Conflicts of Interest: The authors declare that this study received funding from China Three Gorges Renewable (Group) Co., Ltd. The funder was not involved in the study design, collection, analysis, interpretation of data, the writing of this article or the decision to submit it for publication.

Abbreviation

Nomenclature			
Abbreviation	**Definition**		
PRAF	Pontoon raft aquaculture facility	**PV**	Polyvinyl Chloride
CFD	Computational fluid dynamics	**WG**	Wave gauge
Flume tank tests			
T	Wave period, s	D	Monopile diameter, m
H	Wave height, m	H_s	Significant wave height, m
L	Wave length, m		
Wave loads on single-row PRAF			
F	Tension, kN	F_{SB1}	Maximum bridle line tension for the position 1 in a single-row PRAF, kN
ΔF	The reduction in tension, kN ($\Delta F = F_{SN} - F_{SP}$)	F_{SB2}	Maximum bridle line tension for the position 2 in a single-row PRAF, kN
F_{SN}	Maximum mooring line tension without a monopile, kN	F_{SB3}	Maximum bridle line tension for the position 3 in a single-row PRAF, kN
F_{SP}	Maximum mooring line tension with a monopile, kN	F_{SB4}	Maximum bridle line tension for the position 4 in a single-row PRAF, kN
Wave loads on multi-row PRAFs			
F_{TO}	Maximum mooring line tension for the outer raft facility in three-row PRAFs, kN	F_{TB2}	Maximum bridle line tension for the position 2 in three-row PRAFs, kN
F_{TI}	Maximum mooring line tension for the inner raft facility in three-row PRAFs, kN	F_{TB3}	Maximum bridle line tension for the position 3 in three-row PRAF, kN
F_{TB1}	Maximum bridle line tension for the position 1 in three-row PRAF, kN	F_{TB4}	Maximum bridle line tension for the position 4 in three-row PRAF, kN

Wave attenuation

H_i	The incident wave height, m	K_t	Transmission coefficient ($K_t = H_t/H_i$)
H_t	The transmitted wave height, m	K_a	Attenuation coefficient ($K_a = 1 - K_t$)

Appendix A. Validation of the Numerical Flume Model

Appendix A.1. Experimental Setup

To verify the accuracy of the numerical flume model with a wind monopile foundation, we compared the numerical results with the experimental data reported by Mo et al. [49] due to the lack of corresponding experimental data for direct validation. The experimental flume used in Mo's study had a length of 60 m, a width of 5 m, and a depth of 7 m. A steel circular cylinder with a diameter of 0.7 m was placed in the flume to simulate the monopile structure. The experimental conditions selected for the comparison were a depth $h = 4.76$ m, a wave period $T = 4.0$ s, and a wave height $H = 1.2$ m. These conditions were applied to the numerical model to ensure a consistent basis for comparison.

Appendix A.2. Numerical Model Setup

As shown in Figure A1, the numerical flume model was configured to replicate the experimental setup as described in Mo et al. [49]. The model included the same geometric parameters for the flume and monopile structure. The boundary conditions were set to replicate the experimental wave conditions, and the same wave generation and absorption techniques were applied. The numerical simulation was conducted using a fully coupled wave–structure interaction approach, with the governing equations solved using the Reynolds-averaged Navier–Stokes (RANS) method and the SST k-ω turbulence model. Mesh refinement was applied near the monopile to improve the accuracy of the wave elevation results.

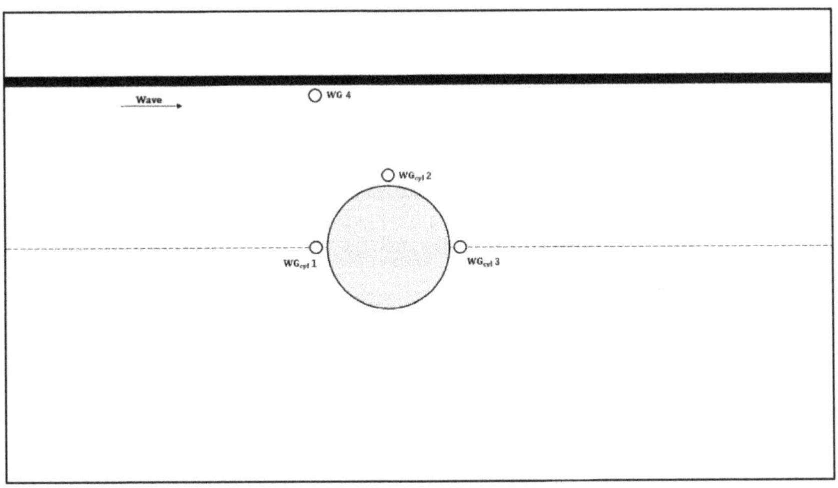

Figure A1. Schematic diagram of four wave-gauge locations around a cylinder in the experiment by Mo et al. (2007) [49].

Appendix A.3. Validation Results

The comparison between the numerical simulation results and the experimental data is presented in Figure A2. The results indicate that the error ranges of the simulation wave heights for WG1 to WG4 were 3.66%, 5.44%, 6.72%, and 8.93%, respectively. The numerical model accurately captured the key hydrodynamic features observed in the experimental study. The wave trough elevations predicted by the numerical model aligned closely with those recorded in the experimental results, indicating that the model successfully reproduced the wave profile around the monopile structure. The successful validation against experimental data from Mo et al. [49] demonstrates that the numerical flume model is applicable for further simulations involving wind farm monopile foundation. The results provide confidence in the accuracy of the numerical predictions presented in the main text.

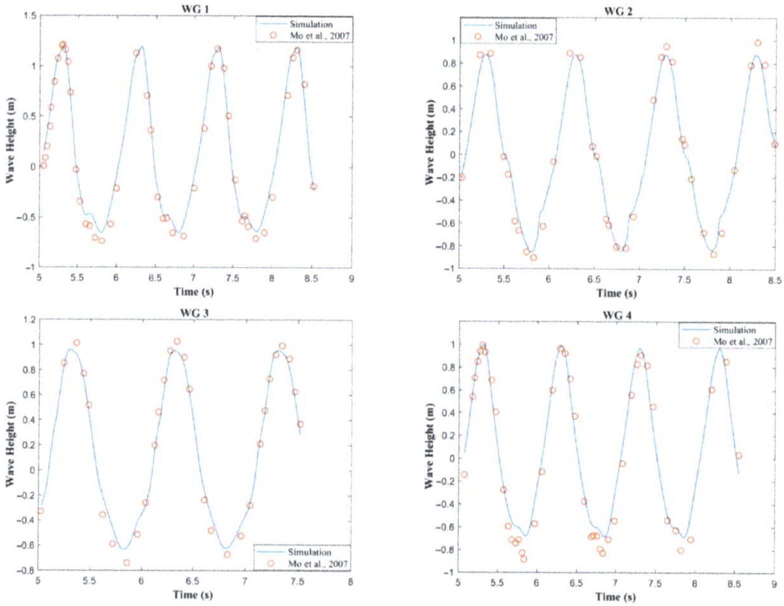

Figure A2. Comparison of numerical results with experimental results by Mo et al. (2007) [49].

Appendix B. The Mesh Sensitivity Analysis

Appendix B.1. Test Setup

A grid sensitivity analysis of the numerical flume was carried out by varying the grid parameters and time step at the encryption of the free surface to determine the optimal grid settings. Case 10 (H = 0.1781 m) was chosen as the incident wave parameter for the grid sensitivity analysis test, and the ratio of the actual wave height (H′) to the theoretical wave height (H) was used as a measure of the grid sensitivity in the test. The grid parameters used for the sensitivity analysis are shown in Table A1, with the number of grids in each wave height range (H/Δz) versus the number of grids in each wavelength range (L/Δx) and time step as the variables, where Δz denotes the minimum grid size in the vertical direction and Δx denotes the minimum grid size in the horizontal direction.

Table A1. The grid parameters used for the sensitivity analysis.

Mesh	H/Δz	L/Δx	Time Step (s)
1	6	100	0.002
2	12		
3	24		
4	36		

Table A1. *Cont.*

Mesh	H/Δz	L/Δx	Time Step (s)
5	24	60	0.002
6		80	
7		100	
8		120	
9	24	100	0.001
10			0.002
11			0.004
12			0.008

Appendix B.2. Test Results

The analysis results are shown in Figures A3 and A4. It can be seen that with the increase in H/Δz, the numerical wave height became closer to the theoretical wave height. When H/Δz > 24, the numerical wave height became stable. Similarly, L/Δx and the time step reached stability when reaching 100 and 0.002 s, respectively. Therefore, H/Δz = 24 and L/Δx = 100 were chosen as the mesh parameters, along with the time step of 0.002 s for subsequent calculations. As shown in Figure A4, compared with the theoretical values, the wave elevation of the numerical simulation was less than 1.1%. The chosen numerical model, mesh size, and time step were very accurate. Therefore, the numerical model could be used to simulate the wave elevation around the monopile.

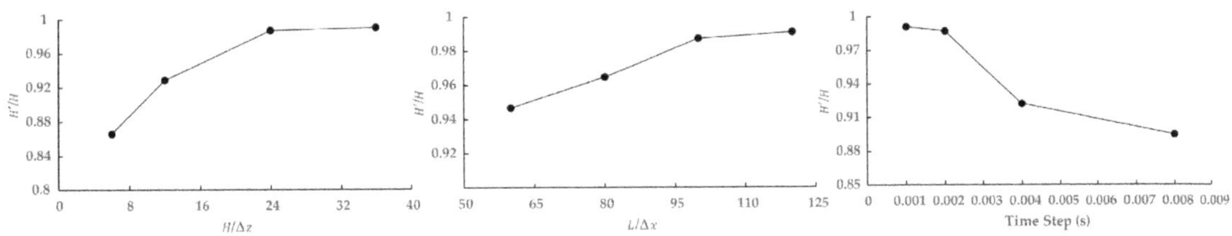

Figure A3. Mesh sensitivity analysis test: mesh number per wavelength L/Δx, mesh number per wave height H/Δz, and time step.

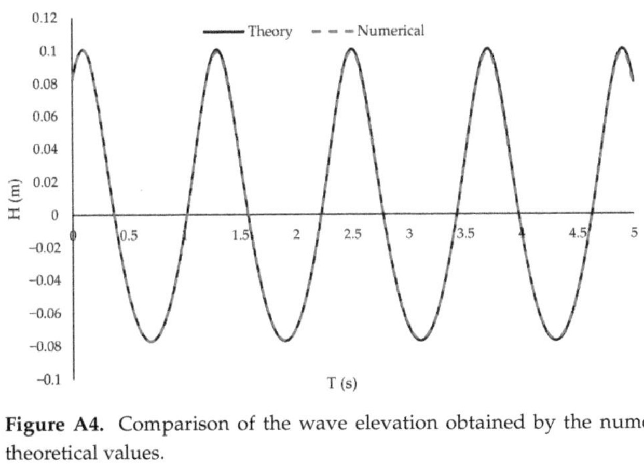

Figure A4. Comparison of the wave elevation obtained by the numerical simulation with the theoretical values.

References

1. Yaghubi, E.; Carboni, S.; Snipe, R.M.J.; Shaw, C.S.; Fyfe, J.J.; Smith, C.M.; Kaur, G.; Tan, S.Y.; Hamilton, D.L. Farmed mussels: A nutritive protein source, rich in omega-3 fatty acids, with a low environmental footprint. *Nutrients* **2021**, *13*, 1124. [CrossRef] [PubMed]
2. Avdelas, L.; Avdic-Mravlje, E.; Marques, A.C.B.; Cano, S.; Capelle, J.J.; Cozzolino, M.; Dennis, J.; Ellis, T.; Polanco, J.M.F.; Guillen, J.; et al. The decline of mussel aquaculture in the European Union: Causes, economic impacts and opportunities. *Rev. Aquac.* **2021**, *13*, 91–118. [CrossRef]
3. Karayücel, S.; Celik, M.Y.; Karayücel, I.; Erik, G. Growth and Production of Raft Cultivated Mediterranean Mussel (Mytilus galloprovincialis Lamarck, 1819) in Sinop, Black Sea. *Turk. J. Fish. Aquat. Sci.* **2010**, *10*, 9–17. [CrossRef]
4. Langan, R. Mussel Culture, Open Ocean Innovations. In *Sustainable Food Production*; Christou, P., Savin, R., Costa-Pierce, B.A., Misztal, I., Whitelaw, C.B.A., Eds.; Springer: New York, NY, USA, 2013. [CrossRef]
5. Maar, M.; Holbach, A.; Boderskov, T.; Thomsen, M.; Buck, B.H.; Kotta, J.; Bruhn, A. Multi-use of offshore wind farms with low-trophic aquaculture can help achieve global sustainability goals. *Commun. Earth Environ.* **2023**, *4*, 447. [CrossRef]
6. Buck, B.H. *Open Ocean Aquaculture und OffshoreWindparks: Eine Machbarkeitsstudie uber Die multifunktionale Nutzung von Offshore-Windparks und Offshore-Marikultur im Raum Nordsee*; Alfred Wegener Institute for Polar and Marine Research: Bremerhaven, Germany, 2002; pp. 136–144.
7. Buck, B.H.; Krause, G.; Rosenthal, H. Extensive open ocean aquaculture development within wind farms in Germany: The prospect of offshore co-management and legal constraints. *Ocean Coast. Manag.* **2004**, *47*, 95–122. [CrossRef]
8. Buck, B.H.; Langan, R. *Aquaculture Perspective of Multi-Use Sites in the Open Ocean: The Untapped Potential for Marine Resources in the Anthropocene*; Springer Nature: Berlin, Germany, 2017. [CrossRef]
9. Alkhabbaz, A.; Hamza, H.; Daabo, A.M.; Yang, H.S.; Yoon, M.; Koprulu, A.; Lee, Y.H. The aero-hydrodynamic interference impact on the NREL 5-MW floating wind turbine experiencing surge motion. *Ocean Eng.* **2024**, *295*, 116970. [CrossRef]
10. Zheng, X.Y.; Lei, Y. Stochastic response analysis for a floating offshore wind turbine integrated with a steel fish farming cage. *Appl. Sci.* **2018**, *8*, 1229. [CrossRef]
11. Zhao, Z.L.; Yu, W.Z.; Xu, T.J. Study of the Interaction between Regular Waves and a Wind-Fish Combined Structure. *Appl. Sci.* **2023**, *13*, 3103. [CrossRef]
12. Zhang, C.L.; Wang, S.M.; Cui, M.C.; Liu, H.; Liu, A.D.; Xu, J.C.; Xie, S.Y. Modeling and dynamic response analysis of a submersible floating offshore wind turbine integrated with an aquaculture cage. *Ocean Eng.* **2022**, *263*, 112338. [CrossRef]
13. Cao, S.G.; Cheng, Y.L.; Duan, J.L.; Fan, X.X. Experimental investigation on the dynamic response of an innovative semi-submersible floating wind turbine with aquaculture cages. *Renew. Energy* **2022**, *200*, 1393–1415. [CrossRef]
14. Yi, Y.; Sun, K.; Liu, Y.Q.; Zhang, J.H.; Jiang, J.; Liu, M.Y.; Ji, R.W. Experimental investigation into the dynamics and power coupling effects of floating semi-submersible wind turbine combined with point-absorber array and aquaculture cage. *Energy* **2024**, *296*, 131220. [CrossRef]
15. Zhai, Y.T.; Zhao, H.S.; Li, X.; Shi, W. Hydrodynamic responses of a barge-type floating offshore wind turbine integrated with an aquaculture cage. *J. Mar. Sci. Eng.* **2022**, *10*, 854. [CrossRef]
16. Li, N.; Zhao, H.S.; Shi, W.; Wang, W.H.; Li, X. Dynamic Analysis of an Integrated Structure Consists of Jacket Offshore Wind Turbine and Aquaculture Cage. In Proceedings of the ASME 2024 43rd International Conference on Offshore Mechanics and Arctic Engineering, Hamburg, Germany, 5–1 June 2022. [CrossRef]
17. Chu, Y.I.; Wang, C.M.; Zhang, H. A frequency domain approach for analyzing motion responses of integrated offshore fish cage and wind turbine under wind and wave actions. *Aquac. Eng.* **2022**, *97*, 102241. [CrossRef]
18. Gong, S.K.; Gao, J.L.; Yan, M.Y.; Song, Z.W.; Shi, H.B. Effects of floater motion on wave loads during steady-state gap resonance occurring between two non-identical boxes. *Ocean Eng.* **2025**, *323*, 120649. [CrossRef]
19. Gong, S.K.; Gao, J.L.; Song, Z.W.; Shi, H.B.; Liu, Y.Y. Hydrodynamics of fluid resonance in a narrow gap between two boxes with different breadths. *Ocean Eng.* **2024**, *311*, 118986. [CrossRef]
20. Gao, J.L.; Mi, C.L.; Song, Z.W.; Liu, Y.Y. Transient gap resonance between two closely-spaced boxes triggered by nonlinear focused wave groups. *Ocean Eng.* **2024**, *305*, 117938. [CrossRef]
21. Mi, C.L.; Gao, J.L.; Song, Z.W.; Yan, M.Y. Gap resonance between a stationary box and a vertical wall induced by transient focused wave groups. *China Ocean Eng.* **2025**, 1–14. [CrossRef]
22. Mi, C.L.; Gao, J.L.; Song, Z.W.; Liu, Y.Y. Hydrodynamic wave forces on two side-by-side barges subjected to nonlinear focused wave groups. *Ocean Eng.* **2025**, *317*, 120056. [CrossRef]
23. Zhong, W.J.; Zhang, X.M.; Wan, D.C. Hydrodynamic characteristics of a 15 MW semi-submersible floating offshore wind turbine in freak waves. *Ocean Eng.* **2023**, *283*, 11594. [CrossRef]
24. Tang, Y.; Shi, W.; You, J.K.; Michailides, C. Effects of nonlinear wave loads on large monopile offshore wind turbines with and without ice-breaking cone configuration. *J. Mar. Sci. Technol.* **2021**, *26*, 37–53. [CrossRef]

25. Wang, J.; Wang, S.Q.; Jiang, Q.D.; Xu, Y.X.; Shi, W.C. Effect of different raft shapes on hydrodynamic characteristics of the attenuator-type wave energy converter. *China Ocean Eng.* **2023**, *37*, 645–659. [CrossRef]
26. Wang, K.; Li, N.; Wang, Z.H.; Song, G.J.; Du, J.; Song, L.; Jiang, H.Z.; Wu, J.H. The impact of floating raft aquaculture on the hydrodynamic environment of an open sea area in Liaoning Province, China. *Water* **2022**, *14*, 3125. [CrossRef]
27. Tomasicchio, G.R.; D'Alessandro, F.; Avossa, A.M.; Riefolo, L.; Musci, E.; Ricciardelli, F.; Vicinanza, D. Experimental modelling of the dynamic behaviour of a spar buoy wind turbine. *Renew. Energy* **2018**, *127*, 412–432. [CrossRef]
28. Lauria, A.; Loprieno, P.; Rizzo, F.; Severini, A.; Foti, D.; Leone, E.; Francone, A.; Tomasicchio, G. On the effects of wind and operating conditions on mooring line tensions for floating offshore wind turbine. *Appl. Ocean Res.* **2024**, *152*, 104197. [CrossRef]
29. Ai, Q.; Cao, J.C.; Jin, L.; Chen, D.M.; Tang, Y.L.; Zhao, F.F.; Sheng, H.X.; Wang, X.X. Experimental Study of the Hydrodynamic Characteristics of the Raft Aquaculture Facilities Around a Windfarm Monopile. In Proceedings of the ASME 2024 43rd International Conference on Offshore Mechanics and Arctic Engineering, Singapore, Singapore, 9–14 June 2024. [CrossRef]
30. Wang, X.X.; Swift, M.R.; Dewhurst, T.; Tsukrov, I.; Celikkol, B.; Newell, C. Dynamics of submersible mussel rafts in waves and current. *China Ocean Eng.* **2015**, *29*, 431–444. [CrossRef]
31. STAR-CCM+. User Manual. Version 19.02.012-R8, Siemens Industry Software Inc. 2024. Available online: https://www.plm.automation.siemens.com/global/zh/products/simulation-test/cae-simulation.html (accessed on 1 June 2024).
32. Menter, F.R. Two-equation eddy-viscosity turbulence modeling for engineering applications. *AIAA J.* **1994**, *32*, 1598–1605. [CrossRef]
33. Hirt, C.W.; Nichols, B.D. Volume of fluid (VOF) method for the dynamics of free boundaries. *J. Comput. Phys.* **1981**, *39*, 201–225. [CrossRef]
34. Fenton, J.D. A fifth-order Stokes theory for steady waves. *J. Waterw. Port Coast. Ocean Eng.* **1985**, *111*, 216–234. [CrossRef]
35. Zhang, T.Z.; Zhao, W.H.; Zhou, C.; Zhao, F.F.; Wan, R. Design and modeling of an eco-friendly anchored fish aggregating device with artificial reef subjected to wave and current. *Ocean Eng.* **2024**, *313*, 3. [CrossRef]
36. Kim, J.; O'Sullivan, J.; Read, A. Ringing analysis of a vertical cylinder by Euler overlay method. In Proceedings of the ASME 2012 31st International Conference on Offshore Mechanics and Arctic Engineering, Rio de Janeiro, Brazil, 1–6 July 2012. [CrossRef]
37. Yang, H.S.; Alkhabbaz, A.; Tongphong, W.; Lee, Y.H. Cross-comparison analysis of environmental load components in extreme conditions for pontoon-connected semi-submersible FOWT using CFD and potential-based tools. *Ocean Eng.* **2024**, *304*, 117248. [CrossRef]
38. Daabo, A.M.; Alkhabbaz, A.; Ibrahim, S.S.; Hamzah, H.; Hassan, A.; Basem, A.; Easa, H.; Pavlovic, S. Thirteen vital factors for micro-scale radial turbine vane's design of geo-solar-powered Brayton cycle applications. *Energy Convers. Manag.* **2024**, *315*, 118774. [CrossRef]
39. Wang, X.X.; Zhao, F.F.; Tang, Y.L.; Huang, L.Y.; Wan, R.; Cheng, H. Numerical Analysis of Submersible Mussel Raft for Exposed Marine Environment. In Proceedings of the ASME 2017 36th International Conference on Offshore Mechanics and Arctic Engineering, 25–30 June 2017. [CrossRef]
40. Wang, X.X.; Xie, J.Y.; Luo, Y.; Wang, X.; Guo, G.B.; You, X.X. Experimental Investigation of the Hydrodynamic Characteristics of Longline Aquaculture Facilities under Current and Wave Conditions. *Fishes* **2023**, *8*, 204. [CrossRef]
41. Liu, H.F.; Bi, C.W.; Zhao, Y.P. Experimental and numerical study of the hydrodynamic characteristics of a semisubmersible aquaculture facility in waves. *Ocean Eng.* **2020**, *214*, 107714. [CrossRef]
42. de Leon, S.P.; Bettencourt, J.H.; Kjerstad, N. Simulation of irregular waves in an offshore wind farm with a spectral wave model. *Cont. Shelf Res.* **2011**, *31*, 1541–1557. [CrossRef]
43. Corvaro, S.; Crivellini, A.; Marini, F.; Cimarelli, A.; Capitanelli, L.; Mancinelli, A. Experimental and numerical analysis of the hydrodynamics around a vertical cylinder in waves. *J. Mar. Sci. Eng.* **2019**, *7*, 453. [CrossRef]
44. Masoudi, E.; Marshall, A. Diffraction of waves by multi-pontoon rectangular floating breakwaters. *Ocean Eng.* **2024**, *310*, 118789. [CrossRef]
45. Li, C.; Zhang, H.J.; Zhang, H.; Sun, B.; Yang, S.L. Wave-attenuation and hydrodynamic properties of twin pontoon floating breakwater with kelp. *Appl. Ocean Res.* **2022**, *124*, 103213. [CrossRef]
46. Miles, J.; Martin, T.; Goddard, L. Current and wave effects around windfarm monopile foundations. *Coast. Eng.* **2017**, *121*, 167–178. [CrossRef]
47. Martinelli, L.; Ruol, P.; Zanuttigh, B. Wave basin experiments on floating breakwaters with different layouts. *Appl. Ocean Res.* **2008**, *30*, 199–207. [CrossRef]
48. Liao, P. Study on Hydrodynamic Behaviors of "Tian" shaped Shellfish Raft Cultivate Facilities. Master's Thesis, Dalian University of Technology, Dalian, China, 2017.

49. Mo, W.H.; Irschik, K.; Oumeraci, H.; Liu, P.L.F. A 3D numerical model for computing non-breaking wave forces on slender piles. *J. Eng. Math.* **2007**, *58*, 19–30. [CrossRef]
50. Wynants, M.; Schendel, A.; Welzel, M.; Kerpen, N.B.; Schlurmann, T. Current-distorted wave diffraction patterns in the near-field of a monopile in coastal seas. *Ocean Eng.* **2024**, *302*, 117432. [CrossRef]

MDPI AG
Grosspeteranlage 5
4052 Basel
Switzerland
Tel.: +41 61 683 77 34

Journal of Marine Science and Engineering Editorial Office
E-mail: jmse@mdpi.com
www.mdpi.com/journal/jmse

www.ingramcontent.com/pod-product-compliance
Lightning Source LLC
Chambersburg PA
CBHW061843210926
51850CB00006B/955